Dismantling Green Colonialism

"A must-read thought-provoking book for every researcher, policymaker and activist working on climate, energy, development and social justice issues in the Arab region. This volume educates and empowers its readers to think about the roots of the problems in clear, systematic, and transformative ways. A significant contribution to the literature on just transition, greenwashing, neocolonialism, extractivism and neoliberalism."
—Fadhel Kaboub, President of the Global Institute
for Sustainable Prosperity

"A brave and timely book that offers hope for our planet. These essays from the Arab world analyse the complexity of the environmental issues at play in the region and offer an optimistic, global, democratic vision of transformative sustainability centred around climate justice."
—Ahdaf Soueif, novelist and political and cultural commentator

"For anyone committed to putting the Just into Just Transition this is a vital intervention that connects the past to the present and challenges us not only to reimagine the future, but to stand with those on the frontlines fighting for it."
—Asad Rehman, War on Want, UK

"A much-needed decolonized examination of the climate crisis for all sacrificial zones. A focus on the situation in North Africa, an area of intense contestations pitching the peoples against the relentless push by fossil fuel speculators and other forces of neoliberalism is both welcome and a clear warning that must not be ignored."
—Nnimmo Bassey, author of *To Cook a Continent: Destructive Extraction and the Climate Change Crisis in Africa*

"This groundbreaking volume by scholars deeply embedded in the region's political and knowledge production milieus, offers a timely, indeed acute, analysis of what a just transition might mean for the region. The authors examine in theoretically and empirically rich essays contestations over the Sahara, greenwashing Israel's colonisation of Palestine, agricultural and mineral extractivism, green capitalism and finance and a range of other urgently pivotal subjects."

—Laleh Khalili, author of *Sinews of War and Trade: Shipping and Capitalism in the Arabian Peninsula*

"Demonstrates that the climate crisis – along with mainstream responses to it - is playing out along colonial lines. It's time to face up to this reality and build an anti-colonial struggle in response."

—Jason Hickel, economic anthropologist and author of *Less is More*

"This book is crucial for those seeking alternative visions and policies to the complete disaster currently being produced by capitalism, and to capitalism's failing global and local projects to deal with an issue that is a question of life and death. Despite the multi-dimensional crisis that the Arab region – and the whole world – is going through […] the Arab region remains largely absent from the intensifying debate over the future."

—Wael Gamal, Egyptian writer and researcher in political economy

"Just as the science is telling us loud and clear that the current situation of climate deterioration may be our last chance "before it is too late", so the research and knowledge presented in this book, including its practical and feasible recommendations (which are directed to people rather than to the indifferent, comprador regimes in the Arab region), serves as wake-up call, reminding us of the urgent need to act before it is too late."

—Nahla Chahal, Professor of political sociology, Editor-in-Chief of *As-Safir Al-Arabi*

"[T]his book serves as a crucial link in the collective efforts and common priorities of climate experts and climate justice advocates in Arab countries who, moreover, refuse the new colonialism that is disguised in some agendas around addressing climate change and harnessing renewable energies. I hope this book can be a catalyst that will prompt governments and civil society organizations and institutions to pursue climate justice and achieve energy democracy in North Africa."

—Houcine Rhili, development specialist, Tunisia

Dismantling Green Colonialism

Energy and Climate Justice in the Arab Region

Edited by
Hamza Hamouchene and Katie Sandwell

PLUTO PRESS

First published 2023 by Pluto Press
New Wing, Somerset House, Strand, London WC2R 1LA
and Pluto Press, Inc.
1930 Village Center Circle, 3-834, Las Vegas, NV 89134

www.plutobooks.com

tni

transnationalinstitute

British Library Cataloguing in Publication Data
A catalogue record for this book is available from the British Library

ISBN   978 0 7453 4921 3   Paperback
ISBN   978 0 7453 4919 0   PDF
ISBN   978 0 7453 4920 6   EPUB

This book is printed on paper suitable for recycling and made from fully
managed and sustained forest sources. Logging, pulping and manufacturing processes are expected to conform to the environmental standards of the country of
origin.

Typeset by Stanford DTP Services, Northampton, England

Simultaneously printed in the United Kingdom and United States of America

Contents

Tables and Figures   ix
Acknowledgements   xi
Abbreviations   xiii

**Introduction: Just in Time – The Urgent Need for a Just Transition
in the Arab Region**
Hamza Hamouchene and Katie Sandwell   1
   North Africa and West Asia as a key node in global fossil capitalism   6
   The colonial gaze and environmental orientalism   9
   What is "just transition"?   11
   Why this book? Why now?   14
   Summary of the book's chapters   16
   In guise of a conclusion   20

**PART I: ENERGY COLONIALISM, UNEQUAL EXCHANGE AND
GREEN EXTRACTIVISM**

1. The Energy Transition in North Africa: Neocolonialism Again!   29
Hamza Hamouchene
   "Green colonialism" and "green grabbing"   29
   Energy transition, dispossession and grabbing in Morocco   31
   Green colonialism and occupation in Western Sahara   34
   Which energy transition in Algeria? Drill baby drill!   35
   Privatization of energy for export   38
   Hydrogen: the new energy frontier in Africa   38
   Desertec 3.0 – or jumping on the green hydrogen bandwagon   40
   Conclusion   43

**2. An Unjust Transition: Energy, Colonialism and Extractivism in
Occupied Western Sahara**   49
Joanna Allan, Hamza Lakhal and Mahmoud Lemaadel
   A brief history of the Western Sahara conflict   50
   Extractivism in occupied Western Sahara   52

Powering the occupation: how energy does diplomatic work for the
   Moroccan regime   53
Powering oppression: Saharawi perspectives of the energy system
   in occupied Western Sahara   55
What would a Saharawi-led "just transition" look like? Inspiration
   and questions from the camps   58
Conclusion   61

3. Arab–Israeli Eco-Normalization: Greenwashing Settler
Colonialism in Palestine and the Jawlan   67
Manal Shqair
Eco-normalization projects   68
Prosperity Blue: Israel quenches parched Jordan   69
Eco-normalization: a violent onslaught on the Palestinians' right to
   self-determination   77
Eco-sumud: A vision for a just transition in Palestine   79
Conclusion   82

4. What Can an Old Mine Tell Us about a Just Energy Transition?
Lessons from Social Mobilization across Mining and Renewable
Energy in Morocco   88
Karen Rignall
A just transition in Morocco   89
Diverse forms of extraction in southeastern Morocco   91
Similar actors and financial interests in conventional extraction
   and renewable energy   92
Land conflict and resource politics in the legal and bureaucratic
   context   95
Why taxes matter, or an argument for reparations   99
Social mobilization and shared political claims across extraction
   and renewable energy   102
Conclusion   104

5. Towards a Just Agricultural Transition in North Africa   109
Saker El Nour
Agricultural policy transformations in North Africa   112
Agroecological and regenerative agriculture as vehicles for a
   just transition in North Africa   121
Conclusion   126

6. The Electricity Crisis in Sudan: Between Quick Fixes and
Opportunities for a Sustainable Energy Transition 134
Razaz H. Basheir and Mohamed Salah Abdelrahman
The crisis 134
Supply 134
Consumption 143
The World Bank report 144
Conclusion 149

PART II: NEOLIBERAL ADJUSTMENTS, PRIVATISATION OF ENERGY
AND THE ROLE OF INTERNATIONAL FINANCIAL INSTITUTIONS

7. International Finance and the Commodification of Electricity
in Egypt 157
Mohamed Gad
After decades of prevarication, liberalization 158
The origins and development of liberalization 164
Outcomes of opening up to external finance 166
The social impacts of electricity liberalization 168
Conclusion 171

8. The Energy Sector in Jordan: Crises Caused by Dysfunctional
and Unjust Policies 175
Asmaa Mohammad Amin
Before the gas stoppage crisis 177
After the crisis 182
The future of the energy sector in Jordan 195

9. Renewable Energy in Tunisia: An Unjust Transition 201
Chafik Ben Rouine and Flavie Roche
The renewable energy law: a turning point in Tunisia's energy
transition 203
Impacts of the current energy transition: a fair shift for Tunisia's
development and people's rights? 209
Conclusion 214

10. The Moroccan Energy Sector: A Permanent Dependence 220
Jawad Moustakbal
The energy sector: from colonial control to neoliberal measures 220

Renewable energy in Morocco: a "green" neoliberalism   222
Energy governance in Morocco   224
Some avenues for a just energy transition in Morocco   229

PART III: FOSSIL CAPITALISM AND CHALLENGES TO A JUST
TRANSITION

**11. A Transition to Where? The Gulf Arab States and the New
"East-East" Axis of World Oil**   235
Adam Hanieh
From the Seven Sisters to OPEC   237
China, world oil and the Gulf's political economy   239
Refining and petrochemicals   241
New "East-East" interdependencies   244
Confronting the climate emergency: Taking the Middle East
   seriously   246

**12. The Challenges of the Energy Transition in Fossil Fuel
Exporting Countries: The Case of Algeria**   252
Imane Boukhatem
The need for an energy transition in Algeria   252
Algeria's climate and energy policy   256
The renewable energy sector in Algeria   258
Challenges of and barriers to the energy transition in Algeria   260
Algeria's urgent need for a just energy transition   266
Conclusion   270

**13. Unjust transitions: The Gulf States' Role in the
"Sustainability Shift" in the Middle East and North Africa**   275
Christian Henderson
A green energy shift in the Gulf?   277
A new market   280
A region of inequality   283
The Gulf and the just transition   285

About the Contributors   291
Index   295

Tables and Figures

TABLES

5.1   Selected economic, social and demographic indicators shaping
      agriculture in North Africa   110
5.2   Selected practices of eco-regenerative agriculture in North Africa 123
5.3   Examples of initiatives supporting eco-regenerative agriculture
      in North Africa   125

FIGURES

6.1   A comparison between the five-year plan of the Government
      of Sudan and the one proposed by the World Bank   146
7.1   A comparison between Egypt and OECD countries in terms
      of electric power transmission and distribution losses
      (in percentage of output) – World Bank   161
7.2   Egypt's total energy supply by source in Tera Joule (TJ) –
      International Energy Agency (IEA)   162
7.3   Petroleum exports and imports (US $ billions)   163
7.4   Fuel subsidy and electricity subsidy (Egyptian pound billions) 164
7.5   Peak load and nominal capacity (1,000 megawatts)   167
7.6   Financial statements of EEHC (Egyptian pound billions)   167
8.1   Annual government losses during the gas stoppage period
      (US $ billions). The disruption of Egyptian gas flows to Jordan
      since 2011 led to the accumulation of losses for the state-owned
      NEPCO   180
8.2   The cost of producing electric energy from 2003 to 2018
      (US cents per kWh)   181
8.3   Energy mix from 2010 to 2020 (in percentage)   182
8.4   The cost of producing the kilowatt-hour (kWh) in US cents   182
8.5   The energy mix according to the Energy Strategy 2020-30   184
8.6   Average cost of various energy sources in US cents per kWh
      (2017-20)   185

8.7   Comparison between costs of direct bids for renewable energy
       in Jordan with the average cost of renewable energy globally
       (in US cents per kWh)   188
8.8   Comparison of the average total electric energy cost for 2021
       with solar energy costs of the suspended third and fourth phases
       (in US cents per kWh)   189
8.9   The total installed capacity of electric power generation projects
       from renewable energy sources until 2021 (in MW)   190
8.10   Government revenues from selling energy to major subscribers
       (US $ millions)   191
8.11   The amount of energy sold to major subscribers (gigawatt-hours
       GWh)   192
8.12   Renewable energy investments in Jordan 2015-20
       (US $ millions)   194
8.13   The volume of investments in the renewable energy sector
       (US $ millions)   195
9.1   Percentage of projects obtained by companies in tenders between
       2017 and 2019 under the concession regime according to
       nationality   211
9.2   Share (in percentage) in the total power (MW) of electricity
       production projects obtained through the authorisation regime
       by companies according to their nationality   211
9.3   Share (in percentage) in the total power (MW) of the electricity
       production projects obtained through the concession regime by
       companies according to their nationality   211
10.1   Distribution of electricity production by source of energy (in
       percentage)   226
10.2   Distribution of electricity production by type of producer
       (in percentage)   226
10.3   Typical financial set-up of Masen-led projects   228
12.1   Algeria's primary energy production in terms of quantity
       (in percentage)   257
12.2   Algeria's final energy consumption in terms of quantity
       (in percentage)   257
12.3   Algeria's primary energy exports in terms of quantity
       (in percentage)   258

Acknowledgements

This book started long before we realized it would eventually become a book. It is the product of years of reflection and work. During this time, we accumulated a great intellectual and personal debt to innumerable friends, comrades and colleagues who, through their encouragement and sustained critique, enriched this book, sharpened its arguments and made its content much clearer. It is not possible to mention all of them here, but we will mention some.

First of all, we thank our wonderful colleagues at the Transnational Institute (TNI). In particular, Jenny Franco, Lyda Fernanda Forero, Daniel Chavez, Zoe Brent, Lavinia Steinfort, Pietje Vervest, Mads Barbesgaard, Sylvia Kay and Carsten Pedersen have been a constant source of inspiration and support. Our interactions and exchanges with them have been very fruitful and contributed greatly to the book's conception and development.

In addition, we are especially grateful to each of the contributing authors, the translators (Meriam Mabrouk and Nellie Epinat) and the various reviewers, without whom this book would not exist. Special thanks go to Thomas Claes from the Friedrich Ebert Stiftung – MENA office, who believed in this project from the start and has supported it generously. We are also immensely thankful to Ashley Inglis, who has been a brilliant and meticulous copyeditor and who made improvements to the text and its flow beyond what we expected. We owe a deep debt of gratitude, too, to Ouafa Haddioui, for her help in bringing the manuscript to its present shape.

Finally, thanks to our editor at Pluto Press, David Shulman, who has put up with all sorts of delays, and who supported the idea of the book from the very beginning.

This book draws, and we hope builds, on the rich body of ongoing discussions and debates around the just and sustainable transitions we would like to see in the Arab region and beyond. We dedicate it to the countless inspiring comrades with whom our paths have crossed, from environmental defenders, trade unionists and activists to researchers, journalists and scholars, who continue to struggle in different ways to build a more just world. A special thought goes to Mohad Gasmi, an inspiring environmental defender and a tireless social justice activist, who is currently a political prisoner in Algeria. We dedicate this book to him and to the many others in the region who never cease fighting for justice and against oppression in all its forms.

Abbreviations

ADNOC: Abu Dhabi National Oil Company
AFD: Agence Française de Développement (French Agency for
　　Development)
AfDB: African Development Bank
ANCC: National Agency for Climate Change
ANME: National Agency for Energy Management
APMM: Association pour la Promotion de la Médiation au Maroc
BNC: Palestinian National Committee of the Boycott, Divestment and
　　Sanctions movement
BOT: build-operate-transfer
BRI: Belt and Road Initiative
CDER: Centre for the Development of Renewable Energy
CEGCO: Central Electricity Generation Company
CEO: Corporate Europe Observatory
CO_2: carbon dioxide
COP: United Nation Climate Conference of the Parties
CSP: concentrated solar power
CSPRON: Committee for the Protection of Natural Resources in Western
　　Sahara
CSR: corporate social responsibility
DCFTA: Deep and Comprehensive Free Trade Agreement
DIC: Dubai International Capital
Dii: Desertec Industrial Initiative
DRC: Democratic Republic of Congo
ECT: Energy Charter Treaty
EDCO: Electricity Distribution Company
EEHC: Egyptian Electricity Holding Company
EEM: Énergie Électrique du Maroc
EETC: Egyptian Electricity Transmission Company
EgyptERA: Egyptian Electric Utility and Consumer Protection Regulatory
　　Agency
EIB: European Investment Bank
EMRC: Energy and Minerals Regulatory Commission

Enefit: Estonian National Energy Company
EPC-F: Engineering, Procurement, Construction and Finance
ESC: Economic and Social Council
ESMAP: Energy Sector Management Assistance Programme
ETAP: Tunisian Enterprise of Petroleum Activities
EU: European Union
FAO: Food and Agriculture Organization
FPI: food price index
FGEG: General Federation of Electricity and Gas
GDP: gross domestic product
GHG: greenhouse gas
GIZ: German Agency for International Cooperation
GW: gigawatt
IFIs: international financial institutions
ILO: International Labour Organization
IMF: International Monetary Fund
IPCC: Intergovernmental Panel on Climate Change
IPO: initial public offering
IPPs: independent power producers
IRENA: International Renewable Energy Agency
ISDS: Investor–State Dispute Settlement
JNF: Jewish National Fund
JREEEF: Jordan Renewable Energy and Energy Efficiency Fund
JV: joint venture
KEC: Kingdom for Energy Investments Company
KfW: Kreditanstalt für Wiederaufbau
kWh: kilowatt hour
LNG: liquefied natural gas
Lydec: Lyonnaise des Eaux Casablanca
Masen: Moroccan Agency for Sustainable Energy
MENA: Middle East and North Africa
MW: Megawatt
NATO: North Atlantic Treaty Organization
NDC: nationally determined contribution
NEPCO: National Electric Power Company
NGOs: non-governmental organizations
NMGP: Nigeria-Morocco Gas Pipeline
NOCs: national oil companies
OCP: Office Chérifien des Phosphates

ONE: Office National de l'Électricité
OPEC: Organization of Petroleum Exporting Countries
PAM: Authenticity and Modernity Party
PMA: Petroleum and Mines Authority
PNEREE: the National Programme for the Development of Renewable
    Energy and Energy Efficiency
POLISARIO: Popular Front for the Liberation of Saguia El Hamra and Río
    de Oro
PPA: power purchase agreements
PPPs: public–private partnerships
PPS: Parti du Progrès et du Socialisme
PST: Plan Solaire Tunisien
PV: photovoltaic
SACE: Italian Export Credit Agency
SADR: Sahrawi Arab Democratic Republic
SAMIR: Société Anonyme Marocaine de l'Industrie du Raffinage
SAP: Structural Adjustment Programme
SHAEMS: Société Algérienne des Énergies Renouvelables
SMD: Société Marocaine de Distribution d'eau, de gaz et d'électricité
SMI: Société Métallurgique d'Imiter
Sodea: Société de développement agricole
SOEs: state owned enterprises
Sogeta: Société de gestion des terres agricoles
STEG: Tunisian Company of Electricity and Gas
TWh: terawatt hours
UAE: United Arab Emirates
UGTT: General Union of Tunisian Workers
UMT: Union Marocaine des Travailleurs
UNFCCC: United Nations Framework Convention on Climate Change
UNFP: Union Nationale des Forces Populaires
USAID: United States Agency for International Development
VAT: value-added tax

Introduction
Just in Time – The Urgent Need for a Just Transition in the Arab Region

Hamza Hamouchene and Katie Sandwell

The reality of climate breakdown is already visible in the Arab region,[1] undermining the ecological and socioeconomic basis of life. Countries such as Algeria, Tunisia, Morocco, Saudi Arabia, Iraq, Jordan and Egypt are experiencing recurrent severe heat waves and prolonged droughts, with catastrophic impacts on agriculture and small-scale farmers.[2] Ranked as one of the world's five most vulnerable nations to climate change and desertification, Iraq was hit in 2022 by many sandstorms that shut down much of the country, with thousands of people admitted to hospital because of respiratory problems. The country's environment ministry has warned that over the next two decades Iraq could endure an average of 272 days of sandstorms a year, rising to above 300 by 2050.[3] In the summer of 2021, Algeria was struck by unprecedented and devastating wildfires; Kuwait experienced a suffocating heat wave, registering the highest temperature on Earth that year, at well over 50°C; and the United Arab Emirates (UAE), Yemen, Oman, Syria, Iraq and Egypt all experienced devastating floods, while southern Morocco struggled with terrible droughts for the third year in a row. In the years ahead, the Intergovernmental Panel on Climate Change (IPCC) projects that the Mediterranean and Gulf regions will see an intensification of extreme weather events, such as wildfires and flooding, and further increases in aridity and droughts.[4]

"It's now or never, if we want to limit global warming to 1.5°C." That's the warning from the IPCC working group behind the 2022 comprehensive review of climate science. The review warns that world temperature is set to rise by 1.5°C within the next two decades and that only the most drastic cuts in carbon emissions, starting immediately, can prevent an environmental and climatic disaster. Since these reviews are conducted every six to seven years, this can be seen as the last warning from the IPCC before the world is set irrevocably on a path to climate breakdown, with terrifying consequences. As United Nations Secretary-General António Guterres declared when the report was released: "In concrete terms, [this level of global heating] means major cities under water, unprecedented heatwaves, terrifying storms, widespread water shortages, and the extinction of one million species of plants and animals."

The impacts of these changes are felt disproportionately by marginalized people, including small-scale farmers, agro-pastoralists, agricultural labourers and fisherfolk. Already, people in the Arab world are being forced off their lands by stronger and more frequent droughts and winter storms, expanding deserts and rising sea levels.[5] Crops are failing and water supplies are dwindling, deeply impacting food production in a region that is chronically dependent on food imports.[6] As the effects of climate change are increasingly felt, there is a huge pressure on already scarce water supplies due to changes in rainfall and seawater intrusion into groundwater reserves, as well as groundwater overuse. According to an article in the *Lancet*, this will place most Arab countries under the absolute water-poverty level of 500 m^3 per person per year by 2050.[7]

Climate scientists are predicting that the climate in large parts of the Middle East and North Africa (MENA) could change in such a manner that the very survival of its inhabitants will be in jeopardy.[8] In North Africa, for example, those whose lives will be changed the most by climate change include small-scale farmers in the Nile Delta and rural areas in Morocco and Tunisia, the fisherfolk of Jerba and Kerkennah in Tunisia, the inhabitants of In Salah in Algeria, the Saharawi refugees in the Tindouf camps in Algeria, and the millions living in informal settlements in Cairo, Khartoum, Tunis and Casablanca. Elsewhere in the Arab region, small-scale farmers and fisherfolk in occupied Palestine, internally displaced people and refugees in Iraq, Syria, Lebanon, Yemen and Jordan, and hyper-exploited migrant workers in the UAE and Qatar will face the violence of the climate crisis with little protection as they are frequently housed in squalid conditions, denied routine medical care and face malnutrition.

The climate crisis was not an inevitable fact: it has been, and continues to be, driven by the choice to keep burning fossil fuels – a choice made predominantly by corporations and governments in the northern hemisphere, together with national ruling classes, including in the Arab region. Energy and climate plans in that part of the world are shaped by authoritarian regimes and their backers in Riyadh, Brussels and Washington, DC. Rich local elites collaborate with multinational corporations and international financial institutions, such as the World Bank, the International Monetary

Fund (IMF) and the European Bank for Reconstruction and Development (EBRD). Despite all of their promises, the actions of these institutions show that they are enemies of climate justice and of humanity's very survival.

Every year, the world's political leaders, advisers, media and corporate lobbyists gather for another United Nations Conference of the Parties (COP) on the issue of climate change. But despite the threat facing the planet, governments continue to allow carbon emissions to rise and the crisis to escalate. After three decades of what the Swedish environmental activist Greta Thunberg has called "blah blah blah", it has become evident that these climate talks are bankrupt and are failing. They have been hijacked by corporate power and private interests that promote profit-making false solutions, like carbon trading, so-called "net-zero" and "nature-based solutions", instead of forcing industrialized nations and multinationals to reduce carbon emissions and leave fossil fuels in the ground.[9]

With COP28 being held in Dubai, UAE, in 2023, the Arab region will have hosted the climate talks five times since their inception in 1995: COP7 (2001) and COP22 (2016) in Marrakech, Morocco; COP18 (2012) in Doha, Qatar; and COP27 (2022) in Sharm el-Sheikh, Egypt. In recent years, and especially since the 2015 Paris Agreement walked back from the (already grossly inadequate) binding targets established in the Kyoto Accord to allow countries to independently determine their own emission-reduction targets, scepticism about the ability of the United Nations Framework Convention on Climate Change (UNFCCC) to tackle the most urgent challenge facing humanity has grown. COPs attract massive media attention but tend not to achieve major breakthroughs. COP27, held in Sharm el-Sheikh in 2022, achieved an agreement on Payment for Loss and Damage that has been lauded by some as an important step in making richer countries accountable for the damage caused by climate change in the global South.[10] However, as the agreement lacks clear funding and enforcement mechanisms, critics worry it will meet with the same fate as the broken promise (first made in COP15 in Copenhagen in 2009) to provide $100 billion in climate finance per year by 2020. That promise was never fully realised, with assistance often taking the form of interest-bearing loans instead.[11] As for COP28, the UAE's appointment of Sultan al-Jaber, Chief Executive Officer of the Abu Dhabi National Oil Company, to preside over the talks seems to many activists and observers to symbolize the deep commitment to continued oil extraction, regardless of the cost, which has characterized negotiations to date.

Middle Eastern and North African states, with their national oil and gas companies, alongside the big oil majors, are doing their best to maintain their operations, and even expand and profit from the remaining fossil fuels they possess. President Abdelfattah al-Sisi's Egypt is aspiring to become a major energy hub in the region, exporting its surplus electricity and mobilizing various energy sources, such as offshore gas, oil, renewable energies and hydrogen, to satisfy the European Union's (EU's) energy needs. This is, of course, inextricable from the ongoing efforts at political and economic normalization with the colonial state of Israel. The Algerian regime, for its part, is also benefiting from the oil price bonanza and taking advantage of the EU's scramble for alternatives to Russian gas in order to expand its fossil fuel operations and plans. The Gulf countries, such as Saudi Arabia, UAE and Qatar, are no different. The ruling classes across the region have been talking about the "after oil" era for decades, and successive governments have paid lip service to the transition to renewable energies for years without taking concrete action, apart from some grandiose and unrealistic plans and projects, such as the proposed, and controversial, futuristic megacity of Neom in Saudi Arabia. For these ruling classes, the iterations of the COP process represent a golden opportunity to advance their greenwashing agenda, as well as their efforts to attract and capture funds and finances for various energy projects and purportedly "green" plans.

Egypt's hosting of the 2022 COP was controversial in view of its government's record of repression and its efforts to prevent access to the summit by environmental groups and climate activists. In fact, the Sharm el-Sheikh COP27 was one of the most exclusionary conferences in history, with a substantially diminished space for the activism, dissidence, discussions, debates, new connections, networking, collective strategies and mobilizations needed to generate pressure on global decision-makers to deliver on their promises and promote real solutions to the unfolding climate emergency. The choice of Egypt as the host in 2022, and of UAE as the host for 2023 is not innocent and is a clear indication that the COP process as a whole is becoming more undemocratic and exclusionary. Moreover, the context of the intensification of geopolitical rivalries unleashed by the war in Ukraine is not amenable to cooperation between major powers and provides yet another pretext for continuing the global addiction to fossil fuels. Indeed, it could be the final nail in the coffin of global climate talks.

Humanity's survival depends on both leaving fossil fuels in the ground and adapting to the already changing climate, while moving towards renewable energies, sustainable levels of energy use and other social transformations. Billions will be spent on trying to adapt – finding new water sources, restructuring agriculture and changing the crops that are grown, building sea walls to keep the saltwater out, changing the shape and style of cities – and on trying to shift to green sources of energy by building the required infrastructure and investing in green jobs and technology. But whose interest will this adaptation and energy transition serve, and who will be expected to bear the heaviest costs of the climate crisis and of the responses to it?

The same greedy and authoritarian power structures that have contributed to climate change are now shaping the response to it. Their main goal is to protect private interests and to make even greater profits. While the international financial institutions, such as the World Bank and the IMF, and governments in the northern hemisphere and their agencies, such as the United States Agency for International Development (USAID), the EU and the German Agency for International Cooperation (GIZ), are all now articulating the need for a climate transition, their vision is of a capitalist, and often corporate-led transition, not one led by and for working people. While the voices of civil society organizations and social movements in the Arab region are largely not heard when it comes to the implications of this transition and the need for just and democratic alternatives, the aforementioned institutions and governments speak loudly, organizing events and publishing reports in all the countries of the Arab region. These actors don't shy away from highlighting the dangers of a warmer world, and even argue for urgent action, including using more renewable energy and developing adaptation plans. However, their analysis of climate change and the necessary transition remains limited – and even dangerous, as it threatens to reproduce the patterns of dispossession and resource plunder that characterize the prevailing fossil fuel regime.

The vision of the future that is pushed by these powerful actors is one where economies are subjugated to private profit, including through further privatization of water, land, resources, energy – and even the atmosphere. The latest stage in this development includes the public–private partnerships (PPPs) now being implemented in every sector in the Arab region, including in renewable energies. The drive towards the privatization of energy and corporate control of the energy transition is global and is not unique to the region, but the dynamic is quite advanced here and has so far been met with only limited resistance. Morocco is already advancing along this path, and so is Tunisia. In Tunisia, a major push is underway to expand the privatization of the country's renewable energy sector and to give huge incentives to foreign investors to produce green energy in the country, including for export. Tunisian law – modified in 2019 – even allows for the use of agricultural land for renewable projects in a country that suffers from acute food dependency,[12] a dependency that was starkly revealed during the COVID-19 pandemic and that is evident once again at the time of writing, as war rages in Ukraine.

As developments like this take place across the region, they highlight the importance of asking: "Energy for what and for whom?" and "Who is the energy transition intended to serve?" The supposedly "green economy" and the broader mainstream vision of so-called "sustainable development" are being presented by international financial institutions, corporations and governments as a new paradigm. But in reality they are merely an extension of the existing logics of capital accumulation, commodification and financialization, including of the natural world.

NORTH AFRICA AND WEST ASIA AS A KEY NODE IN GLOBAL FOSSIL CAPITALISM

North Africa and West Asia/the MENA region must be understood within the context of the larger capitalist world market, which is characterized by the concurrent rise of new zones of accumulation and growth in some parts of the world and the relative decline of long-established centres of power in North America and Europe. Not only does the region today play a major role in mediating new global networks of trade, logistics, infrastructure and finance,[13] but it is also a key nodal point in the global fossil fuel regime and plays an integral role in keeping fossil capitalism intact through its oil and gas supplies. In fact, the region remains the central axis of world hydrocarbon markets, with a total share of global oil production standing at around 35 percent in 2021.[14] Historically, these supplies fuelled a major shift in the global energy regime during the mid-twentieth century, with oil and gas replacing coal as the primary fuel for global transportation, manufacturing and industrial production.[15] More recently, the resources of the Middle East have been essential in regard to meeting the increased demand for oil and gas fuelled by the rise of China, heralding a key structural shift in the global political economy over the last two decades based on closer ties between the Middle East and East Asia. All of this has positioned Middle East oil producers as indisputable protagonists in climate change debates and any future transition away from fossil fuels.[16]

The historical, political and geophysical realities of the Arab region mean that both the effects of, and the solutions to, the climate crisis there will be distinct from those in other contexts. From the mid-nineteenth century to the second half of the twentieth century, the region was forcibly integrated into the global capitalist economy in a subordinate position: colonial/imperial powers influenced or forced the countries of the region to structure their economies around the extraction and export of resources – usually provided cheaply and in raw form – coupled with the import of high-value industrial goods. The result was a large-scale transfer of wealth to the imperial centres at the expense of local development and ecosystems.[17] The persistence till today of such unequal and asymmetric relations (which some call unequal economic/ecological exchange, or ecological imperialism)[18] preserves the role of Arab countries as exporters of natural resources, such as oil and gas, and primary commodities that are heavily dependent on water and land, such as monoculture cash crops. This entrenches an outward-looking extractivist economy, thereby exacerbating food dependency and the ecological crisis, and it also maintains relations of imperialist domination and neocolonial hierarchies.[19] However, it is important to avoid the tendency to see the region as an undifferentiated whole, but rather to be aware of its inherent unevenness and deep inequalities. A closer look reveals the underlying role of the Gulf states[20] in this configuration, as a semi-periphery – or even as a sub-imperialist – force.[21] Not only are the Gulf states significantly richer than their other Arab neighbours, but they also participate in the capture and syphoning off of surplus value at the regional level, reproducing core–periphery-like relations of extraction, marginalization and accumulation by dispossession. In this regard, the work of Adam Hanieh (one of the contributors to this book) is enlightening in terms of how economic liberalization in the Middle East over recent decades (through various structural adjustment packages in the 1990s and 2000s) has been closely bound up with the internationalization of Gulf capital throughout the wider region.[22] Gulf capitalists now dominate key economic sectors of many neighbouring countries, including real estate and urban development, agribusiness, telecommunications, retail, logistics, and banking and finance.

Crucial questions, therefore, need to be raised when talking about addressing climate change and transitioning towards renewable energies in the region. What would a just response to climate change look like here? Would it mean the freedom to move across, and to open the borders within, the region, and to open the borders with Europe? Would it mean the payment of climate debt, restitution, and redistribution – by Western governments, multinational corporations, and the rich local elites nationally and regionally? Would it mean a radical break with the capitalist system? What should happen to the fossil fuel resources in the region that are currently being extracted by national companies and foreign corporations? Who should control and own the region's renewable energy? What does adapting to a changing climate mean here, and who will shape and benefit from these adaptations? And who are the key agents and actors that will fight for meaningful change and radical transformation?

While governments all over the world are beginning to take climate change seriously, they often see it through a "climate security" lens[23] – bolstering defences against rising sea levels and extreme weather events, and too often also shoring up their defences against the "threat" of climate refugees and activists, and against renegotiations of global power. The securitization and militarization of the climate response in the Middle East is itself a potential challenge and threat to the climate justice agenda, given that the region plays a pivotal role in the global development of coercive technologies, techniques and doctrines. This role extends beyond the region's status as the world's single largest export market for weapons and military hardware, and includes its crucial involvement in the testing of new security technologies, including emerging forms of surveillance and population control. Several authors have drawn attention to the intricate international networks that support the region's arms trade and surveillance industry, including the flow of "war on terror" logics, military technologies, personnel, training manuals, cross-border operations, police forces, and private military and security companies.[24] All of these factors combine to make the Middle East an important hub in the global spread of new norms of militarism and securitization. Moreover, the dynamics of warfare in the region itself are also significantly shaped by these global ties, as are the various ways that militaries have been assimilated into political and economic systems on both a national and regional level.[25]

It is of the utmost importance and urgency to start looking at the issue of climate change through a justice lens rather than a security one. Seeing the future through the frame of "security" subordinates our struggles to a conceptual and imaginative framework that ultimately re-empowers the state's repressive power and securitizes and militarizes the response. More tanks and guns, higher walls and more militarized borders will not solve the climate crisis. At best, they will allow the rich to survive in comfort while the rest of the world pays the price for climate inaction. We need to break with the system of capitalist exploitation of people and the planet that has given rise to the climate crisis, not arm and entrench it.

THE COLONIAL GAZE AND ENVIRONMENTAL ORIENTALISM

Just as economic subjugation and imperialist domination have undermined the political and economic autonomy of the Arab region, knowledge production about, and representations of, the Arab peoples and their environments have equally been used by colonial powers to legitimize their colonial project and imperial goals. Such strategies of domination continue today, as countries in the region are being recast (once again) as objects of development (sustainable or otherwise), echoing the colonial *mission civilisatrice* (civilizing mission).

Rejecting the theses of French colonial historians regarding the supposed "historical backwardness" and "the state of being frozen in time" of the Berbers/Amazigh, Arabs and Muslims, and of their civilizations, the Moroccan historian and philosopher Abdallah Laroui argues that the reality of the indigenous populations in the Maghreb, or Arab west, in its multiple facets (political, economic, cultural, environmental, etc.), and at various historical times, has been deliberately misrepresented in order to advance a false and essentialist narrative that serves a colonial agenda of subjugation, domination and expansion.[26] The American geographer Diana K. Davis concurs and argues that Anglo-European environmental imaginaries in the nineteenth century represented the environment in the Arab world most often as "alien, exotic, fantastic or abnormal, and frequently as degraded in some way". She aptly uses Edward Said's concept of orientalism[27] as a framework to interpret early Western representations of the Middle Eastern and North African environment as displaying a form of "environmental orientalism". This representation of the environment was narrated by those who became the imperial powers, primarily Britain and France, as a "strange and defective" one, as compared to Europe's "normal and productive" environment. This implied the need for some kind of intervention "to improve, restore, normalize and repair" it.[28]

This deceptive representation of presumed environmental degradation and ecological disaster was used by colonial authorities to justify all sorts of dispossession, as well as policies designed to control the populations of the region and their environments. In North Africa (and later in the Mashriq, or Arab east), the French constructed an environmental narrative of degradation in order to implement "dramatic economic, social, political and environmental changes".[29] According to this perspective, the natives and their environments warranted the "blessings" of the *mission civilisatrice* and required the attention of the white man.

Narratives are always the product of their historical moment and are never innocent, and therefore one always needs to ask: in whose benefit do knowledge production, representations and narratives work? One glaring contemporary example is the current representation of the North African Sahara as a vast, empty, dead land that is sparsely populated and thus constituting a golden opportunity to provide Europeans with cheap energy so they can continue their extravagant consumerist lifestyle and excessive energy consumption. This false narrative ignores issues of ownership and sovereignty, while masking ongoing global hegemonic relations that facilitate the draining of resources, privatization of common land and resources, and dispossession of communities. As in many places where working people's lives and livelihoods are invisible or "illegible"[30] to colonizing states, "there is no vacant land" in North Africa.[31] Even when sparsely populated, traditional landscapes and territories are embedded in cultures and communities, and people's rights and sovereignty must be respected in any socio-ecological transformation.

It is crucial to analyse the mechanisms by which "the other" is dehumanized and how the power of representing and constructing imaginaries about them (and their environments) is used to entrench structures of power, domination and dispossession. In this regard, the process Said describes in *Orientalism* of "disregarding, essentializing, [and] denuding the humanity" of another culture, people or geographical region continues today to be employed to justify violence towards "the other" and towards nature. This violence takes the form of displacing populations, grabbing land and resources, making people pay for the social and environmental costs of extractive and renewable projects, bombing, massacring, letting people drown in the Mediterranean and destroying the earth in the name of progress. Naomi Klein put this eloquently in her 2016 Edward Said Lecture,[32] in which she described a white-supremacist/racist culture that is increasingly evident in parts of Europe and the United States: "A culture that places so little value on black and brown lives, that it is willing to let human beings disappear beneath the waves, or set themselves on fire in detention centres, will also be willing to let the countries where black and brown people live disappear beneath the waves, or desiccate in the arid heat." Such a "culture" won't blink an eye when it places catastrophic socio-environmental costs onto the poor in these countries.

Resisting and dismantling the orientalist and neocolonial environmental narrative about the Arab region will both enable and require building visions of collective climate action, social justice, and socio-ecological transformation that are rooted in the experiences, analyses and emancipatory visions of the African and Arab regions and beyond.

WHAT IS "JUST TRANSITION"?

As outlined above, discussions of climate action are often narrow and technocratic, neoliberal and market-based, top-down and implicitly focused on preserving the structures of racist, imperialist and patriarchal capitalism. Against this backdrop, which at best largely ignores questions of power and justice, the concept of "just transition" has emerged as a framework that places justice at the centre of the discussion. This approach recognizes that, in the words of Eduardo Galeano, "the rights of human beings and the rights of nature are two names for the same dignity".[33] Where did the idea of just transition come from, and what might it offer to the project of developing grounded, bottom-up and non-imperialist visions of emancipation and climate action in the context of the Arab region?

The origin of the concept of just transition is usually traced back to the US in the 1970s, when pathbreaking alliances between labour unions and environmental justice and indigenous movements emerged to fight for environmental justice in the context of polluting industries. In the face of environmental regulations which were being implemented for the first time or tightened during that decade, companies claimed that policies to protect the environment would require them to lay off workers. Unions and communities rallied against this attempt to divide and conquer, arguing that workers and communities – especially black, brown and indigenous communities, who were (and remain) the most impacted by polluting industries – had a shared interest in a liveable environment, and in decent, safe and fairly paid work.

Over the decades that followed, the concept of just transition was taken up, explored and elaborated by a range of different movements, initially in the US and Canada, but subsequently also around the world, and especially in South America and South Africa. Labour and environmental justice movements, working with indigenous nations, women's movements, youth, students and other groups, have built coalitions and shared visions of what a just transition would look like: transformative solutions to the climate crisis that tackle its underlying causes, and that put human rights, ecological regeneration and people's sovereignty at the centre.

As the framework gained in popularity, corporations and governments increasingly tried to advance their own visions of just transition, but these lacked class analysis and denied the need for radical transformation. With the inclusion of the term "just transition" in the preamble of the Paris Agreement – a hard-won victory for global labour and climate justice movements – this cooptation intensified. Today, just transition is not a single concept but a field of contestation, a space where struggles about what responses to the climate crisis are possible and necessary are playing out. The term does not automatically imply progressive or emancipatory politics, and many actors use it to describe and defend proposals which are basically business as usual or intensified green extractivism. Nonetheless, far more than rhetoric about "sustainable development" or the "green economy", the idea of just transition still provides a space that movements can use to insist on the primacy of justice in all climate solutions. Despite attempts at cooptation, the centrality of *justice* in the term itself is an important strength of the concept of just transition.

Just transition proposals being advanced by progressive social movements are driven by a conviction that the people who bear the heaviest costs of the current system should not be the ones who pay the costs of a transition to a sustainable or regenerative society, and at the same time, should be the leading actors in shaping such a transition. Different movement dynamics have explored different dimensions of this, seeking to better understand the costs of the current system, the possibilities for transformation, and the likely costs of proposed alternatives. From feminist and indigenous perspectives to regional and national programmes, movements are advancing their own definitions of both *justice* and *transition* in their diverse contexts.[34]

A meeting of environmental justice and labour movements from three continents that took place in Amsterdam in 2019 (which, incidentally, laid the foundations for the present book) sought to identify key characteristics of just transition: (1) just transition looks different in different places; (2) just transition is a class issue; (3) just transition is a gender issue; (4) just transition is an anti-racist framework; (5) just transition is about more than just climate; and (6) just transition is about democracy.[35]

While not claiming to be an exhaustive definition or final set of permanent principles, this analysis lays out the contours of a position that recognizes that discussions of just transition must respond to the reality of unequal development caused by imperialism and colonialism; that just transition must include radical shifts that increase the power of working people in all their diversity (see below) and reduce the power of capital and governing elites; that environmental issues cannot be addressed without addressing the racist, sexist and other oppressive structures of the capitalist economy; that the environmental crisis is much broader than just the climate crisis, encompassing loss of habitats and biodiversity, and a fundamental breakdown in human relationships with the "natural world"; and that a just transition cannot be achieved without transformations in political and economic power towards greater democratization.

A second important strength of just transition is its history as a tool or framework for unifying diverse movements by overcoming differences and potential divisions. As mentioned above, the term emerged originally as a response to the "divide and conquer" tactics of businesses resisting environmental regulation. These tactics continue to be used as corporations push for policies that protect profits regardless of the costs for communities, workers and the planet; and that pit different regions and different kinds of working people against each other. International climate justice movements, national and regional coalitions, and local alliances around the world recognize that virtually all of us benefit from a liveable and flourishing environment, and suffer when wealth and power are concentrated in the hands of a tiny elite who count on being able to protect themselves from the worst effects of the climate crisis. Yet building shared campaigns and common visions, cultivating trust and solidarity, and developing and fighting for shared proposals is slow and politically challenging work – but necessary, as any shortcuts that try to side-step this process are likely to compromise the justice that must be at the heart of any just transition. The concept of just transition, and the growing body of experiences of working and campaigning with it around the globe, can help to provide some guides on this difficult path.

The concept of just transition has been shaped partially by labour movements, so the question of decent work remains central to many articulated proposals for just transition. This is of particular importance for the MENA region, which the International Trade Union Confederation has dubbed the worst in the world for workers' rights, with systematic violations across the area.[36] Millions of non-citizen migrant workers (from both in and outside of the region) are also located there. In the Gulf Arab states, for example, more than half of the labour force is made up of non-citizens, with more migrants working in those states than in any other region in the global South.[37] At the same time, across the Arab world, youth unemployment is almost twice the global average[38] and in North Africa about two thirds of workers are employed in the informal sector.[39]

In this context, what does it mean to talk about decent work, and how should we understand working people? Inspired by the Guyanese historian and political activist Walter Rodney's political mobilizations of "working people", Tanzanian scholar Issa Shivji has argued that "under neoliberalism, primitive accumulation assumes new forms and becomes generalized in almost all sectors of the economy, including the so-called informal sector. The producer self-exploits him or herself just to survive while subsidising capital."[40] Following from this, he argues that we need a new understanding of working people that recognizes the common exploitation faced by organized industrial workers; informal, precarious, temporary or migrant workers; unpaid or underpaid workers (usually women) doing domestic, care and social reproductive work; and nominally self-employed or independent small-scale peasant farmers, pastoralists and fisherfolk working directly for their own survival.

Today, the vast majority of humanity, regardless of the kind of work they do, are giving up some part of their essential daily consumption, their human rights, or their ability to live a dignified life in order to keep propping up the super-profits of transnational corporations. Whether this is the case because their food, health, energy and care systems have been privatized, putting the full burden of care on the family unit; because they have lost or are at risk of losing access to their traditional lands, territories or fishing grounds; or because they are unable to find work and must struggle to make ends meet in an informal economy where they have no political means to demand a living wage, the effects are the same. It is no coincidence that this precarious and exploited majority is also the group most at risk from climate change, and least able to protect themselves from its effects.

Taken together with the concept of just transition, we can use this definition of "working people" when developing our vision of who should be in control of the energy transition, and the response to the climate crisis more generally. Together, these concepts provide a basis for asking what justice in climate action would look like, and what concrete steps we need to take to achieve it in different contexts. It is these questions that this book seeks to answer. It does so by drawing together the diverse perspectives of many different kinds of working people across the Arab region and by illuminating some of the possibilities for building alliances and coalitions.

WHY THIS BOOK? WHY NOW?

Much writing on climate change, the ecological crisis and the energy transition in the Arab region is dominated by, or reproduces the perspectives of, international neoliberal institutions. The analyses put forward by these institutions are biased and do not include questions of class, race, gender, justice, power or colonial history. Their proposed solutions and prescriptions are market-based, top-down, and do not address the root causes of the climate, ecological, food and energy crises. The knowledge produced by such institutions is profoundly disempowering and overlooks questions of oppression and resistance, focusing largely on the advice of "experts", to the exclusion of voices "from below".

This book is one attempt to remedy that. It is a collection of essays by writers mainly from the MENA region focusing on dimensions of the energy transition and how to make this process equitable and just. The chapters cover a wide range of countries, from Morocco, Western Sahara, Algeria and Tunisia to Egypt and Sudan, and from Jordan and Palestine to the UAE, Saudi Arabia and Qatar. The book also includes contributions with a regional perspective: on the agricultural transitions and the neocolonial scramble for various energy sources (including green hydrogen) in North Africa, as well as the challenges and contradictions of the energy transition in the Gulf.

Until now, no widely available collection of writings by critical North African and Middle Eastern researchers or activists on a just energy transition has been published in Arabic, English or French – either as a book or an online resource. While important books on various green new deals[41] and on the necessary energy transition are gaining increasing attention, writings by critical authors from the global South remain marginal, including in the Arab region. Given the utmost importance of challenging eurocentrism and the need for a class-conscious approach to climate change mitigation and adaptation (including an urgent move towards renewable energies), as well as the importance of critically reflecting on, and challenging, the role the governments and elites of the region play in the current fossil energy regime, we think there is a massive gap to fill.

This book adopts an explicit *justice* lens. It aims to expose policies and practices that protect political elites, multinational corporations and authoritarian/military regimes. It seeks to recognize and contribute to processes of knowledge production and resistance against extractivism, land/resource grabbing and neocolonial agendas, and aims to support a move towards transformative sustainability from the ground up, based on the assumption that this offers the greatest potential for dealing with environmental, food, energy and social crises.

To our knowledge, this is the first collection of essays to directly tackle the question of the energy transition in the Arab region using a justice lens and a just transition framework. The book strives to make an important contribution to evolving global discussions on climate action, and just transition more generally, by interrogating what these processes will mean in the unique circumstances of different countries in the Arab region that are characterized by (a) authoritarian regimes, (b) oil export-dependent economies, (c) histories of colonization and imperialism, and (d) potentially immense green energy resources. Because a just transition entails planetary transformation, and since the Arab region will be one critical locus of that change, we strongly believe that the relevance of the book is not only regional, but global. The book aims to do the following:

- To advance a deeper analysis of where we are now in terms of energy transition in the Arab region. This is important because having a better understanding of the current situation, the actors involved and the potential winners and losers, is crucial for any efforts to bring about a just transition;
- To increase the emphasis on structural critiques in *green* transition debates, by placing at the centre the voices of activists, scholars and writers from the Arab region;
- To highlight the urgency of the climate crisis in the Arab region and to push back against the entrenchment of extractivism and energy colonialism, emphasizing the need for holistic analyses and structural change;
- To counteract the dominant neoliberal/neocolonial discourse on the *green* transition that is promoted by various international actors in the region;
- To overcome the dominance of a *security* discourse. The book avoids demands framed around security, like climate security, food security or energy security. Instead, it promotes notions of justice, sovereignty and decoloniality;
- To support progressive forces/movements/grassroots groups in the Arab region to articulate a localized, democratic and public response to the urgently needed energy transition, incorporating political, economic, social, class and environmental analyses.

SUMMARY OF THE BOOK'S CHAPTERS

The book's chapters are divided into three sections:

Part I, titled Energy Colonialism, Unequal Exchange and Green Extractivism, focuses on the ongoing (neo)colonial dynamics of appropriating different natural resources (including fossil fuel and renewable energy), as well as the persistence of extractivist approaches and practices of plunder and the externalizing of socio-environmental costs and placing them on peripheral and oppressed populations.

In his chapter, Hamza Hamouchene shows how renewable energy engineering projects tend to present climate change as a problem that is common to the whole planet, without questioning the capitalist and productivist energy model, or the historical responsibilities of the industrialized West. As he argues, in the Maghreb this tends to translate into "green colonialism", rather than into the search for an energy transition that benefits working people. He takes as an example the new green hydrogen hype and argues that green hydrogen projects constitute neocolonial schemes of plunder and dispossession.

Joanna Allan, Hamza Lakhal and Mahmoud Lemaadel, in highlighting how extractivism operates today in the part of Western Sahara currently occupied by Morocco, focus principally on renewable energy developments. Morocco is widely celebrated on the international stage for its commitments to the so-called "green energy transition", but the writers offer a different story that emphasizes the voices of the Saharawi population, and they argue that current renewable energy projects in Western Sahara simply sustain and "greenwash" colonialism, undermining a just transition that could truly benefit local communities.

In her chapter, Manal Shqair shines a light on Arab eco-normalization with the State of Israel. She presents eco-normalization as the use of "environmentalism" to greenwash and normalize Israeli oppression, and the environmental injustices resulting from it in the Arab region and beyond. She investigates how eco-normalization undermines the Palestinian anti-colonial struggle and hampers a just agricultural and energy transition in Palestine, which is inextricably linked to the Palestinian struggle for self-determination. She introduces the concept of eco-*sumud* (eco-steadfastness) in the face of Israeli oppression, and its role in countering the greenwashing function of eco-normalization.

Karen Rignall shows how solar energy is embedded in a long history of extraction in Morocco and reveals some of the striking continuities between fossil fuel commodity chains and those of renewable energies in the country. These continuities raise questions about how to work towards a just transition not only in Morocco, but also in other countries around the world that are seeing a surge in renewable energy projects, often in areas with long histories of mining. She considers how to advocate for new forms of energy that do not reproduce the same economic and political inequalities inherent in carbon-fuelled capitalism.

Sakr El Nour, in his chapter on the needed just agricultural transition in North Africa, argues that countries in the region are subjected to unequal exchange with the global North, particularly the EU, through trade agreements that enable the North to benefit from North African agricultural products at preferential rates. He contends that North Africa needs to recast its agricultural, environmental, food and energy policies. He convincingly advocates for alternatives that are locally centred and that are able to flourish autonomously, independent of European interests.

Mohamed Salah and Razaz Basheir, in their contribution on the electricity crisis in Sudan, chart the evolution of the energy sector in the country since colonial times, and attribute Sudan's uneven development to policies from that era and to their continuation in the post-colonial period. They put forward a critique of hydro-electric projects in Sudan in terms of their socioeconomic and environmental costs, deepening of existing inequalities, and negative impacts on livelihoods. They also challenge the World Bank's agenda of liberalizing and privatizing the energy sector in the country and show how these plans would only pauperize more people and limit access to energy. This chapter provides a segue to the second section of the book.

Part II, titled Neoliberal Adjustments, Privatization of Energy and the Role of International Financial Institutions, puts a spotlight on neoliberal global political economic structures that facilitate the continued exploitation of peripheral ecologies and the scramble for resources, in the name of the energy transition. The chapters in this section document the various dynamics of privatization and liberalization of the energy sector, and the ensuing economic and social impacts of such policies, while highlighting the crucial role of international financial institutions, such as the World Bank and the IMF, in pushing such an agenda.

Mohamed Gad documents Egypt's response to massive power outages in 2014 in terms of the liberalization of electricity production and the move away from subsidizing electricity prices for a wide range of income groups. He debunks the World Bank's claim that the liberalization of electricity prices ended subsidies to the rich and redirected resources towards the poor. Instead, he shows how it paved the way for the entry of international finance, at the expense of the poorest – radically transforming a basic service into a commodity.

Asmaa Mohammad Amin delves into the various policies that have generated successive crises in the Jordanian energy sector. She shows how the disruption of gas supplies from Egypt between 2011 and 2013 revealed that such policies, starting with the privatization and liberalization agenda pushed by the World Bank and IMF, were not only short-sighted, but also not fit for purpose. She also challenges the celebratory view of Jordan as one of the leaders of renewable energy in the region and argues that beyond the shiny statistics lies the bleak fact that huge profits have been syphoned off by the private sector while the state has continued to register significant losses. This in turn has exacerbated the country's debt burden and increased its dependence on external lenders, at the expense of the most vulnerable in society.

In their contribution on Tunisia, Chafik Ben Rouine and Flavie Roche show how the country's energy transition plan relies heavily on privatization and foreign funding, while neglecting democratic decision-making, placing the country firmly in the global neoliberal scheme for the development of renewable energy. They argue that instead of chasing private profits, a just transition in Tunisia would give households and communities the means to produce their own electricity, which would reduce dependency and promote the development of local industry and the creation of decent jobs.

Jawad Moustakbal, in his chapter on the energy sector in Morocco, asks a number of very important questions: Who benefits from, who pays the price for, and who decides on Morocco's so-called energy transition? He argues that public-private partnerships guarantee high profits for private corporations, while the poor have to pay ever higher prices for energy. He asserts that there will be no just transition as long as Morocco's energy sector remains under the control of foreign transnational companies and a local ruling elite that is allowed to plunder the state and generate as much profit as it wishes.

Because a just transition will look different in different contexts, Part III, titled Fossil Capitalism and Challenges to a Just Transition, touches on the difficulties and contradictions of the energy transition in fossil fuel-exporting countries. The chapters in this final section reveal the extent of the challenges, regionally and globally, while warning about the dangers of a continued addiction to fossil fuels.

Adam Hanieh argues that the rise of the Gulf states needs to be understood in the light of the significant changes that have taken place in global capitalism over the last two decades. Key to this is a new hydrocarbon axis linking the oil and gas reserves of the Middle East with the production networks of China and Asia, which serve to locate the Gulf at the core of contemporary "fossil capitalism". For him, the character of any green transition, both in the Middle East and globally, will be significantly determined through the actions and policies of these states. He thus argues that without understanding the changes to the control and structure of the oil industry – and strategizing effectively around them – it will be impossible to build successful campaigns to halt and reverse the effects of anthropogenic climate change.

In her chapter on Algeria, Imane Boukhatem contends that the country faces a triple challenge in its energy sector: economic dependence on hydrocarbon revenues, growing domestic electricity demand, and long-term fossil fuel export agreements. She highlights the opportunities, challenges and potential injustices facing the green energy transition in Algeria, and argues that Algeria must rapidly transform its energy sector, with a core focus on social justice. She lists several socioeconomic, institutional and policy obstacles that need to be overcome to achieve a just transition in Algeria.

In his chapter, Christian Henderson challenges some reductionist assumptions about the Gulf put forward by various mainstream reports and analyses, which portray the Gulf states as being simply victims of climate change and facing demise due to the potential decline in demand for oil and gas. According to him, rather than being powerless producers and passive actors in the politics of climate change, the Gulf countries are working to ensure they remain at the centre of the global energy regime. This entails the formulation of a dualistic policy: one that allows them to benefit from both fossil fuels and renewable energies.

IN GUISE OF A CONCLUSION

Through these essays, the authors aim to initiate a deeper discussion of what just transition means in the context of the Arab region. The dynamics are complex and obviously differ across the countries and sub-regions, yet many shared challenges and questions also emerge from these explorations: Whose needs and rights should be prioritized in an energy transition? What model of energy production and extraction can deliver energy to all working people in the region? How are countries in the global North and international financial institutions pushing the region into shouldering the burden of the energy transition, and what would a more just solution look like? What role should states play in driving a just transition, and what are the possibilities for a democratic reclaiming of state power for this goal? What alliances of working people, environmental justice movements and other political actors within the region are possible and necessary, and what role can international solidarity and resistance play in supporting these?

It is increasingly clear that a just transition for the Arab region will require not only a recognition of the historic responsibility of the industrialized West in causing global warming, but also of the role of emerging economic powers, including the Gulf states, in perpetuating a destructive global economic order. It will also need to acknowledge the role of power in shaping both how climate change is caused, and who carries the burden of its impacts, and of "solutions" to the crisis. Climate justice and a just transition will mean breaking with "business as usual" approaches that protect global political elites, multinational corporations and non-democratic regimes, and adopting a radical social and ecological transformation and adaptation process. The imperatives of justice and pragmatism are increasingly converging on the need for climate reparations or debts to be (re)paid to countries in the global South by the rich North. This must take the form not of loans and additional debts, but of transfers of wealth and technology, cancelling current odious debts, halting illicit capital flows, dismantling neocolonial trade and investment agreements, such as the Energy Charter Treaty,[42] and stopping the ongoing plunder of resources. The financing of the transition needs to take into account the current, ongoing and future loss and damage, which is occurring disproportionately in the South. At the same time, inequalities exist not only between North and South, but also within all countries of the world, including those in the Arab region. This being so, there is a need to consider how a programme of climate reparations can be combined with the creation of a just, democratic and equitable energy system within these countries.

These questions are increasingly urgent. International negotiations on climate action are stagnating while at the same time climate change is accelerating, with its effects increasingly deadly and undeniable. This book is intended as a tool for activists, both in the Arab region and around the world, to help them to continue to pose critical questions and to build coalitions, alliances and popular power in support of their own solutions for a just transition.

Obviously, this collective book has some lacunas – things that are not addressed, such as the impact of ongoing wars and conflicts (and the resulting devastating cross-border displacement of populations)[43] on questions of just transitions in countries like Iraq, Libya, Syria and Yemen. This is partly due to our own limitations. Nonetheless, although we would not pretend, or seek, to be fully comprehensive when discussing such a vast region, we hope we offer here an important glimpse, and contribute to the emerging study of energy transitions through a political economy lens which investigates the relationships between fossil fuel industries, the renewable energy sector, regional elites and international capital. Ultimately, the goal is to articulate and explore concepts and political ideas that can help to guide and galvanize transformative grassroots-led change in the region. We hope that this collection will spark more and deeper conversations and explorations about the role of the Arab region in a global just transition.

NOTES

1. We note briefly here the various ways the book's editors and contributors refer to the region that is the focus of this volume. Some use "Middle East" or 'Middle East and North Africa (MENA)', others refer to the "Arab region" or "Arab world", while others (beyond this book) go for the less used coinages "North Africa and West Asia (NAWA)" or "West Asia and North Africa (WANA)". Our own view is that if we are committed to advancing counter-hegemonic narratives that challenge structures of power, and to decolonizing concepts and names, it is only fitting to call into question the colonial designation "Middle East" – a construct of, and designed to sit in opposition to, the West; part of the legacy of orientalism, of creating an "other". We are sympathetic to the use of "Arab region", but without its reductionist ethnic connotations. We acknowledge that this naming can arouse feelings of exclusion and oppression among some. No naming is perfect, and each has its own limits. In our view, without trying to efface the rich shared cultural and political legacies in the region, a reference rooted in a geographic identification, such as North Africa and West Asia (NAWA), is a more apt description, but that does not mean we reject the other forms of naming. In fact, we use them interchangeably, while being aware that "Middle East", "MENA" or "NAWA" go beyond the Arab region to include countries such as Iran and Turkey. Finally, we refer to the various sub-regions, using the terms "North Africa", the "Maghreb", "the Gulf" and the "Mashriq".

2. At the time of writing this introduction (April 2023), Tunisia is witnessing its worst drought on record, forcing the authorities to cut water supplies to citizens at night during the holy month of Ramadan. Other tight restrictions on water usage include a ban on the use of potable water for irrigating farmland and green spaces, and for cleaning public areas and cars.

3. Staff and agencies, "Ninth Sandstorm in Less than Two Months Shuts Down Much Of Iraq," *The Guardian*, May, 24, 2022, https://tinyurl.com/zbe576p7; "Thousands Hospitalized as Latest Sandstorm Brings Iraq to Standstill," *France24*, May 16, 2022, https://tinyurl.com/5n848e3w.

4. Richard P. Allan et al., *Climate Change 2021: The Physical Science Basis. Contribution of Working Group I to the Sixth Assessment Report of the Intergovernmental Panel on Climate Change*, (Cambridge University Press, Cambridge, United Kingdom, and New York, USA, 2021), DOI:10.1017/9781009157896.

5. Hamza Hamouchene and Mika Minio-Paluello, *The Coming Revolution in North Africa: The Fight for Climate Struggle*, Platform London, Environmental

Justice North Africa, Rosa Luxemburg Stiftung and Ritimo, April 2015, https://tinyurl.com/4urr3pzs.

6. Ali Amouzai and Sylvia Kay, *Towards a Just Recovery from the Covid-19 Crisis: The Urgent Struggle for Food Sovereignty in North Africa*, Transnational Institute, 2021, https://tinyurl.com/4pzjj59f.

7. Abbas El-Zein, Samer Jabbour et al., "Health and Ecological Sustainability in the Arab World: A Matter of Survival," *The Lancet* 383, No. 9915 (2014): 458-476, DOI: 10.1016/S0140-6736(13)62338-7.

8. Jos Lelieveld et al., "Strongly Increasing Heat Extremes in the Middle East and North Africa (MENA) in the 21st Century," *Climatic Change* 137, (2016): 245-60, https://doi.org/10.1007/s10584-016-1665-6.

9. Friends of the Earth International, *Chasing Carbon Unicorns: The Deception of Carbon Markets and "Net Zero"*, 2021, https://tinyurl.com/9y4rnbm3; Corporate Accountability, *Not Zero: How "Net Zero" Targets Disguise Climate Inaction*, 2020, https://tinyurl.com/5dap2tva.

10. Dorothy Grace Guerrero, "The Historic Loss and Damage Victory at COP27 was Thanks to the Unity of Developing Countries and Civil Society," *Global Justice Now*, 2022, https://tinyurl.com/49t2bf2y.

11. See Kate Abnett, "Rich Countries Failed to Meet Their Climate Funding Goal," *Reuters*, July 29, 2022, https://tinyurl.com/4pc97ccn; Zia Weise, "Rich Countries Broke Climate Finance Promise, Says OECD," *Politico*, July 29, 2022, www.politico.eu/article/oecd-rich-countries-broke-climate-finance-promise.

12. The Working Group on Food Sovereignty – Tunisia, "filahtuna, ghadhawna, siadatuna tahlil lilsiyasat alfalahiat altuwnisiat ealaa daw' mafhum alsiyadat alghidhayiya, [Our Food, Our Farming, Our Sovereignty: Analysis of Tunisian Policies in Light of Food Sovereignty]," 2019, https://tinyurl.com/2ja4v4du.

13. Laleh Khalili, *Sinews of War and Trade: Shipping and Capitalism in the Arabian Peninsula* (Verso, London/New York, 2021).

14. BP, *BP Statistical Review of World Energy 2022*, 71st Edition, 2022, https://tinyurl.com/29kcuvb9.

15. Timothy Mitchell, *Carbon Democracy: Political Power in the Age of Oil*, (Verso, London/New York, 2013).

16. Adam Hanieh, *Money, Markets, and Monarchies: The Gulf Cooperation Council and the Political Economy of the Contemporary Middle East* (Cambridge University Press, Cambridge, 2018), https://doi.org/10.1017/9781108614443.

17. Samir Amin, *Delinking: Towards a Polycentric World* (Zed Books, London, 1990); Amin, *The Implosion of Capitalism* (Pluto Press, London, 2013). See also Walter Rodney, *How Europe Underdeveloped Africa* (Pambazuka Press, London, 2012); Eduardo Galeano, *Open Veins of Latin America.* (Monthly Review Press, New York, 1973).

18. Amin, *Unequal Development* (Monthly Review Press, New York, 1976). See also Immanuel Wallerstein, *World-Systems Analysis: An Introduction* (Duke University Press, Durham, 2004). More analysis can be found in John Bellamy Foster and Hannah Holleman , "The Theory of Unequal Ecological Exchange: A Marx-Odum Dialectic," *The Journal of Peasant Studies* 41, No. 2 (2014),

pp. 199-233; Mariko Lin Frame, "The Neoliberalization of (African) Nature as the Current Phase of Ecological Imperialism," *Capitalism Nature Socialism* 27, No. 1 (2016), pp. 87-105, https://doi.org/10.1080/10455752.2015.1135973.

19. Hamouchene, *Extractivism and Resistance in North Africa*, Transnational Institute, 2019, www.tni.org/en/ExtractivismNorthAfrica; Layla Riahi and Hamouchene, *Deep and Comprehensive Dependency: How a Trade Agreement with the EU Could Devastate the Tunisian Economy*, Transnational Institute, 2020, www.tni.org/en/deep-and-comprehensive-dependency.

20. We refer here to the political-economic union of the Gulf Cooperation Council (GCC) that includes Bahrain, Kuwait, Oman, Qatar, Saudi Arabia and the United Arab Emirates.

21. On sub-imperialism, see Patrick Bond, "Bankrupt Africa: Imperialism, Sub-Imperialism and the Politics of Finance," *Historical Materialism* 12, No. 4 (2004), pp. 145-72, https://doi.org/10.1163/1569206043505211.

22. Hanieh, *Lineages of Revolt: Issues of Contemporary Capitalism in the Middle East* (Haymarket Books, Chicago, 2013), ISBN: 9781608463251; *Money, Markets, and Monarchies*, op.cit.

23. Nick Buxton, *A Primer on Climate Security: The Dangers of Militarising the Climate Crisis*, Transnational Institute, 2021. www.tni.org/en/publication/primer-on-climate-security.

24. See Hanieh, (2021) "Space, Scale, and Region: Thinking Through the New Dynamics of the Middle East," *Arab Council of Social Sciences, Working Paper* (draft version); Rafeef Ziadah, *Saudi-UAE Interventions: Arms, Aid and Counter-Revolution*, Transnational Institute, October 27, 2021, https://tinyurl.com/3ju4vuju; Khalili, "Gendered Practices of Counterinsurgency," *Review of International Studies* 37, no. 4(2011): 1471-1491, https://doi.org/10.1017/S0260 21051000121X.

25. Yezid Sayigh, *Owners of the Republic: An Anatomy of Egypt's Military Economy* (Carnegie Middle East Center, Beirut; and Carnegie Endowment for International Peace, Washington DC, 2019), https://tinyurl.com/3duy5ha7.

26. Abdallah Laroui, *The History of the Maghrib: An Interpretive Essay* (Princeton Legacy Library, Princeton, 1977).

27. Edward Said, *Orientalism* (Penguin, London, 1977).

28. Diana K. Davis, "Imperialism, Orientalism, and The Environment in the Middle East: History, Policy, Power and Practice," in Davis and Edmund Burke (eds.), *Environmental Imaginaries of the Middle East and North Africa* (Ohio University Press, Athens and Ohio, 2011).

29. *Ibid.*

30. James C. Scott, *Seeing Like a State: How Certain Schemes to Improve the Human Condition Have Failed* (Yale University Press, New Haven, 1998).

31. Oliver Springate-Baginski, *'There is No Vacant Land:' A Primer on Defending Myanmar's Customary Tenure Systems*, Transnational Institute, 2019, www.tni.org/en/publication/there-is-no-vacant-land.

32. Naomi Klein, "Let Them Drown: The Violence of Othering in a Warming World," *London Review of Books* 38, No. 11 (2016), www.lrb.co.uk/the-paper/v38/n11/naomi-klein/let-them-drown.

33. Movement Generation Justice and Ecology Project, *From Banks and Tanks to Cooperation and Caring: A Strategic Framework for a Just Transition*, https://tinyurl.com/2p8ferj9.
34. Friends of the Earth International, *If it's Not Feminist, it's Not Just*, 2021, https://tinyurl.com/4j9dd8u9; Indigenous Environmental Network, *Indigenous Principles of Just Transition*, 2017, https://tinyurl.com/2zxfe65a; Trade Union Confederation of the Americas, *PLADA – Development Platform of the Americas*, 2014, https://tinyurl.com/4wuv66d2; Brian Ashley, Dick Forslund et al., *One Million Climate Jobs* (Alternative Information and Development Centre, Cape Town, 2016), https://tinyurl.com/5ydj4yhf.
35. Transnational Institute et al., *Just Transition: How Environmental Justice Organizations and Trade Unions are Coming Together for Social and Environmental Transformation*, February 11, 2020 https://www.tni.org/en/justtransition.
36. International Trade Union Confederation, *2020 ITUC Global Rights Index: The World's Worst Countries for Workers*, 2020, https://tinyurl.com/mryav896.
37. Abdulhadi Khalaf, Omar Al-Shehabi and Hanieh Khalaf (eds.), *Transit States: Labour, Migration and Citizenship in the Gulf* (Pluto Press, London, 2014).
38. Gilbert Achcar, *The People Want: A Radical Exploration of the Arab Uprising* (Saqi Books, London, 2013).
39. Roberto Cardarelli et al., *Informality, Development, and the Business Cycle in North Africa*, Departmental Paper No. 2022/011 (International Monetary Fund, Washington, DC 2022), https://tinyurl.com/yhuzbt6m.
40. Issa G. Shivji, "The Concept of 'Working People'," *Agrarian South Journal of Political Economy* 6, No. 1 (2017): pp. 1-13, https://doi.org/10.1177/22779760 17721318.
41. "Green New Deal (GND) proposals call for public policy to address climate change along with achieving other social aims like job creation and reducing economic inequality. The name refers back to the New Deal, a set of social and economic reforms and public works projects undertaken by US President Franklin D. Roosevelt in response to the Great Depression." – *Wikipedia* (https://en.wikipedia.org/wiki/Green_New_Deal). For more information, see Green New Deal UK's website: www.greennewdealuk.org.
42. Pia Eberhardt, Cecilia Olivet and Lavinia Steinfort, *The Ever-Expanding Energy Charter Treaty and the Power it Gives Corporations to Halt the Energy Transition*, Transnational Institute, 2018, www.tni.org/en/energy-charter-dirty-secrets.
43. The region has recently witnessed the largest forced displacement since the Second World War. See Shanta Devarajan and Lili Mottaghi, *MENA Economic Monitor: The Economics of Post-Conflict Reconstruction in MENA* (World Bank, Washington, DC, April 2017), p. 26, doi: 10.1596/978-1-4648-1085-5.

PART I

Energy Colonialism, Unequal Exchange and Green Extractivism

1

The Energy Transition in North Africa: Neocolonialism Again!

Hamza Hamouchene

The COVID-19 pandemic, which contributed to the global multidimensional crisis we are living through, demonstrates that what we are experiencing now is a taste of the worst things to come if we don't take the necessary measures to implement just solutions to the unfolding climate crisis.

The impacts of climate change in the MENA region are already a reality, and they are undermining the socioeconomic and ecological basis of life in the region. These impacts are felt disproportionately by the marginalized in society, especially small-scale farmers, agropastoralists and fisherfolk. People are being forced off their lands due to stronger and more frequent droughts and winter storms, and due to the growth of deserts and rises in sea levels. Moreover, crops are failing and water supplies are dwindling, which is having a large impact on food production.[1]

Addressing this global climate crisis requires a rapid and drastic reduction of greenhouse gas emissions. At the same time, we know that the current economic system is undermining the life support systems of the planet and will eventually collapse. Therefore, a transition towards renewable energies has become urgent. However, it is very possible that this transition, if and when it comes, will maintain the same practices of dispossession and exploitation that currently prevail, reproducing injustices and deepening socioeconomic exclusion. Thus, before talking about "green" projects, it is appropriate to question the frameworks and design choices applied in an energy transition, to shine some light on transitions that would be unjust, and on some problematic aspects of renewable energy that have been sidelined by the mainstream narrative.

"GREEN COLONIALISM" AND "GREEN GRABBING"

The Sahara is usually described as a vast empty land that is sparsely populated, and as representing an Eldorado of renewable energy, thus constituting a golden opportunity to provide Europe with energy so it can continue its extravagant consumerist lifestyle and excessive energy consumption. However, this deceptive narrative overlooks questions of ownership and sovereignty and masks ongoing global relations of hegemony and domination that facilitate the plunder of resources, the privatization of commons and the dispossession of communities, thus consolidating undemocratic and exclusionary ways of governing the energy transition.

Several examples from the North African region show how energy colonialism and extractivist practices are reproduced even in transitions to renewable energy, in what can be described as "green colonialism" or "green grabbing". If what really matters to us is not just any kind of transition but rather a just transition that will benefit the impoverished and marginalized in society, instead of deepening their socioeconomic exclusion, these examples raise some serious concerns.

Before delving into some of these examples, I would like to provide some short definitions of the terms "green colonialism" and "green grabbing". Green colonialism, or "renewable energy colonialism" can be defined as the extension of the colonial relations of plunder and dispossession (as well as the dehumanization of the other) to the green era of renewable energies, with the accompanying displacement of socio-environmental costs onto peripheral countries and communities, prioritizing the energy needs of one region of the world over another. Basically, the same system is in place, but with a different source of energy, moving from fossil fuels to green energy, while the same global energy-intensive production and consumption patterns are maintained and the same political, economic and social structures that generate inequality, impoverishment and dispossession remain untouched.

Scholars and activists have coined another useful concept: "green grabbing". This refers to cases where the dynamics of land grabs take place within a supposedly green agenda.[2] In other words, land and resources are appropriated for purportedly environmental ends. This ranges from certain conservation projects that dispossess indigenous communities of their land and territories, to the confiscation of communal land in order to produce biofuels, and to the installation of big solar plants/wind farms on agropastoralists' land without their proper consent.

The current uneven transition to renewable energies, which is happening mainly in the global North, is predicated on the ongoing extraction of base minerals and rare earth metals (such as cobalt, lithium, copper, nickel and graphite) that are used for manufacturing solar panels, wind turbines, blades and batteries. Where will these resources come from? The answer is from countries such as the Democratic Republic of Congo (DRC), Bolivia, Chile, Indonesia and Morocco, where environmental destruction and workers' exploitation will continue and even intensify.

Colonialism – if it ever formally ended – is continuing in other forms and at various levels, including in the economic sphere. This is what some scholars and activists call neocolonialism or re-colonization. The economies of the peripheries/the global South have been placed in a subordinate position within a profoundly unjust global division of labour: on the one hand, as providers of cheap natural resources and a reservoir of cheap labour, and, on the other hand, as a market for industrialized/high-technology economies.[3] This situation has been imposed and shaped by colonialism and attempts to break away from it have been defeated so far by the new tools of imperial subjugation: crippling debts, the religion of "free trade", and the Structural Adjustment Programmes (SAPs) imposed by international financial institutions (IFIs), such as the World Bank and the International Monetary Fund (IMF).

So, if we are serious about moving beyond fossil fuels, it is crucial to closely examine the linkages between fossil fuels and the wider economy, and to address the power relations within, and the hierarchies of, the international energy system.[4] This means recognizing that countries of the global South are still systematically exploited by a colonial, imperialist economy built around the pillaging of their resources and a massive transfer of wealth from South to North.

ENERGY TRANSITION, DISPOSSESSION AND GRABBING IN MOROCCO

Let's take Morocco as an example, as it has advanced much further in its energy transition than its neighbours. Morocco has set the goal of increasing its share of renewable energy to more than 50 percent by 2030. The Ouarzazate solar power plant, which began operating in 2016, is one element in the country's plan to achieve this goal. The plant has failed to benefit the impoverished communities that surround it: the Amazigh agropastoralists whose lands were used without their consent to install the 3,000 hectare facility.[5] Moreover, the debt of $9 billion relating to loans from the World Bank, the European Investment Bank (EIB) and others for the plant's construction is backed by Moroccan government guarantees, which means potentially more public debt in a country that is already overburdened with debt. The project, which is a public–private partnership (PPP) – a euphemism for the privatization of the profits and the socialization of losses through derisking strategies – has been recording, since its launch in 2016, an annual deficit of around 80 million euros, which is covered by the public purse. Finally, the Ouarzazate plant uses concentrated solar power (CSP), which necessitates extensive use of water in order to cool down the system and clean the solar panels. In a semi-arid region like Ouarzazate, diverting water use from drinking and agriculture is outrageous.

Another example of an unjust energy transition is the Noor Midelt project, which constitutes Phase II of Morocco's solar power plan. It is planned to provide more energy capacity than the Ouarzazate plant and will be one of the biggest solar projects in the world to combine both CSP and photovoltaic (PV) technologies. The Noor Midelt facilities will be operated by the French entity EDF Renewables, the Emirati entity Masdar and the Moroccan conglomerate Green of Africa, in partnership with the Moroccan Agency for Sustainable Energy (Masen), for a period of 25 years. The project has accumulated around $4 billion in debt so far, including more than $2 billion from the World Bank, the African Development Bank (AfDB), the EIB, the Agence Française de Développement (AFD) and Kreditanstalt für Wiederaufbau (KfW).[6,7]

Construction of Noor Midelt started in 2019, and commissioning was initially expected in 2022. However, delays have accumulated for various reasons, including the slow pace of progress on the solar plan and the political problems that the head of Masen encountered during 2021, as well as the geopolitical tensions between Morocco and Germany. The Noor Midelt solar complex will be developed on a 4,141 hectare site on the Haute Moulouya Plateau in central Morocco, approximately 20 km northeast of Midelt town. Of this site, a total of 2,714 hectares are managed as communal/collective land by the three ethnic agrarian communities of Ait Oufella, Ait Rahou Ouali and Ait Massoud Ouali, while approximately 1,427 hectares are declared forest land and are currently managed by these communities. This land has been confiscated from its owners through national laws and regulations that allow for expropriation in order to serve the public interest. The expropriation was granted in favour of Masen by an administrative court decision in January 2017, with the court decision publicly disclosed in March 2017.

In an ongoing colonial environmental narrative that labels the lands to be expropriated as marginal and underutilized, and therefore available for investing in green energy, the World Bank, in a study conducted in 2018,[8] stressed that "the sandy and arid terrain allow only for small scrubs to grow, and the land is not suitable for agricultural development due to lack of water". This argument/narrative was also used when promoting the Ouar-zazate plant in the early 2010s. At that time, one person stated: "The project people talk about this as a desert that is not used, but to the people here it is not desert, it is a pasture. It is their territory, and their future is in the land. When you take my land, you take my oxygen."[9]

The World Bank report does not stop there but goes on to assert that "the land acquisition for the project will have no impacts on the livelihood of local communities". However, the transhumant pastoralist tribe of Sidi Ayad, which has been using that land to graze its animals for centuries, begs to differ. In 2019, Hassan El Ghazi, a young shepherd, declared to an activist from the association ATTAC Morocco:

"Our profession is pastoralism, and now this project has occupied our land where we graze our sheep. They do not employ us in the project, but they employ foreigners. The land in which we live has been occupied. They are destroying the houses that we build. We are oppressed, and the Sidi Ayad region is being oppressed. Its children are oppressed, and their rights and the rights of our ancestors have been lost. We are 'illiterates' who do not know how to read and write… The children you see did not go to school and there are many others. Roads and paths are cut off… In the end, we are invisible and we do not exist for them. We demand that officials pay attention to our situation and our regions. We do not exist with such policies, and it is better to die, it is better to die!"[10]

In this context of dispossession, misery, underdevelopment and social injustice, the people of Sidi Ayad have been voicing their discontent since 2017 through a series of protests. In February 2019, they carried out an open sit-in, which led to the arrest of Said Oba Mimoun,[11] a member of the Union of Small Farmers and Forest Workers. He was sentenced to 12 months in jail.

Mostepha Abou Kbir, another trade unionist who has been supporting the struggle of the Sidi Ayad tribe, has described how the land was enclosed without the approval of the local communities, who have endured decades of socioeconomic exclusion. The land has now been fenced off and no-one is allowed to approach it. Abou Kbir contrasts the mega-development projects of the Moroccan state with the fact that basic infrastructure is inexistent in Sidi Ayad. Moreover, he points to another dimension of the enclosure and resource grab, which is the exhaustion of water resources in the Drâa-Taf-ilalet region for the sake of these gargantuan projects (the Midelt solar plant will be fed from the nearby Hassan II dam) from which the communities complain they do not benefit.[12] In a challenging context in which owners of small herds are being driven out of the sector, with wealth concentrated in fewer and fewer hands, along with the commoditization of the livestock market and chronic droughts, the Midelt solar project is set to exacerbate the threat to the livelihoods of these pastoralist communities and to worsen their predicament.

It is not only the communities of Sidi Ayad that have voiced concerns about the Midelt project. Women from the Soulaliyate movement have also demanded their right to access land in the Drâa-Tafilalet region and have demanded appropriate compensation for the loss of their ancestral land, on which the solar plant has been built. The term "Soulaliyate women" refers to tribal women in Morocco who live on collective land. The Soulaliyate women's movement, which began in the early 2000s, arose in the context of intense commodification and privatization of collective lands.[13] At the time, tribal women began demanding equal rights and an equal share when plans were made to privatize or divide up their lands. Despite intimidation, arrests and even sieges by the authorities, the movement has spread nationwide and women from different regions now rally behind the Soulaliyate movement's banner of equality and justice.

Despite all these concerns and injustices, the Midelt project is going ahead, protected by the monarchy and its tools of repression and propaganda. It seems that there is no end in sight for the logic of externalizing socio-ecological costs and displacing them through space and time, which is characteristic of capitalism's extractivist drive.

GREEN COLONIALISM AND OCCUPATION IN WESTERN SAHARA

While some of the projects in Morocco, such as the Ouarzazate and Midelt solar plants, certainly qualify as "green grabbing" – the appropriation of land and resources for purportedly environmental ends – similar renewable projects (solar and wind) that are being implemented, or will be implemented, in the occupied territories of Western Sahara can simply be labelled "green colonialism", as they are carried out in spite of the Saharawis and on their occupied land.

At present, there are three operational wind farms in occupied Western Sahara; a fourth is under construction in Boujdour, while several others are in the planning stage. Combined, these wind farms will have a capacity of over 1,000 MW. These wind farms are part of the portfolio of Nareva, the wind energy company that is owned by the holding company of the

Moroccan royal family. Ninety five percent of the energy that the Moroccan state-owned Office Chérifien des Phosphates (OCP) needs to exploit from Western Sahara's phosphate reserves in Bou Craa is produced by windmills. This renewable energy is generated by 22 Siemens wind turbines at the 50 MW Foum el Oued farm, which has been operational since 2013.[14]

In November 2016, at the time of the COP22 climate talks, Saudi Arabia's ACWA Power signed an agreement with Masen to develop and operate a group of three solar PV power stations generating 170 MW of electricity. However, two of these power stations (operational today), producing 100 MW of elecricity, are not located in Morocco but rather inside the occupied territory of Western Sahara (El Aaiún and Boujdour). Plans have also been made for a third solar plant at El Argoub, near Dakhla.[15]

It is clear that these renewable energy projects are being used to entrench the occupation by deepening Morocco's ties to the occupied territories, with the obvious complicity of foreign capital and companies (more details and analysis are in the second chapter of this book).

WHICH ENERGY TRANSITION IN ALGERIA? DRILL BABY DRILL!

The Algerian ruling classes have been talking about the "after oil" era for decades and successive governments have paid lip-service to the transition to renewable energies for years, without taking any concrete action. In fact, there have been significant delays in the implementation of current renewable energy plans, which in my view reflects a lack of any serious or coherent vision for the transition. Announcements and declarations from officials follow each other, while promises remain merely ink on paper. For example, the recent tender for the deployment of 1 GW of solar capacity has been delayed for more than two years. Algeria's plans to deploy 15 GW of solar energy generation capacity by 2030 are simply not realistic when we realize that the country had around 423 MW of total installed solar capacity at the end of 2021, according to the International Renewable Energy Agency (IRENA).[16] All sources combined, the installed capacity of renewable energy does not currently exceed 500 MW. This is a far cry from the 22 GW planned for 2030 that was announced in 2011. The Ministry of Energy Transition and Renewable Energies, which came into being in June 2020, has reduced these targets to four GW by 2024 and to 15 GW by 2035. However, even this is overly optimistic.

In a nutshell, Algeria needs to move fast towards renewable energies as the day will come when the country's European clients will stop importing its fossil fuels for energy purposes. The European Union (EU) is expanding and accelerating its energy transition, a pattern that has been rendered urgent by the Russian invasion of Ukraine. In the short term, the EU will obviously continue importing gas and will intensify its efforts to diversify its sources, but in the long term it will do its best to move away from fossil fuels. This will be an existential threat for countries like Algeria, if they continue to remain dependent on oil and gas. Therefore, the urgent move towards the production of renewable energies (primarily for the local market) is not only the right thing to do ecologically, but is also a strategic and a survival imperative.

However, the general tendency in the country in the last few years has been to move towards more liberalization of the economy and to extend more concessions to the private sector and foreign investors. The cases of the budget laws of 2020-21 and the new Hydrocarbon Law are illuminating in this respect. The new Hydrocarbon Law is very friendly to multinationals and offers more incentives and concessions for them to invest in Algeria. It also opens the way to destructive projects, such as the exploitation of shale gas in the Sahara and offshore resources in the Mediterranean.

Regarding the budget laws of 2020-21, they reopened the door to international borrowing, and imposed harsh austerity measures by lifting various subsidies and cutting public spending. In the name of encouraging foreign direct investment, they exempted multinationals from tariffs and taxes and increased their share in the national economy by removing the 51/49 percent investment rule that limits the part of foreign investment in any project to 49 percent, undermining national sovereignty still further.[17] Now it is the turn of the renewable energy sector. This is definitely not a decision that is likely to ensure sovereignty in this strategic sector that will grow in importance in the coming years!

While certain Western governments portray themselves as pro-environment by banning fracking within their borders and by setting carbon emissions-reduction targets, they simultaneously offer diplomatic support to their multinationals to exploit shale resources in their former colonies, as France did with Total in Algeria in 2013.[18] If that's not energy colonialism and environmental racism, I don't know what it is!

In the context of the war in Ukraine and the EU's attempts to cut reliance on Russian gas, we see once again that EU energy security comes above everything else. We are seeing more gas lock-in, more extractivism,[19] more path dependency and a halt to the green transition where those extractive projects are taking place. That's exactly what has happened in the case of

Italy and Algeria agreeing to boost gas supplies to Italy. In fact, Algeria's national hydrocarbons company Sonatrach and the Italian ENI will pump an additional nine billion cubic metres from 2023/24.[20] The EU will also receive liquefied natural gas (LNG) shipments from Egypt, Israel, Qatar and the United States.

Some North African economies heavily dependent on fossil fuels will be hugely impacted when Europe significantly reduces its fossil fuel imports from this region in the coming decades. Therefore, a serious discussion and public debate need to take place regarding the necessary and urgent transition to renewables, while phasing out fossil fuels. New projects for the exploration and exploitation of fossil fuels should be considered out of the question and this cannot be disconnected from questions of democratization and popular sovereignty over land, water and other natural resources. In kleptocratic military dictatorships like Algeria and Egypt (where COP27 took place), how can people decide and shape their future without demilitarizing and democratizing their countries and societies? Moreover, there is a need to consciously build alliances between labour movements and other social and environmental justice movements and organizations. We need to find a way of involving workers in the oil industry in discussions around the transition and green jobs as the transition won't take place without them. It is therefore of paramount importance to start engaging with the trade unions around these issues.

In Algeria and in other countries in North Africa and the global South, the energy transition needs to be a sovereign project, primarily looking inwards and directed towards satisfying local needs first, before embarking on any export initiatives. We cannot continue in the old ways of producing for Europe and obeying its diktats, including its desire to wean itself off its dependency on Russian gas by diversifying its energy sources. The priority now is to decarbonize North African economies by reaching 70 to 80 percent renewables in the energy mix before even starting to think about exporting to the EU.

On top of this, it needs to be borne in mind that countries like Algeria that have been locked into a predatory form of an extractivist model of development have neither the financial means nor the sufficient know-how to carry out a rapid energy transition. In this respect, some financial compensation must be put on the table to keep the oil in the ground, and monopolies on green technology and knowledge must be ended, and these resources must be made available to countries and communities in the global South.

PRIVATIZATION OF ENERGY FOR EXPORT

The drive towards the privatization of energy and corporate control of the energy transition is global. Morocco is already on this path, and so is Tunisia. Right now, there is a strong push towards privatization of the Tunisian renewable energy sector and providing huge incentives to foreign investors to produce green energy in the country, including for export. The 2015-12 law and its amendment in 2019 enables the use of agricultural land for renewable energy projects in a country that already suffers from acute food dependency, laid bare by the COVID-19 pandemic and with the war in Ukraine. In this context, one must ask, energy transition for whom?

In 2017, the company TuNur applied to build a 4.5 GW solar plant in the Tunisian desert in order to deliver enough electricity via submarine cables to power two million European homes. This still unrealized project is a partnership between UK-based Nur Energy and a group of Maltese and Tunisian investors in the oil and gas sector.[21] Until recently, TuNur was openly describing itself as a solar energy export project linking the Sahara and Europe. Given that Tunisia depends on Algeria for some of its energy needs, it is outrageous that such projects are turning to exports rather than producing energy for domestic use.

The same goes for another project proposed in 2021 by a former Tesco chief executive officer,[22] in partnership with the Saudi ACWA Power, which aims to connect southern Morocco to the United Kingdom through underwater cables that will channel electricity over 3,800 km. Once again, the same relations of extraction and the same practices of land grabbing are maintained while people in the region are not even self-sufficient in energy. These big renewable projects, while proclaiming their good intentions, end up sugar-coating brutal exploitation and robbery. It seems that a familiar colonial scheme is being rolled out in front of our eyes: the unrestricted flow of cheap natural resources (including solar energy) from the global South to the rich North, while fortress Europe builds walls and fences to prevent human beings from reaching its shores!

HYDROGEN: THE NEW ENERGY FRONTIER IN AFRICA

As the world seeks to switch to renewable energy amid a growing climate crisis, hydrogen has been presented as a "clean" alternative fuel. Most current hydrogen production is the result of extraction from fossil fuels, leading to large carbon emissions (grey hydrogen). This process can be made cleaner (blue hydrogen), for example, through carbon capture technology. However, the cleanest form of hydrogen extraction uses electrolysers to split water molecules, a process that can be powered by electricity from renewable energy sources (clean or green hydrogen).

In recent years, in response to heavy lobbying from various interest groups, the EU has embraced the idea of a hydrogen transition as a centrepiece of its climate response, introducing in 2020 its hydrogen strategy within the framework of the European Green Deal. The plan proposes shifting to "green" hydrogen by 2050 through local production and by establishing a steady supply from Africa.[23] This strategy was inspired by ideas put forward by the trade body and lobby group Hydrogen Europe, which has set out the 2x40 GW Green Hydrogen Initiative. Under this initiative, by 2030 the EU will have in place 40 GW of domestic renewable hydrogen electrolyser capacity and will import a further 40 GW from electrolysers in neighbouring areas, including the deserts of North Africa, using existing natural gas pipelines that already connect Algeria to Europe.[24]

It is worth saying here that the drive for green hydrogen and the push for a hydrogen economy has already gained support from major European oil and gas companies, which see it as a back door to the continuation of their operations, with hydrogen being extracted from fossil gas (the production of grey and blue hydrogen). It is thus becoming clear that the fossil fuel industry wants to preserve the existing natural gas and pipeline infrastructure.[25]

In Africa and elsewhere, fossil fuel companies continue to use the same exploitative economic structures set up during the colonial era to extract local resources and to transfer wealth out of the continent. They are also keen on preserving the political *status quo* in African countries so they can continue to benefit from lucrative relations with corrupt elites and authoritarian leaders. This allows them to engage in labour exploitation, environmental degradation and violence against local communities with impunity.

In the context of the war in Ukraine, replacing fossil gas with hydrogen from renewables has become a key plank of REPowerEU, the European Commission's plan to end dependence on Russian gas.[26] Commission Vice-President Frans Timmermans told the European Parliament in May 2022: "I strongly believe in green hydrogen as the driving force of our energy system of the future. [...] and I also strongly believe that Europe is never going to be capable to produce its own hydrogen in sufficient quantities."[27]

As well as shifting gas suppliers from President Vladimir Putin's Russia to other authoritarian regimes, like those ruling Algeria, Azerbaijan, Egypt and Qatar, or to the settler-colonial apartheid state of Israel, and building more ports and pipelines to import and transport gas, the European Commission has quadrupled its hydrogen target from five million tonnes by 2030 to 20 million tonnes, with half of that to be imported primarily from North Africa, though other countries are also on the target list, including Namibia, South Africa, DRC, Chile and Saudi Arabia. Recent studies have shown that these targets are unrealistic from a cost and energy perspective, and are already leading to more fossil fuel exploitation.[28]

Within Europe, Germany is at the forefront of green hydrogen efforts in Africa. It is working with the DRC, Morocco and South Africa to develop "decarbonized fuel" generated from renewable energy for export to Europe, and is exploring other potential areas/countries that are particularly suited to green hydrogen production. In 2020, the Moroccan government entered into a partnership with Germany to develop the first green hydrogen plant on the continent.[29] As always, Morocco, boasting one of the most neoliberal(ized) economies in the region, garners praise for its business-friendly environment, openness to foreign capital and its "leadership" in the renewable energy sector. According to some estimates, the country can take up to four percent of the global Power-to-X market (production of green molecules) by 2030, given its "exceptional renewable resources and its successful track record in deploying large scale renewable plants".[30]

All of this is happening at the expense of energy access and energy transitions in these African countries. If these developments are not stopped, the green transition will be derailed in the name of the EU's energy security and its efforts to meet its climate targets. Moreover, the EU plans for renewable hydrogen strategy (RePowerEU) are not simply about emissions, but are part of a broader move to reposition itself and its corporations as global players within a green high-tech economy,[31] competing with other powers such as China.

DESERTEC 3.0 – OR JUMPING ON THE GREEN HYDROGEN BANDWAGON

In 2009, the Desertec project, an ambitious initiative that aimed to power Europe from Saharan solar plants and wind farms stretching across the Middle East and North Africa (MENA) region, was launched by a coalition of European industrial firms and financial institutions based on the idea that a tiny surface area of the desert can provide around 20 percent of Europe's electricity by 2050 via special high-voltage direct current transmission cables.

After a number of years of hype, the Desertec venture ultimately stalled amid criticisms of its astronomical costs and its neocolonial connotations.[32] However, attempts to revive it as Desertec 2.0, with a focus on the local market for renewable energy, followed, and the project was eventually reborn as Desertec 3.0, which aims to satisfy Europe's demand for hydrogen – seen as a "clean" energy alternative to fossil fuels. In early 2020, Desertec Industrial Initiative (Dii) launched the MENA Hydrogen Alliance, which brings together private and public sector actors, as well as science and academia, to initiate green hydrogen economies and produce hydrogen for export.[33] Two of Dii's partners are the French energy giant Total, and the Dutch oil major Shell.

The Desertec 3.0 proposal,[34] which advocates for a European energy system based on 50 percent renewable electricity and 50 percent green hydrogen by 2050, starts with the assumption that "due to its limited size and population density, Europe will not be able to produce all its renewable energy in Europe itself". The new Desertec proposal attempts to distance itself from the focus on exports in the initiative's early days, by adding the dimension of local development of a clean energy system. However, the export agenda cannot be underestimated or ignored as the Desertec 3.0 manifesto points out that "…over and beyond catering for domestic demand, most North African countries have huge potential in terms of land and resources to produce green hydrogen for export".

If that was not convincing enough for the political and business elites on both sides of the Mediterranean, the Desertec crew have other tricks up their sleeves. The document goes on: "Furthermore, a joint European-North African renewable energy and hydrogen approach would create economic development, future-oriented jobs and social stability in North-African countries, potentially reducing the number of economic migrants from the region to Europe." It is not clear if this is a desperate aggressive sales strategy but it seems clear that this Desertec vision lends itself to the agenda of consolidating fortress Europe and expanding an inhuman regime of border imperialism, while trying to tap into the cheap energy potential of North Africa, which also relies on undervalued and disciplined labour.

Desertec is thus presented as one solution to Europe's energy transition: an opportunity for economic development in North Africa and a brake on migration from the South. Being an apolitical technofix, it promises to overcome these problems without making any fundamental changes: basically, maintaining the *status quo* and the contradictions of the global system that have led to these problems in the first place. Technofixes of this kind embrace the obsession with endless economic growth, repackaged in the oxymoron "green growth", and give the illusion of an endless availability of energy and resources, thus indirectly perpetuating consumerist lifestyles and energy intensive productivism. This will do nothing to bring our socioeconomic system within the planet's limits in time to avert a climate and ecological breakdown.

Big engineering-focused "solutions" like Desertec tend to present climate change as a shared problem that has no political or socioeconomic context. This perspective hides the historical responsibility of the industrialized West, the problems of the capitalist energy model, and the different vulnerabilities experienced by countries of the North compared to those of the South. Moreover, using language such as "mutual cooperation" and "for the benefit of both", presents the Euro-Mediterranean region as a unified community – in other words: we are all friends now, fighting against a common foe! – and masks the real enemy of African and Arab people, which is neo-colonial structures of power that exploit them and plunder their resources.

Furthermore, pushing for the use of the current gas pipeline infrastructure effectively advocates for a mere switching of the energy source, while maintaining the existing authoritarian political dynamics and leaving intact the current hierarchies within the international order. The fact that Desertec is encouraging the use of pipelines from Algeria and Libya (including through Tunisia and Morocco) raises the question of the future of the populations in these two fossil fuel-rich countries. What will happen when Europe stops importing gas from them (in a context in which 13 percent of the gas consumed in Europe is from North Africa)? What about the ongoing chaos and destabilization caused by the North Atlantic Treaty Organization (NATO) intervention in Libya? Will Algerians' aspirations for democracy and sovereignty – well expressed in the 2019-21 uprising against the country's military dictatorship – be considered in this equation? Or is it simply another remake of the *status quo* where hydrogen simply replaces gas? Perhaps there is nothing new under the sun after all.

To add insult to injury, the Desertec manifesto points out that "in an initial phase (between 2030-2035), a substantial hydrogen volume can be produced by converting natural gas to hydrogen, whereby the CO_2 is stored in empty gas/oil fields (blue hydrogen)". First of all, carbon capture and storage technologies are still expensive and unreliable. Second, there is a big risk that the captured carbon dioxide (CO_2) will be used for enhanced oil recovery, as is currently the case around the world. Notwithstanding this, storing CO_2 in empty gas fields in North Africa, alongside the use of the rare water resources there to produce hydrogen, and the potential pollution from desalination, would be yet another example of dumping waste in the global South and displacing environmental costs from North to South (the creation of sacrifice zones): a strategy of imperialist capital in which environmental racism is wedded to energy colonialism.

Last but not least, huge upfront investment will be needed in order to establish the infrastructure required to produce and transport green hydrogen. Given previous experiences of implementing such high-cost and capital-intensive projects (such as the Ouarzazate solar plant), the investment could well end up amassing yet more debt for the receiving country, deepening the dependence on multilateral lending and foreign assistance.

If these plans do go ahead, they will represent the latest neocolonial resource grab, at a time when renewable resources should be used for local energy needs and to meet local climate targets, rather than helping the EU safeguard its energy security and deliver its climate strategy.

CONCLUSION

What seems to unite all the aforementioned "green" projects and the hype around them is a deeply erroneous assumption that any move towards renewable energy is to be welcomed, and that any shift from fossil fuels, regardless of how it is carried out, is worthwhile. One needs to say it clearly: the climate crisis we are currently facing is not attributable to fossil fuels *per se*, but rather to their unsustainable and destructive use in order to fuel the capitalist machine. In other words, capitalism is the culprit, and if we are serious in our endeavours to tackle the climate crisis (which is only one facet of the multidimensional crisis of capitalism), we cannot elude the issue of radically changing our ways of producing and distributing things, our consumption patterns, and fundamental issues of equity and justice. It follows from this that a mere shift from fossil fuels to renewable energy sources, while at the same time remaining within the capitalist framework of commodifying and privatizing nature for the profit of the few, will not solve the problem we face. In fact, if we continue down this path, we will end up only exacerbating the problem, or creating another set of problems, relating to issues of ownership of land and natural resources.

A green and just transition must fundamentally transform the global economic system, which is not fit for purpose at either the social, ecological or even biological level (as revealed by the COVID-19 pandemic). It must put an end to the colonial relations that still enslave and dispossess people. We must always ask: Who owns what? Who does what? Who gets what? Who wins and who loses? And whose interests are being served? Because if we don't ask these questions we will go straight to a green colonialism, with an acceleration of extraction and exploitation, in the service of a so-called common "green agenda".

The fight for climate justice and a just transition needs to take into account the differences in responsibilities and vulnerabilities between North and South. Ecological and climate debt must be paid to countries in the global South, which happen to be the hardest hit by global warming and which have been locked by global capitalism into a system of predatory extractivism. In a global context of forced liberalization and the push for unjust trade deals, as well as an imperial scramble for influence and energy resources, the green transition and talk about sustainability must not become a shiny façade for neocolonial schemes of plunder and domination.

Furthermore, while there is always talk about the lack of technology expertise where renewable projects are installed in the global South, it is usually not asked why this is the case in the first place. Isn't this lack attributable to the monopolizing of technology and the existing intellectual property regime (the cruelty of which was revealed during the pandemic)? Isn't it because of all the Structural Adjustment Programmes (SAPs) that have been imposed, which have hollowed out public services and scientific research? Technology transfer must be a cornerstone of any just energy transition, otherwise nations of the global South will always remain dependent.

In this context, just transition is a framework for a fair shift to an economy that is ecologically sustainable, equitable and just for all its members. A just transition means a transition from an economic system that is built around the excessive extraction of resources and the exploitation of people, to one that is structured around the restoration and regeneration of territories and people's rights and dignity. A robust and radical vision of just transition sees environmental destruction, capitalist extraction, imperialist violence, inequality, exploitation, and marginalization along the axes of race, class and gender, and as simultaneous effects of one global system which must be transformed.[35] Seen in this light, "solutions" which try to address a single dimension, such as the environmental catastrophe, in isolation from the social, cultural and economic structures which give rise to it, will inevitably remain "false solutions".[36]

A just transition will obviously look different in different places. It is indeed better to speak of transitions in the plural, in recognition of this reality. We must be sensitive to the fact that massive global and historical inequalities, and their continuation at present, are part of what must be transformed in order to bring about a just and sustainable society. Thus, a just transition may mean very different things in different places. What might work in Europe will not necessarily be applicable in Africa. What might work in Egypt might not work in South Africa. And what might work in urban areas in Morocco may not be good for rural areas there. And, perhaps, a transition in a fossil fuel-rich country like Algeria will look different to one in other countries that are less endowed with such resources. Thus, we need to be imaginative and to have a decentralized approach, and we need to seek guidance from local populations themselves.

The concept of just transition draws on concepts like energy democracy and energy sovereignty to elaborate a vision of a world where people have access to, and control over, the resources they need to lead dignified lives, and have a political role in making decisions about how those resources are used, and by whom. This transition must be under the control of communities and their democratically elected representatives. It cannot be left to the private sector and to companies. Active participation in the decision-making and shaping of transitions is crucial.

Finally, just transition is not just about energy. In this regard, the way we do agriculture must also be transformed. Industrial agriculture and farming or agribusiness is another locus where imperialist domination and climate change intersect. Not only is it one of the drivers of climate change, but it also keeps so many countries in the South prisoners of an unsustainable and destructive agrarian model, a model that is based on the exporting of a few cash crops and the exhaustion of land and the rare water resources in arid and semi-arid regions, such as Egypt, Tunisia and Morocco (and Algeria to a growing extent).

In many ways, the climate crisis and the needed green transition offer us a chance to reshape politics. Coping with the dramatic transformation will require a break with existing militarist, colonial and neoliberal projects. Therefore, the struggle for a just transition and climate justice must be fiercely democratic. It must involve the communities who are most affected, and it must be geared towards providing for the needs of all. It means building a future in which everybody has enough energy, and a clean and safe environment: a future with an eco-socialist horizon that is in harmony with the revolutionary demands of the African and Arab uprisings: popular sovereignty, bread, freedom and social justice.

NOTES

1. Hamza Hamouchene and Mika Minio-Paluello, *The Coming Revolution in North Africa: The Fight For Climate Struggle*, Platform London, Environmental Justice North Africa, Rosa Luxemburg Stiftung and Ritimo, April 2015, https://tinyurl.com/4urr3pzs.

2. James Fairhead, Melissa Leach and Ian Scoones, "Green Grabbing: A New Appropriation of Nature?" *Journal of Peasant Studies* 39, No. 2 (2012), pp. 237-61.

3. Walter Rodney, *How Europe Underdeveloped Africa*, (Pambazuka Press, London, 2012); Samir Amin, *Accumulation on a World Scale*, (Monthly Review Press, New York, 1974); Eduardo Galeano, *Open Veins of Latin America*. (Monthly Review Press, New York, 1973); John Bellamy Foster and Robert McChesney, *The Endless Crisis: How Monopoly Finance Capital Produces Stagnation and Upheaval from the USA to China*. (Monthly Review Press, New York, 2012); Michael Brie, "A Critical Reception of Accumulation of Capital," in Judith Dellheim and Frieder Otto Wolf (eds.), *Rosa Luxemburg: A Permanent Challenge for Political Economy*, (Palgrave Macmillan, London, 2016), pp. 261-303.

4. Adam Hanieh, "When Oil Markets Go Viral," *Verso Blog*, 2020, www.versobooks.com/blogs/4651-when-oil-markets-go-viral.

5. Hamouchene, "The Ouarzazate Solar Plant in Morocco: Triumphal 'Green' Capitalism and The Privatization Of Nature," *Jadaliyya*, March 23, 2016, www.jadaliyya.com/Details/33115.

6. In English, the AFD is the French Development Agency, the KfW is the German state-owned investment and development bank.

7. "Noor Midelt Solar Power Project, Morocco," Projects, NS Energy, https://tinyurl.com/4h8rd8jw.

8. The World Bank, *Morocco - Noor Solar Power Project: Additional Financing*, Project Appraisal Document PAD2642, 2018, https://tinyurl.com/4w9paund.

9. Karen Rignall, "Theorizing Sovereignty in Empty Land: Contested Global Landscapes," *Land Deal Politics Initiative*, 2012, https://tinyurl.com/2ffuejp7.

10. "Oh land," produced by ATTAC Maroc and War On Want, Documentary Film, July 8, 2020, www.youtube.com/watch?v=rQouJbUXEPw.

11. ATTAC Morocco, "alhuriyat alfawriat lilmunadil alniqabii saeid awba mimunila litajrim muqawamat alfalaahin alsighar difaean ean haqihim fi 'aradiihim watharawatihim [Immediate Freedom for Trade Union Activist Said Oba Mimoun: Stop Criminalizing the Resistance of Small Farmers Defending Their Right to Their Land and Wealth]," Statement, April, 17, 2019, https://tinyurl.com/sh7rkuv9.

12. *Ibid.*

13. ATTAC Morocco, "The Soulaliyate Movement: Moroccan Women Fighting Land Dispossession," *War On Want*, August 14, 2020, https://tinyurl.com/5eermtxt.

14. Western Sahara Resource Watch (WSRW), "Dirty Green Energy on Occupied Land," April 14, 2022, https://wsrw.org/en/news/renewable-energy.

15. WSRW, "Concrete Plans for Third Solar Plant in Occupied Western Sahara," August 24, 2020, https://tinyurl.com/3yjaa6sa.

16. IRENA, *Renewable Capacity Statistics 2021*, April 2021, https://tinyurl.com/28p56e7s.
17. Hamouchene, *The New Algerian Revolution and Black Lives Matter – a Fanonian Perspective*, Transnational Institute, October 27, 2021, https://longreads.tni.org/the-new-algerian-revolution-and-black-lives-matter.
18. Olivier Petitjean and Sophie Chapelle, "Shale gas: How Algerians Rallied Against the Regime and Foreign Oil Companies," June 2, 2016, *Multinationals Observatory*, https://tinyurl.com/3rscs9wv.
19. Hamouchene, *Extractivism and Resistance in North Africa*, Transnational Institute, November 20, 2019, https://www.tni.org/en/publication/extractivism-and-resistance-in-north-africa.
20. "Eni, Sonatrach Sign Deal to Boost Algeria Gas Exports to Italy," *Reuters*, May 26, 2022, https://tinyurl.com/2s35auxv.
21. Arthur Neslen, "Huge Tunisian Solar Park Hopes to Provide Saharan Power to Europe," *The Guardian*, September 6, 2017, https://tinyurl.com/yckkhtbb.
22. Rachel Morison, "Ex-Tesco CEO Wants $22 Billion Power Link From Morocco to U.K.," *Bloomberg*, September 26, 2021, https://tinyurl.com/yvbamv4a.
23. European Commission, *Communication from the Commission to the European Parliament, the Council, the European Economic and Social Committee and the Committee of the Regions*, July 8, 2020, https://tinyurl.com/47basuhn.
24. John Parnell, "European Union Sets Gigawatt-Scale Targets for Green Hydrogen," *Greentech Media*, July 9, 2020, https://tinyurl.com/4xxwc5tm.
25. Corporate Europe Observatory (CEO), *The Hydrogen Hype: Gas Industry Fairy Tale or Climate Horror Story?*, 2020, https://corporateeurope.org/en/hydrogen-hype.
26. "REPowerEU: Affordable, Secure and Sustainable Energy for Europe," *European Commission*, 2022, https://tinyurl.com/mpj5nxjc.
27. Leigh Collins, "Europe is Never Going to be Capable of Producing its Own Hydrogen in Sufficient Quantities: EU climate chief," *Recharge News*, May 4, 2022, https://tinyurl.com/4ch3p2tr.
28. Michael Barnard, *Assessing EU Plans to Import Hydrogen from North Africa*, Transnational Institute and Corporate Europe Observatory, May 17, 2022, https://tinyurl.com/swyzvsc7.
29. Clifford Chance, "Focus on hydrogen: A New Energy Frontier for Africa," January 21, 2021, https://tinyurl.com/3fmev2hv.
30. Eichhammer Wolfgang, Oberle Stella et al., *Study on the Opportunities of "POWER-TO-X" in Morocco: 10 hypotheses for discussion*, Fraunhofer Institute for Systems and Innovation Research ISI, February 2019, https://tinyurl.com/3p2fmush.
31. CEO, *Hydrogen from North Africa – A Neocolonial Resource Grab*, May 17, 2022. https://tinyurl.com/2d3r78nb.
32. Hamouchene, "Desertec: The Renewable Energy Grab?" *New Internationalist*, Mach 1, 2015, https://newint.org/features/2015/03/01/desertec-long.
33. "MENA Hydrogen Alliance," Dii Desert Energy, https://dii-desertenergy.org/mena-hydrogen-alliance.

34. Dii Desert Energy, *A North Africa-Europe Hydrogen Manifesto*, 2019, https://tinyurl.com/4c9ksct3.

35. Transnational Institute et al., *Just Transition: How Environmental Justice Organizations and Trade Unions are Coming Together for Social and Environmental Transformation*, February 11, 2020 https://www.tni.org/en/justtransition.

36. *Ibid.*

2

An Unjust Transition: Energy, Colonialism and Extractivism in Occupied Western Sahara

Joanna Allan, Hamza Lakhal and Mahmoud Lemaadel

The multiple ecological crises provoked by human activities are linked to and exacerbate the other political, social and economic challenges currently faced by North Africa.[1] In Western Sahara, these challenges and crises are shaped by its continued condition as a colony. This chapter aims to contribute to conversations on a just transition – that is, a transition to "thriving economies that provide dignified, productive and ecologically sustainable livelihoods; democratic governance and ecological resilience" – in Western Sahara.[2] It discusses how extractivism operates in the part of Western Sahara currently occupied by Morocco. The bulk of the analysis focuses on renewable energy developments, because Morocco is widely celebrated on the international stage for its commitments to the so-called "green energy transition".[3] The story told here is different. It aims to raise the voices of the Saharawi population, who are indigenous to Western Sahara. Precisely because renewable energy developments undermine Saharawi self-determination and further (perceived and actual) inequalities between indigenous Saharawis and Moroccans, such developments undermine a just transition.

Below, after giving a brief history of the Western Sahara conflict, the authors, first, identify forms of extractivism in occupied Western Sahara and map who contributes to, and profits from, extractive industries there. While the primary focus of this chapter is on energy developments, it also shines a light on related forms of extractivism, including phosphate extraction, fishing, and sand and agricultural industries. The authors situate their research on extractivism in occupied Western Sahara in wider academic and activist conversations on energy and colonialism globally. The chapter also makes the case for why renewable developments in the occupied territory should be considered forms of extractivism.

Secondly, the authors go on to argue that energy (potentially) produced in occupied Western Sahara contributes to the diplomacy of the Moroccan regime abroad, furthering its colonial hold on occupied Western Sahara. Finally, the chapter asks what a Saharawi just transition would look like. For inspiration, the authors turn to the Saharawi refugee camps and government-in-exile located near Tindouf, Algeria. A small sample of Saharawi initiatives there are analysed in terms of how they might relate to, or inform, a just transition.

A BRIEF HISTORY OF THE WESTERN SAHARA CONFLICT

The Spanish colonization of Western Sahara started in 1884, after the Berlin Conference at which the European states divided up Africa among themselves, with Western Sahara becoming a Spanish possession. At first, the Spanish presence in the so-called "Spanish Sahara" was limited to fishing in the coastal waters and trading with Saharawi tribes. However, the discovery of phosphates, oil and other mineral deposits in the 1940s encouraged Spain to extend its hold over the territory politically, socially and economically.[4]

At the start of the 1960s, a new era of decolonization began, with the Declaration on the Granting of Independence to Colonial Countries and Peoples adopted by the United Nations General Assembly in 1960.[5] Spanish Sahara (Western Sahara) was included in the United Nations list of non-self-governing territories to be decolonized in 1963. Around this time, organized mass movements for Saharawi independence emerged, the first of which was the Vanguard Organization for the Liberation of the Sahara, which was formed in 1968 by Mohamed Sidi Brahim Bassiri. Later, after Spain disappeared Bassiri,[6] a group of young students and members of the Vanguard Organization formed the Popular Front for the Liberation of Saguia El Hamra and Río de Oro (Polisario) in 1973. In the same year, they launched an armed struggle against the Spanish.[7]

Since Morocco's independence in 1956, and with expansionist ambitions, the Moroccan regime has expressed its dream of a "Greater Morocco", which would encompass Western Sahara, Mauritania and parts of Algeria and Mali.[8] Thus, when Spain signalled its plan to hold a self-determination referendum for Saharawis in 1974, Morocco and Mauritania again expressed their own cases for territorial sovereignty over Western Sahara. The two states' claims, that prior to Spain's colonization, Western Sahara had belonged to Greater Morocco and Greater Mauritania, were heard by the International Court of Justice. The latter rejected these claims in an advisory opinion and urged the application of United Nations Resolution 1514 (XV), allowing for the self-determination of indigenous Saharawis.[9] Spain, however, signed an illegal tripartite agreement with Morocco and Mauritania, which divided Western Sahara between the two African countries and gave Spain a 35 percent share of profits from Western Sahara's phosphates reserves, as well as continued access to Western Sahara's fisheries.[10]

In October 1975, Morocco and Mauritania invaded Western Sahara.[11] Tens of thousands of Saharawis fled to refugee camps in neighbouring Algeria, and some of them were bombed with napalm en route.[12] In 1976, Polisario, based in the camps, declared the establishment of the Saharawi Arab Democratic Republic (SADR) in exile. This would be the headquarters of Polisario's armed struggle against Morocco and Mauritania until a United Nations-brokered ceasefire in 1991, which was agreed based on the promise that a self-determination referendum on independence for the Saharawis would be held. This referendum never took place, resulting in a stagnant diplomatic process that stretched on until November 2020 (see below).

Mauritania withdrew from the war in 1979, when it signed a peace treaty with Polisario. Morocco, on the other hand, remains the occupying power of Western Sahara. The United Nations General Assembly has "urge[d] Morocco to join in the peace process and terminate the occupation of the territory of Western Sahara".[13] Polisario currently controls approximately a quarter of the territory of Western Sahara, lying eastwards of the Moroccan-built berm, which is considered the "largest functional military barrier in the world".[14]

Today, some 180,000 Saharawi refugees live on international humanitarian aid in the refugee camps in Algeria, while Morocco continues to pursue settler colonial policies in occupied Western Sahara. Such policies range from forced disappearance and the torture of prisoners of conscience[15] to moving a sizeable Moroccan settler population into the territory (there is no reliable data on the exact proportion of settlers to indigenous Saharawis, but the consensus is that the former today greatly outnumber the latter), as well as cultural appropriation.[16]

The United Nations-brokered ceasefire between Polisario and Morocco which began in 1991 lasted for 29 years and ended on 13 November 2020 after a violent incident. Saharawi civilians had mounted a roadblock at a breach in the military wall near the village of Guerguerat, which is located near the Mauritanian border in a demilitarized buffer zone. Abdelhay Larachi, a Saharawi who helped mount the roadblock, explained: "Our purpose was to close down the illegal breach at Guerguerat [...][it's] a gate through which Morocco passes our plundered natural resources to Mauritania and other countries."[17] Morocco fired on protesters at the site, and Polisario, declaring the ceasefire broken, fired back.

It is no coincidence that the new war was provoked by the Saharawi blocking of the so-called "plunder corridor" at Guerguerat (through which produce from the occupied territory passes on its way to the port of Nouadhibou, from which it is exported globally): extractivism is at the heart of conflict and colonialism in Western Sahara.

EXTRACTIVISM IN OCCUPIED WESTERN SAHARA

Extractivism is a capitalist mode of accumulation through which some regions, usually in the global North, extract the natural resources of other regions, primarily for export.[18] Extractivism has characterized Europe's relationship with the Americas, Africa and Asia since the era of conquest and colonization.[19] Today, in North Africa, extractivism continues in a neocolonial guise.[20] Resources that are extracted range from oil and gas, to precious ore, fish and agricultural goods.[21] Tourism and cultural appropriation are today also widely understood as forms of neocolonial extraction, in that global Southern or indigenous resources, including intellectual or artistic resources, are exploited for the benefit of global Northern populations.[22]

In recent years, it has become increasingly clear that renewable energy projects can also perpetuate or strengthen extractivism. For example, the failed Desertec Industrial Initiative, which aimed to meet approximately 20 percent of Europe's energy needs by 2050 via solar and wind farms built across the Middle East and North Africa, was understood by local activists as a neocolonial capitalist endeavour. Desertec raised concerns about the possible plunder of already scarce water resources, the export of energy to Europe without meeting local energy needs and the colonial language it used to describe the Sahara Desert. The initiative eventually collapsed for financial reasons.[23] Likewise, drawing on research among indigenous communities in Mexico, Alexander Dunlap describes industrial-scale renewable developments as "fossil fuel+", on the basis that such large-scale, corporate-led developments renew and expand the exploitative, capitalist, colonial order of the fossil fuel industry.[24] Renewable energy developments in occupied Western Sahara can be understood as extractivist because they further capitalist modes of accumulation, as well as colonialism and military occupation, and because they use resources in ways that do not benefit or recognize the human rights of local communities.

Apart from one privately-owned wind farm that powers a cement factory, wind energy developments in occupied Western Sahara are all part of the portfolio of a wind energy company called Nareva, which belongs to the Moroccan monarchy's own holding company, Al Mada.[25] Nareva has worked in partnership with the German multinational energy company Siemens (and later Spanish Siemens Gamesa) on all the wind farms that it has developed in occupied Western Sahara. The 200 MW Aftissat farm generates power for industrial users, including the Moroccan state-owned company, the Office Chérifien des Phosphates (OCP).[26] The 50 MW Foum el Oued farm provides 95 percent of the energy needed for running OCP's phosphates mine at Bou Craa.[27] Several more wind farms are planned for occupied Western Sahara, with a combined capacity of over 1,000 MW. There are also plans to expand two existing solar farms in occupied Western Sahara, and to build a third solar farm. Studies exploring the occupied country's geothermal potential are also underway.[28]

While this article focuses on renewable energy developments, it is worth placing such developments within the wider context of extractivism in occupied Western Sahara. Phosphate from the Bou Craa mine – the mining of which involves the draining of precious underground wells – is transported around the world for use in agricultural fertilizers.[29] Industrial-sized greenhouses produce fruits and vegetables for the European Union market.[30] Western Sahara's rich fisheries are also exploited by trawlers from several countries and regions, not least the EU and Russia, using practices that are unsustainable.[31] Locally, several fishing licences have been granted to high-profile figures within the Moroccan *makhzen* (the ruling elite).[32]

Many legal scholars question the legality of such activities, since the resources of an occupied territory cannot legally be exploited without the consent of the people of that territory.[33] In this respect, several international courts have judged claims raised by the government of SADR and by Saharawi solidarity groups.[34]

POWERING THE OCCUPATION: HOW ENERGY DOES DIPLOMATIC WORK FOR THE MOROCCAN REGIME

Energy developments are used to create new forms of dependency outside Morocco on energy that is at least partially sourced in Western Sahara. This arguably creates a diplomatic incentive for other countries to support the occupation. Western Sahara is connected to Morocco's electricity grid via an interconnection in its capital, El Aaiun. A 400 kilovolt (KV) interconnection is now being set up between El Aaiun and Dakhla, a city in the south of Western Sahara:[35] Morocco hopes to connect its grid to the Mauritanian one via Dakhla, with the eventual aim of exporting energy to the West African market.[36] Similarly, at the COP22 climate talks in Marrakech in 2016, Morocco signed a plan to eventually establish energy exports to the European internal market.[37] These plans and agreements represent serious additional obstacles to the self-determination of the Saharawi people. If these interconnections are established, Morocco could create a partial European and West African dependency on energy generated in Western Sahara.

The Moroccan regime also uses the promise of energy to enhance its "soft power", that is the power to persuade or coerce other states to pursue certain policies or take certain actions on the continent.[38] For instance, the Nigeria-Morocco Gas Pipeline (NMGP) is a planned onshore and offshore project that aims to deliver Nigerian gas to West and North Africa, with the potential to supply Europe. NMGP is a huge energy project that has equally huge political implications: while the Nigerian regime has traditionally been a strong supporter of Polisario, its diplomatic stance on the Western Sahara conflict has softened because of this project.[39] This can be seen as a form of energy diplomacy: Morocco makes powerful actors complicit in the occupation and creates alliances for its colonial project through its planned energy system developments.

It is also possible to read Morocco's renewable energy developments in occupied Western Sahara through the lens of greenwashing. To "greenwash" is to deceptively promote a product, policy or action as environmentally friendly. Morocco currently markets itself as "the African leader in the development of renewable energy in Africa".[40] In doing so, it greenwashes its occupation of Western Sahara. The environmental impact of a huge military deployment, of the wall that bisects the country, of phosphate exploitation and of draining freshwater wells to irrigate industrial-sized greenhouses is hidden behind the Moroccan regime's carefully curated "green" image.

Energy developments in occupied Western Sahara bolster a false energy "sovereignty" for Morocco (false because Morocco is not legally the sovereign power of Western Sahara) by making Morocco "energy independent" from other countries in the region, through the expropriation of Western Sahara's resources. As at autumn 2021, Morocco is said to be attempting to hasten the NMGP project, due to Algeria's refusal to continue gas cooperation with Morocco, after cutting diplomatic relations with the kingdom, due in no small part to the Western Sahara conflict.[41] Indeed, in a situation in which the kingdom produces only marginal amounts of its own oil and gas,

Morocco's renewable energy plans are designed to end its reliance on foreign imports of energy. Western Sahara Resource Watch (WSRW) reports that "the energy produced from wind in occupied Western Sahara could constitute 47.20 percent of Morocco's total wind capacity by the year 2030. By that same year, the share of solar power generated in the territory could be between 9.70 percent and 32.64 percent of Morocco's total solar capacity – likely towards the higher end of that range."[42] Morocco is thus seeking to alleviate the energy supply issues it faces through its colonial exploitation of Western Sahara's resources.

POWERING OPPRESSION: SAHARAWI PERSPECTIVES OF THE ENERGY SYSTEM IN OCCUPIED WESTERN SAHARA

The authors gathered data on Saharawi perceptions of the energy system in occupied Western Sahara through participant observation (2015), two focus groups (2019) and 20 in depth, semi-structured interviews (2019-20). The research participants, whose names have been changed, were Saharawis living in occupied El Aaiun or Boujdour, who self-identified as non-activists or as low-profile activists (on the issues of independence, environmentalism and/or human rights).[43] By "energy system", we refer to energy developments, infrastructure, transmission, use and imaginaries (that is, understandings of energy and the meanings attached to energy in any given community). This covers both fossil fuel-powered systems and renewable ones.

The research participants described power outages as "frequent" and gave several explanations as to why this was the case. Dadi said: "[A blackout] happens for political reasons, for example because of late-night demonstrations." Similarly, Hartan explained: "When there is a homecoming of Saharawi political detainees, Moroccan occupying authorities intentionally cut [the power] off in order to screw up the event... I was personally able to see the suffering of media activists and this was when we were trapped during popular demonstrations in conjunction with UN envoy Christopher Ross's visit to occupied El Aaiun... I noticed how their camera batteries had run out so they couldn't monitor the violations..."

Mahmoud reported: "[The energy providers] say [power outages] are due to problems in the grid, but we know that they sometimes cut the power on purpose, when they want to bring secret things to the city or when the young people protest in the streets." As for the "secret things" mentioned by Mahmoud, Fadel reported: "Sometimes they cut [the power] if they bring more soldiers and arms from the airport to the desert, to the berm; they don't want people or activists to know how many arms, tanks and soldiers are entering."[44]

Who is the "they" referred to by Fadel? Is it both the energy provider and the Moroccan state? Or just the latter? The need to ask this question reflects the frequent conflation of the two – energy providers and the Moroccan state – by the research participants. Such a conflation is common in neocolonial contexts and has wide implications for how states are viewed by their citizens. As Idalina Baptista argues, when service providers are perceived as closely associated with a state, the provider-customer relationship becomes understood as a proxy for the state-society relationship.[45] Similarly, Charlotte Lemanski argues that a people's access to public infrastructure shapes their identity as citizens, and their relationship to the state.[46] In Western Sahara, the research participants' experiences of the energy systems deepened the antagonism they felt towards the Moroccan state.

The research participants felt that districts with higher proportions of ethnic Saharawis, such as Maatalla district in El Aaiun city, were prone to more power outages. Some participants were also keen to highlight that the same was true for running water. For example, Ali, aged 31, told us: "These cuts are usual in Maatalla and the other Saharawi suburbs but you can bet the settlers still have their showers."[47] He understood infrastructure – both water and energy in this case – as a tool that is wielded by the colonizer in order to differentiate the settlers from the natives. As in other colonial situations, historic and current, energy infrastructure mediates ethnic segregation.[48] The gender dimensions of power outages should also be taken into account. In Saharawi society, the burden (or pleasure) of childcare and looking after the home falls disproportionately on women and girls. The impact of domestic power outages is therefore gendered. In Mahmoud's words, "[a]s a nomad [a power cut] doesn't affect me, I'm familiar with it. But sometimes we are really in need of electricity, and my wife and kids especially".[49]

All research participants that were connected to the grid felt that their energy bills were "expensive", and in most cases the expense caused significant anxiety. Salka told the authors that she spent over half of her monthly income on energy bills.[50] The research participants also reported that there were several families, especially in the slums of east El Aaiun, who had no electricity at all. Zrug's words are worth quoting extensively, as they reveal the sense of injustice linked to the expensive nature of energy, the importance of popular sovereignty over energy resources, and the wider political issue of natural resource exploitation:

"We are in 2019 and in a few days, we will be in 2020. I know many who do not have electricity at home. A lot of companies have launched big projects of energy power and, not far from these projects, people in El Aaiun are living without electricity... There was a protest in Al Matara neighbourhood concerning the energy and water outages... Wind farms and so on are making the poor poorer and the rich richer. Green energy is being exported out of Western Sahara to other places in Africa and elsewhere. Although this is illegal because it is done by the Moroccan occupation, I feel proud as many elsewhere will be able to use electricity for lighting and other activities. They need electricity, just as I do. I am in favour of benefits for people everywhere and I can compromise my rights for them to produce light for poor people, but under one condition: it has to be for free and not for sale".[51]

Several participants said that energy providers had mischarged them. For example, Mahmoud stated: "They sometimes send us invoices with the wrong amounts. In our house we haven't a lot of machines, so we know how much energy we use." Such mistrust of providers among the research participants was also reflected in the latter's perceptions of who manages and owns energy in occupied Western Sahara. Nguia understood the energy developers to be "foreign companies" with "no humanity". She stated: "The occupying power is letting other countries invest here as a way of getting them to recognize Moroccan sovereignty over Western Sahara." Dadi commented: "These companies contribute to Moroccan colonization and endlessly support its presence." Salka reported: "All profits go to the Moroccan occupation and foreign companies."[52]

All of the research participants voiced a desire to further protest against energy developments, but some were too scared to act on this desire. Those that had attended protests against energy developments in the past reported that they had been beaten by police, and/or had suffered other forms of retribution, including having their social security benefits and/or employment terminated, and/or receiving threats to their relatives and having travel bans imposed on them. While Saharawi-led non-governmental organizations (NGOs) are mostly prohibited from officially registering their existence in occupied Western Sahara, there are two unregistered Saharawi NGOs that have focused their work on campaigning against the exploitation of Western Sahara's natural resources, including in the realm of energy. One is the Saharawi League for Human Rights and Natural Resources, led by Sultana Khaya; the other is the Committee for the Protection of Natural Resources in Western Sahara (CSPRON), whose president is Sidahmed Lemjeyid. Both individuals have suffered serious human rights abuses at the hands of the Moroccan state due to their work: Lemjeyid is currently serving a life sentence in a Moroccan prison,[53] while Khaya is currently under house arrest, having also lost an eye during police torture.[54] Police have recently attempted to rape her; they also raped her sister in the Khaya family home, in retribution for Sultana's activism.[55] This follows a wider, entrenched pattern of gender repercussions against Saharawi activists: the Moroccan state has used gender-based forms of torture against Saharawi political prisoners since 1975, including sexual assault, sexual humiliation and forced sex between prisoners.[56] The energy system in occupied Western Sahara is thus clearly linked to grave and gender-based human rights abuses.

WHAT WOULD A SAHARAWI-LED "JUST TRANSITION" LOOK LIKE? INSPIRATION AND QUESTIONS FROM THE CAMPS

High-level debates on the future of energy systems often fail to engage with indigenous voices.[57] In this section, the authors wish to highlight a handful of Saharawi initiatives that illuminate what a Saharawi just transition might look like. These include low-tech hydroponics for sustainable food production, homes made from re-used plastic, and plans for future towns powered by renewable energy in a free Western Sahara. Nevertheless, we must be aware that such "good practice" cases from the camps are not in themselves a guarantee that the government of an independent Western Sahara would realize a truly just transition in the event of decolonization. Although self-determination is, as we have seen in the previous section, a fundamental component of a Saharawi just transition, it would not guarantee a just transition in and of itself. In this section, then, the authors also wish to highlight questions that would need to be addressed in a future independent Western Sahara in order to ensure a transition away from extractivism towards a just, equitable and regenerative system.

Engineer Taleb Brahim has developed innovative low-tech hydroponics to allow refugee-citizens to grow their own fruits and vegetables, and fodder for their animals. Hydroponics is a type of horticulture that involves growing plants without soil. "Low-tech" here refers to technologies that, according to Brahim, refugee citizens have access to and can afford. This method is designed to be accessible to all, so that even the poorest families can reasonably have access to self-produced, healthy, nutritious food. The hydroponic units recycle water and use naturally produced fertilizers. As

Brahim points out: "If you insist that pesticides and artificial fertilizers are necessary for agriculture, then you will rely on multinationals." Brahim explained that he is driven by an ethic of "sustainability, self-sufficiency and independence for Saharawis". According to Brahim, as far as he knows he is the first person globally to have developed low-tech hydroponics in conditions that are widely considered to be "extreme" in terms of climate and availability of resources. The World Food Programme is now trialling his model in seven other countries with refugee populations, and 1,200 Saharawis in the camps have received the training necessary to allow them to replicate his innovation.[58]

Engineer Tateh Lehbib has created a new construction method that leads to lower household temperatures and higher resistance to winds and floods (traditional houses are made using adobe, which crumbles in the rain). His method relies on cheap materials – recycled water bottles – and can be easily replicated by anyone. The curved dome shape of these buildings keeps interior temperatures lower than in traditional square homes. Especially vulnerable refugees, including the elderly and those with long-term health conditions, have been the first to benefit from Lehbib's new form of housing.[59]

While Brahim and Lehbib have spearheaded innovations that make life in the camps more sustainable, comfortable and healthy, other refugee citizens are looking to the future of the Polisario-controlled zone of Western Sahara. Architect and engineer, Hartan Mohammed Salem Bechri, has designed a future sustainable city or, as he calls it, a "durable, permanent habitat" for humans and their non-human companions (camels and goats), keeping in mind the zone controlled by the Polisario. His design includes areas to house sedentary citizens, as well as zones with amenities for visiting nomads and for animals. The city would be run fully on renewable energy.[60]

Bechri's, Lehbib's and Brahim's innovations speak to a just transition in several ways. A just transition requires an equitable redistribution of resources.[61] Lehbib's and Brahim's innovations reveal a concern for affordability and self-sufficiency. The two engineers have developed ways to ensure the poorest families have access to shelter and healthy food, without reliance on multinationals for raw materials, with their innovations aiming to be economically sustainable (for the families themselves) and environmentally sustainable. Lehbib's designs, although they are just plans at this stage, take into account more than just humans in his vision for a Saharawi future in an independent Western Sahara. Most frameworks for a just transition emphasize the importance of caring for "more than human nature", as well as for human communities. In the Saharawi case, this is in line with nomadic traditions. Traditional ecologically-aware and environmentally-conscious Saharawi practices have been documented back to the eighteenth century at least,[62] while the traditional centrality of, and care for, camels is also well-documented.[63] SADR's indicative Nationally Determined Contribution (NDC) to the Paris Climate Agreements further illustrates its government's intention to contribute to wider, global conversations on addressing the climate crisis and to sustaining these traditional ecologically-aware practices.[64]

More immediately, SADR's Energy Department has plans for rolling out renewable energy in the part of Western Sahara controlled by SADR. The rollout would incentivize a return of refugees to Western Sahara. The department has carried out a scoping study and is looking for funding to pilot some recommendations of the study, which calculates the solar and wind infrastructure that would be needed to power essential public infrastructure, such as hospitals, and takes stock of existing infrastructure, such as communal wells, currently powered by wind turbines, which are used by nomads. The study also looks at options for residential energy. Electrical engineer and co-author of the scoping study, Daddy Mohammed Ali, together with his team, has discussed the option of extensive solar farms. However, they wonder if such a model would be "adaptable enough" for nomadic lifestyles. The team has, therefore, scoped the possibility of providing every Saharawi family with its own portable, independent solar technology. Mohammed Ali explains: "We find that families in the liberated zone often travel, so it's good if they have their independent panel, which they can transport, have their own independent network if you like".[65] Such concern for sustaining non-sedentary lifestyles would be a vital part of a Saharawi just transition, ensuring inclusionary spaces for nomadic practices.

The recent plans for a renewable future set out by the SADR government's Energy Department depart drastically from older plans by the government's Petroleum and Mines Authority (PMA). Through licensing rounds that began in 2005, SADR entered into assurance agreements with four international companies over oil exploration rights in a future independent Western Sahara.[66] The PMA says it has consulted extensively with civil society ahead of launching its licensing round.[67] However, research among Saharawi youth activists found both civil society groups that were supportive of such agreements (on the basis that they challenged Morocco's efforts to exploit petroleum) and those that were critical of such plans on the basis that solar energy is far preferable for environmental reasons.[68] This raises the question of popular sovereignty – integral to any just transition – and how energy-related decisions would be made in a free Western Sahara. Would oil be exploited despite the climate crisis and its disproportionate impact on communities living in hot climates like the Saharawis? Would existing wind and solar farms in occupied Western Sahara be nationalized? A just transition, as well as moving away from fossil fuel extraction, requires democratic, participative decision-making over, and equitable benefit from, energy resources.

On the other hand, there are reassuring aspects in the SADR government's existing energy policy in the camps. For example, when limited opportunities for solar-powered electricity arrived in the camps in the late 1980s (largely via funding from Swiss and Spanish NGOs), the government prioritized three public institutions for electrification: hospitals and pharmacies, primary schools, and women's education and training centres.[69] Arguably, such prioritization reflects SADR's professed dedication to gender equality.[70] As the authors have argued in the previous section, the current energy model in occupied Western Sahara has disproportionately negative impacts on women and girls, due to the frequent power outages and the gender-based oppression of those who oppose the extractivist energy model. A Saharawi just transition, as in other contexts, should, therefore, be a feminist one.[71]

CONCLUSION

The energy system in occupied Western Sahara physically connects Morocco and Western Sahara through transmission lines and cables. As well as providing Morocco with opportunities to greenwash its occupation, Morocco's renewable energy developments in occupied Western Sahara provide it with a false energy "sovereignty", which decreases its energy dependency on neighbours such as Algeria. Furthermore, these developments are used to create new forms of dependency outside Morocco on energy that is at least partially sourced in Western Sahara. These energy developments arguably create a diplomatic incentive for other countries to support the occupation.

For Saharawis, the present energy system in occupied Western Sahara is an oppressive, colonial tool. For Saharawis living in the occupied territory, energy justice is inextricably linked with independence and decolonization. This is also true of Saharawis living in the state-in-exile and refugee camps in Algeria, where innovations based on sustainability, self-sufficiency and self-determination have been trialled. Nevertheless, questions over energy policy in a future free and independent Western Sahara remain. While an end to the Moroccan occupation and full decolonization are integral to a

Saharawi just transition, the SADR government's ability to ensure popular sovereignty over Western Sahara's energy resources will also be of fundamental importance.

NOTES

1. Basem Aly, "5 Key Security Challenges for North Africa in 2019," *Africa Portal,* January 10, 2019, accessed September 29, 2021, https://tinyurl.com/ypw2kn5c.
2. "Just Transition," What We Do, Climate Justice Alliance, accessed September 30, 2021, https://climatejusticealliance.org/just-transition.
3. See, for example, International Renewable Energy Agency (IRENA), "Morocco and IRENA Partner to Boost Renewables and Green Hydrogen Development," press release, June 10, 2021, accessed September 30, 2021, https://tinyurl.com/3y7b79eh.
4. For more information on the Spanish history of resource exploitation in Western Sahara, see: Jesús Martínez-Milán, "La larga puesta en escena de los fosfatos del Sahara Occidental, 1947-1969," *Revista de historia industrial* 26, No. 69 (November 2017), pp. 177-205.
5. United Nations General Assembly, *Declaration on the Granting of Independence to Colonial Countries and Peoples,* A/RES/1514(XV), December 14, 1960, accessed September 28, 2021, https://www.refworld.org/docid/3b00f06e2f.html.
6. Tony Hodges, "The Origins of Saharawi Nationalism," *Third World Quarterly* 5, No. 1 (1983), pp. 28-57, 49.
7. For more on the history of Saharawi nationalism, see Pablo San Martín, *Western Sahara: The Refugee Nation* (University of Wales Press, Cardiff, 2010).
8. San Martin, *Western Sahara,* p. 66.
9. International Court of Justice, *Western Sahara: Advisory Opinion of 16 October 1975,* 1975, accessed September 30, 2021, https://www.icj-cij.org/en/case/61.
10. For more on the Tripartite Agreement, see Chapter 1 of Stephen Zunes and Jacob Mundy, *Western Sahara: War, Nationalism and Conflict Irresolution,* (Syracuse University Press, New York, 2010).
11. Hassan II's so-called "Green March", in which some 350,000 Moroccan citizens descended on Spanish Sahara armed only with Qurans, is often described as "peaceful". Nevertheless, Moroccan troops had been crossing into Spanish Sahara since the preceding summer, and by October 1975 Morocco had launched a "full-scale military invasion that involved several thousand regular troops" (Martin, *Western Sahara,* p. 104). As Zunes and Mundy note, several reputable human rights groups published detailed accounts of extensive attacks against civilian populations and systematic violations of the Geneva Conventions and other laws of war (Zunes and Mundy 2010, p. 114).
12. The Moroccan air force bombed civilian refugee encampments at Guelta Zemmour and Um Draiga (both in Western Sahara) in February 1976, using napalm on four known occasions (Zunes and Munday *Western Sahara,* p. 114).

13. See United Nations General Assembly, *Question of Western Sahara*. A/ RES/34/37, November 21, 1979, accessed September 29, 2021, https://www. refworld.org/docid/3b00f1aa8.html. The vast majority of legal scholars working on the case of Western Sahara also understand Morocco to be the "occupying power". See Joanna Allan and Raquel Ojeda-García, "Natural resource Exploitation in Western Sahara: New Research Directions," *Journal of North African Studies* 27, No. 7 (2021), pp. 4-13, https://tinyurl.com/y2x257u5.

14. Geoffrey Jensen and Lovelace, D. C., *War and Insurgency in the Western Sahara* (US Army War College Press, Carlisle Barracks, PA, 2013), p. 10.

15. See especially Amnesty International, (1996) "Human Rights Violations in Western Sahara," MDE/29/04/96, 1996, accessed September 30, 2021, https:// www.amnesty.org/en/documents/MDE29/004/1996/en/.

16. Robert F. Kennedy Human Rights, Collective of Sahrawi Human Rights Defenders, et al., "Report on the Kingdom of Morocco's Violations of the International Covenant on Economic Social and Cultural Rights in the Western Sahara," 2015, accessed October 2021, https://tinyurl.com/2p8ye556.

17. Telephone interview with Abdelhay Larachi, November 19, 2020.

18. Alberto Acosta, "Extractivism and Neoextractivism: Two Sides of the Same Curse," in M. Lang and D. Mokrani (eds.) *Beyond Development: Alternative Visions from Latin America.* (Rosa Luxemburg Foundation and Transnational Institute, Quito and Amsterdam, 2013).

19. *Ibid.*

20. Hamza Hamouchene, *Extractivism and Resistance in North Africa,* (Transnational Institute, Amsterdam, 2019).

21. *Ibid.*

22. On tourism, see Hamouchene (*Extractivism and Resistance*, p. 4). On cultural appropriation, see: Sunna Juhn and Emily Ratté, "Intellectual Extractivism: The Dispossession of Maya Weaving," *Intercontinental Cry,* 2018, https://tinyurl. com/msbkumbs.

23. Hamouchene, "Desertec: The Renewable Energy grab?," *New Internationalist,* Mach 1, 2015, accessed September 21, 2021, https://newint.org/features/2015/ 03/01/desertec-long.

24. Alexander Dunlap, *Renewing Destruction: Wind Energy Development, Conflict, and Resistance in a Latin American Context,* (Rowman and Littlefield, London, 2019).

25. For more on the links between the Moroccan royal family and energy developments in occupied Western Sahara, see Western Sahara Resource Watch (WSRW), *Greenwashing Occupation: How Morocco's Renewable Energy Projects in Western Sahara Prolong the Conflict over the Last Colony in Africa,* 2021, accessed October 12, 2021, https://tinyurl.com/26nt63kp.

26. WSRW, *Dirty Green Energy on Occupied Land,* 2020, accessed September 22, 2021, https://wsrw.org/en/news/renewable-energy.

27. *Ibid.*

28. *Ibid.*; WSRW, *Greenwashing Occupation.*

29. WSRW, *P for Plunder: Morocco's Exports of Phosphates from Occupied Western Sahara,* 2021, accessed September 23, 2021, https://tinyurl.com/yckpt8by.

30. WSRW and Emmaus Stockholm, *Label and Liability: How the EU Turns a Blind Eye to Falsely Stamped Agricultural Products made by Morocco in Occupied Western Sahara*, 2012, accessed September 23, 2021, https://wsrw.org/files/dated/2012-06-17/wsrw_labelliability_2012.pdf.

31. Saharawi Campaign Against the Plunder (SCAP), *Saharawis: Poor People in a Rich Country*, 2013, accessed September 23, 2021, www.hlrn.org/img/documents/snrw_report_eng2013.pdf.

32. Observatorio de Derechos Humanos y Empresas en el Mediterráneo (ODHE), *Los tentáculos de la ocupación*, 2019, 2013, accessed September 23, 2021, www.odhe.cat/es/los-tentaculos-de-la-ocupacion.

33. Allan and Ojeda-García, "Natural resource exploitation in Western Sahara".

34. *Ibid.*

35. Neimat Khatib et al., "Country Profile: Morocco 2018," *Renewable Energy Solutions for the Mediterranean & Africa RES4MED&Africa*," 2019, accessed October 4, 2021, https://tinyurl.com/3mynh655.

36. Office National de l'Electricité et de l'Eau Potable, "ONEE au Maroc et en Afrique: Activité électricité," 2016, accessed October 4, 2021, https://tinyurl.com/msyee7ef.

37. European commission, *Joint Declaration on the Establishment of a Roadmap for Sustainable Electricity Trade between Morocco and the European Internal Energy Market Between Germany, France, Spain, Portugal and Morocco*, 2016, accessed November 2, 2021, https://tinyurl.com/49btrfxe.

38. Amine Bennis, "Morocco's Contemporary Diplomacy as a Middle Power," *Journal of International Affairs*, August 10, 2019, accessed October 4, 2021, https://tinyurl.com/rnk9u2fj; *North Africa Post*, "Morocco Reaps Diplomatic Gains of Soft Power in Africa," April 10, 2019, accessed October 4, 2021, https://tinyurl.com/y3spch9h.

39. *North Africa Post*, 'Morocco reaps diplomatic gains."

40. Boris Ngounou, "Morocco: Sharing Experience in Renewable Energy with Africa," *Afrik 21*, December 20, 2018, accessed October 4, 2021, https://tinyurl.com/yckpdjpn.

41. Ediallo, P. (5 September 2021). 'Morocco-Algeria Dispute: A Challenge for the Kingdom's Energy Supply'. *Africa Logistics Magazine*, https://www.africalogisticsmagazine.com/?q=en/content/morocco-algeria-dispute-challenge-kingdoms-gas-supply (accessed October 4, 2021); Yahia H. Zoubir, "Why Algeria Cut Diplomatic Ties with Morocco: and Implications for the Future," *The Conversation*, September 12, 2021, accessed October 4, 2021, https://tinyurl.com/m5sj4sfk.

42. WSRW, "Greenwashing Occupation."

43. Joanna Allan, Mahmoud Lemaadel and Hamza Lakhal, "Oppressive Energopolitics in Africa's Last Colony: Energy, Subjectivities, and Resistance," *Antipode*, 54 (August 2021), pp. 44-63, https://doi.org/10.1111/anti.12765.

44. *Ibid.*

45. Idalina Baptista, "Maputo: Fluid Flows of Power and Electricity – Prepayment as Mediator of State-Society Relationships," in Andrés Luque-Ayala and Jonathan Silver (eds.), *Energy, Power and Protest on the Urban Grid, Geog-*

raphies of the Electric City, (Routledge, London, 2016), 112-32, https://doi. org/10.4324/9781315579597.

46. Charlotte Lemanski, "Infrastructural Citizenship: The Everyday Citizenships of Adapting and/or Destroying Public Infrastructure in Cape Town, South Africa," *Transactions of the British Institute of Geographers* 45, 3 (2020), pp. 589-605, https://doi.org/10.1111/tran.12370.

47. Allan et al., "Oppressive Energopolitics."

48. Conor M. Harrison, "The American South: Electricity and Race in Rocky Mount, North Carolina, 1900-1935;" Baptista, "Maputo," in Luque-Ayala and Silver, *Power, and Protest on the Urban Grid.*

49. Interview with Mahmoud (pseudonym), El Aaiun, occupied Western Sahara, May 27, 2019.

50. Allan et al., "Oppressive energopolitics."

51. *Ibid.*

52. *Ibid.*

53. Tone Moe, "Observer Report: The 2017 Trial against Political Prisoners from Western Sahara," *International Observer*, September 21, 2017, accessed June 28, 2021, https://ssrn.com/abstract=3050803.

54. Allan, "Natural Resources and Intifada: Oil, Phosphates, and Resistance to Colonialism in Western Sahara," *Journal of North African Studies 21, 4 (2016),* pp. 645-66, https://doi.org/10.1080/13629387.2016.1174586.

55. Amnesty International, "Saharawi Activist at Risk of Further Assault," MDE 29/4198/2021, May 26, 2021, accessed October 12, 2021, https://tinyurl.com/59637afb.

56. Joanna Allan, *Silenced Resistance: Women, Dictatorships, and Genderwashing in Western Sahara and Equatorial Guinea* (Wisconsin University Press, Madison, 2019).

57. Tristan Loloum, Simone Abram and Nathalie Ortar, "Politicising Energy Anthropology," in Loloum, Abram and Ortar (eds.), *Ethnographies of Power: A Political Anthropology Of Energy,* (Berghahn Books, New York, 2021), https://doi.org/10.3167/9781789209792.

58. Interview with Taleb Brahim, Smara camp, October 11, 2019.

59. *Ibid.*

60. Hartan Mohammed Salem Bechri, *Towards a Nature-Friendly Durable Permanent Habitat in Western Sahara*, Master's dissertation, Hadj Lakhdar University; Allan interview with Bechri, Aaiun camp, October 15, 2019.

61. What We Do, Climate Justice Alliance, "Just Transition", https://climatejustice alliance.org/just-transition/

62. The oldest written documentation that the authors have found is Pierre Raymond de Brisson and Saugnier, *Voyages to the Coast of Africa*, containing an account of their shipwreck on board different vessels, and subsequent slavery, and interesting details of the manners of the Arabs of the desert and of the slaves as carried on in Senegal and Galam with an accurate map of Africa, (G.G.J. and J. Robinson, London, 1792), p. 35. Shipwrecked Mr Saugnier, "abducted" at Boujdour (modern day Western Sahara) by "wandering Arabs", marvels at

the Saharawis' ecologically-aware practices, such as their insistence on only using dead wood for kindling and never any live shrubs.

63. Gabriele Volpato and Patricia Howard, "The Material and Cultural Recovery of Camels and Camel Husbandry among Sahrawi Refugees of Western Sahara," *Pastoralism* 4, No. 7 (2014), https://doi.org/10.1186/s13570-014-0007-4.

64. WSRW, "Saharawi Government Launches Climate Plan," November 8, 2021, https://tinyurl.com/3mmy94kd.

65. Interview with Daddy Mohammed Ali, October 9, 2019.

66. Fadel Kamal, "The Role of Natural Resources in the Building of an Independent Western Sahara," *Global Change, Peace & Security* 27, 3 (2015), pp. 345-59, https://doi.org/10.1080/14781158.2015.1080235.

67. Randi Irwin, "Derivative States: Property Rights and Claims-Making in a Contested Territory," doctoral thesis, Doctor of Philosophy, (New School for Social Research, New York, 2019), p. 79.

68. *Ibid.*

69. Focus group with director and several civil servants of the SADR Energy Department, Rabouni camp, Tindouf, October 7, 2019.

70. For more on official Saharawi nationalist discourses on gender equality, see Joanna Allan, "Imagining Saharawi Women: The Question of Gender in Polisario Discourse," *Journal of North African Studies* 15, 2 (2010), pp. 189-202.

71. For more on the need for a just transition to be feminist, see Lavinia Steinfort, *Ecofeminism: Fueling the Journey to Energy Democracy*, Transnational Institute, 2018, https://tinyurl.com/mw33n9z.

3

Arab–Israeli Eco-Normalization: Greenwashing Settler Colonialism in Palestine and the Jawlan

Manal Shqair

Israel has portrayed pre-1948 Palestine as an empty, parched desert, and has suggested that after the establishment of the State of Israel that parched desert became a blooming oasis.[1] For Israel and its supporters, what surrounds that oasis is a fearsome, degraded and arid Middle East that is sinking in primitiveness and backwardness.[2] Israel's green image, which is set in contrast to a savage and undemocratic Middle East, has been central to its efforts to greenwash its settler colonial and apartheid structure. Israel uses its expertise in agribusiness, afforestation, water solutions and renewable energy technology as constituents of its greenwashing efforts and narrative globally.[3]

The assertion of the environmental superiority of Israel over the rest of the Middle East (and North Africa) was reinforced after it signed the Abraham Accords with the United Arab Emirates (UAE), Bahrain, Morocco and Sudan in 2020. The Abraham Accords are a US-brokered normalization deal that also seeks to reinforce (already existing) normalizing relations with other Arab countries which are not officially part of the agreement, including those that have not yet formalized their longstanding relations with Israel, such as Saudi Arabia and Oman, and those that have, such as Egypt and Jordan. The coalition of these Arab states formed under the umbrella of the Abraham Accords has vowed to increase their collaboration with Israel on issues related to security, the economy, health, culture and the environment, among others.[4] In the last two years, under the deal, Israel and these normalizing Arab states have signed a number of memorandums of understanding (MoUs) to jointly implement environmental projects concerning renewable energy, agribusiness and water.[5]

The Palestinian National Committee of the Boycott, Divestment and Sanctions movement (BNC), which is working to end international complicity with Israeli oppression, defines normalization as "participation in any project, initiative or activity, local or international, that brings together (on the same platform) Palestinian (and/or Arabs) and Israelis (individuals and institutions)".[6] The BNC elaborates that acts of normalization do not meet the conditions set by the BNC concerning the Palestinian right to self-determination, dismantling Israel's three-layered system of oppression (settler colonialism, apartheid and military occupation) and the right of Palestinian refugees to return to their homes as enshrined in United Nations Resolution 194.[7] Israel uses normalization to naturalize its apartheid settler colonialism. In this vein, the Palestinian think tank Al-Shabaka observes that so-called environmentally friendly collaborative projects between Israel and Arab states are a form of eco-normalization.[8] Eco-normalization is presented in this chapter as the use of "environmentalism" to greenwash and normalize Israeli oppression, and the environmental injustices resulting from it in the Arab region and beyond.

This chapter investigates eco-normalization both in Palestine and in the Jawlan (the occupied Syrian Golan Heights),[9] and pursues two questions: 1) How does eco-normalization as a tool of greenwashing undermine the Palestinian anti-colonial struggle? 2) In what ways does eco-normalization hamper a just agricultural and energy transition in Palestine, which is inextricably linked to the Palestinian struggle for self-determination? The chapter also introduces the concept of eco-*sumud* (eco-steadfastness) in the face of Israeli oppression, and its role in countering the greenwashing function of eco-normalization.

ECO-NORMALIZATION PROJECTS

On 8 November 2022, during the 27th United Nation Climate Conference of the Parties (COP27) in Sharm El Sheikh in Egypt,[10] Jordan and Israel signed a UAE-brokered MoU to continue a feasibility study on two interlinked projects, called Prosperity Blue and Prosperity Green, which together constitute Project Prosperity. According to the terms of the agreement, Jordan will buy 200 million cubic metres of water annually from an Israeli water desalination station, which will be established on the Mediterranean coast (Prosperity Blue). The water desalination station will use power produced by a 600 MW solar photovoltaic plant that will be constructed in Jordan (Prosperity Green) by Masdar, a UAE state-owned renewable energy company.

The parties to the agreement intend to submit more concrete plans regarding the implementation of the projects at COP28, to be held in the UAE in 2023.[11]

The idea of Project Prosperity was first proposed by Eco-Peace Middle East, an Israeli-Jordanian-Palestinian non-governmental organization that promotes environmental normalization between the three parties, within the framework of the "Green Blue Deal for the Middle East", an initiative that claims to address water and energy issues in Israel, Palestine and Jordan. Although Palestine is a party to this deal, it has no role in Project Prosperity.[12]

In another development, a few months ahead of COP27, in August 2022, Jordan joined Morocco, the UAE, Saudi Arabia, Egypt, Bahrain and Oman in signing an MoU with two Israeli energy companies to implement renewable energy projects in these countries. Enlight Green Energy (ENLT) and NewMed Energy (henceforth ENLT-NewMed), the two Israeli companies involved in this enormous energy project, will initiate, finance, construct, develop and operate renewable energy plants on Arab lands. These "green" energy projects will include wind and solar energy production, and energy storage. While ENLT specializes in renewable energy projects, NewMed is a natural gas and oil company, and both, particularly NewMed, play a key role in strengthening normalization ties with Arab states through both fossil fuel-based and green energy deals.[13]

PROSPERITY BLUE: ISRAEL QUENCHES PARCHED JORDAN

A decades-long water crisis in Jordan has deepened in recent years. Mainstream media suggest the reason for this is the increasing number of Syrian and Iraqi refugees Jordan hosts, in addition to the climate crisis. Indeed, the influx of refugees who have fled imperialist wars waged against their countries has rendered Jordan incapable of meeting the rising demand for water.[14] However, placing the blame solely on Syrian and Iraqi refugees for the worsening water shortage without highlighting the root cause of that shortage – Israeli usurpation of Jordan's water – is racist and xenophobic. It also deflects attention from Israel's role in making Jordan a parched country. For decades, Israel has depleted Jordan's water resources to achieve economic and political gains in the region. The greenwashing framings of Prosperity Blue in Israeli and Western media outlets absolve Israel of its responsibility for the water crisis in Jordan.[15]

Following the signing of the MoU for Project Prosperity in 2022, *Times of Israel* commented that "Jordan is one of the world's most water-deficient nations. The nation [...] faces dire water prospects as its population expands and temperatures rise. Israel is also a hot, dry country, but its advanced desalination technology has opened opportunities for selling freshwater."[16] This statement reflects the core of Israel's greenwashing narrative of environmental benevolence and stewardship.[17] Israel has always depicted itself as a dry country which, despite this, and unlike its Arab neighbours, has developed the technology needed to efficiently manage its scarce water resources and mitigate the climate crisis. In the last two decades, Israel has exalted its advanced water technology and celebrated its success in water desalination.[18] According to this narrative, as an "environmental altruist", Israel always seeks to put its technology at the service of its parched neighbour, Jordan, even during times of tension between the two countries. This view is reflected in a 2021 comment published in *The Hill*, apropos of Project Prosperity: "Israel and Jordan have a long history of collaboration on water, even amid political tensions. Ever since the 1994 Israel-Jordan Peace Treaty, Israel has been storing some of the kingdom's Jordan River allocations in the Sea of Galilee and discharging the supplies as needed."[19] This is false. Israel has not been "storing" some of the kingdom's "allocations" in the Sea of Galilee. Rather, it has been plundering Jordan's share of water from the Jordan and Yarmouk rivers, against the expressed will of Jordan (and this was especially so in the past). Nor has Israel discharged "the supplies as needed"; rather, it continues to hoard Jordan's usurped water.[20]

Historically, the Jordan River was one of Jordan's main sources of water and also supplied water to the rest of *Bilad al-Sham* (the Levant region): Palestine, Syria and Lebanon. After the creation of Israel in 1948, this changed dramatically. In the 1950s, the Jewish National Fund (JNF), an Israeli parastatal organization, drained Lake Hula and its surrounding swamps in the north of historic Palestine (present-day Israel and the occupied Palestinian territories).[21] The Israeli government claimed this was necessary to increase farmland as part of the nascent state's efforts to "make the desert bloom". The project not only failed to expand "productive" agricultural land for newly arrived Jewish settlers from Europe, but it also caused substantial environmental damage, destroying the natural habitats of numerous animal and plant species.[22] It also severely affected the quality of water flowing to the Sea of Galilee (Lake Tiberias), one of the largest sources of freshwater in the country. Furthermore, the deteriorated water quality in the Sea of Galilee disrupted the flow of water in the Jordan River.[23]

In the same period, Mekorot, the Israeli national water company, started the construction of Israel's national water carrier, which was built to divert the water of the Jordan River from the West Bank and Jordan to serve Israeli settlers along the coast and Jewish settlements in the Naqab (Negev) Desert.[24] Following the Israeli occupation of the rest of Palestine (the West Bank, including East Jerusalem, and the Gaza Strip – together, the occupied Palestinian territories) in 1967, Israeli plundering of water from the Jordan intensified. The Jordan River, and especially its lower part, is now no more than a creek full of dirt and sewage water.[25]

The lower Jordan River is particularly affected by Israeli practices because it is fed by the outlet of the Sea of Galilee and the Yarmouk River. The Yarmouk is equally negatively affected by the Israeli plunder of Palestinian and Arab water resources. It has its source in Syria and is the largest tributary to the Jordan River. Syria, Jordan and occupied Palestine/Israel are the three co-riparian parties which draw on the Yarmouk, with Syria retaining the largest share and Jordan and Israel sharing the rest. Palestinians are totally denied access to the Yarmouk by Israel. Before 1967, Israel had limited access to the river, but after the occupation of the Jawlan in that year it extended its direct territorial control by a further two miles of the river and began to exploit more of its water. The Israel-Jordan peace agreement signed in 1994 maintained Israeli control over the amount of water Jordan can utilize from the river. Israel has forced Jordan to accept the construction of infrastructure that ensures Israeli capture of the excess flows of the Yarmouk.[26]

The infrastructure of dams and wells built by Syria (whose share in the Jordan River Israel totally denies) ensures that most of the Yarmouk is utilized by Syrians. Syria claims that this infrastructure was built so as to limit Israel's exploitation of the Yarmouk, especially as the latter refuses to withdraw from the Jawlan. This, along with Israeli exploitation of the Yarmouk, Jordan's acceptance of an unfair agreement, and a myriad other bureaucratic issues, has decreased the amount of water from the Yarmouk that Jordan can access. The deterioration of the river flow due to the decline in precipitation will reduce Jordan's share even more in the upcoming years.[27]

It is thus clear that the innocent and benevolent rhetoric behind Prosperity Blue hides Israel's role in looting Palestinian and Arab water. Instead of appropriating and commodifying water, in the form of selling it to Jordan, Israel should give the usurped water it continues to hoard back to Jordan. Far from doing so, through Prosperity Blue Israel denies its responsibility for water scarcity in Jordan and claims to be offering a solution, portraying itself as an environmental steward and a regional waterpower.

Vanquishing the (Arab) heart of darkness: An Israeli greenwashing trope

The two renewable energy projects on the eco-normalization agenda, Prosperity Green and ENLT-NewMed, bolster the image of Israel as a hub for creative renewable energy technologies. In upraising Israel in this regard, the mainstream narrative omits that its innovations in the energy sector are predicated on (green) energy colonialism in Palestine and the Jawlan. Energy colonialism refers to foreign companies and countries plundering and exploiting the resources and land of countries and communities in the global South to generate energy for their use and benefit. Perpetuating the North/South dichotomy, energy colonialism also wreaks havoc on the socioeconomic life of local populations in the South, along with their environments. Green energy colonialism includes the appropriation and plunder of green sources of energy while maintaining the same political, economic and social structures of power asymmetry between the North and the South. Energy colonialism is ingrained in the colonial capitalist paradigm of power, exploitation, dehumanization and otherness, and persists decades after many parts of the world entered the post-colonial era.[28] In Palestine and the Jawlan, energy colonialism, including through green sources of energy, is one facet of Israeli settler colonialism. Israel employs it as a means, among others, to dispossess and ghettoize Palestinians and Jawlanis (the 26,000 Syrians currently living in the Israeli-occupied Jawlan) in ever-smaller enclaves, while expanding Israeli-Jewish supremacy on their land. Both Prosperity Green and ENLT-NewMed can also be seen as colonialist energy projects that enable Israel to continue its settler colonial project and geopolitical power in the Middle East and North Africa, under cover of a greenwashing narrative.

Prosperity Green

According to the terms of Prosperity Green, Jordan will sell to Israel for $180 million per year *all* the electricity generated from the solar farm to be built on its land. The proceeds will be split between the Jordanian government and Masdar, the Emirati firm that will build the solar farm. The rationale is that Israel will not need to use its energy to operate the water desalination station that will supply Jordan with 200 million cubic metres of water annually. This is part of the Israeli goal of strengthening both its energy and water desalination sectors. Water desalination, which Israel seeks to rely on as its main source of water by 2030, is energy-intensive, accounting for 3.4 percent of its energy consumption.[29] Israel is thus seeking to increase its access to alternative sources of energy, with Prosperity Green offering one such source.[30]

The deal does not allow Jordan, whose imports of fossil gas account for 75 percent of its energy sources, to receive energy from the project and to leverage its own energy sector.[31] Thus, while the country's solar energy will be extracted, its heavy reliance on imported fossil gas will remain. Jordan will continue to receive gas from Israel, which since 2020, after the infamous 2014 gas agreement was struck between the two countries, has become a major exporter of fossil gas to the country. According to the $10 billion deal, Leviathan, a natural gas field in the Mediterranean over which Israel exercises control, will supply Jordan with 60 billion cubic metres of gas over 15 years.[32] Thus, Jordan will remain hostage to imports of natural gas (particularly from Israel), while it exports its own green energy in order to receive desalinated water from Israel![33]

In the way it is designed to empower Israel's renewable energy sector while maintaining Jordan's reliance on Israeli fossil energy sources, Prosperity Green is a form of energy colonialism – or, more specifically, green colonialism. This is clear from the fact that the solar farm will be built in Jordan rather than in Israel. Consider this 2021 quote from *Axios*, an American news website: "The logic was that Israel needs renewable energy but lacks the land for massive solar farms, which Jordan has."[34] This is echoed by Karine Elharrar, previously Israeli energy minister: "Jordan, which has abundance of open spaces and sunlight, will help advance the State of Israel's transition to green energy and to achieve the ambitious goals we have set, and Israel, which has an excellent desalination technology, will help tackle Jordan's water shortage."[35] This hierarchical categorisation of the land, where the desert is perceived as inferior to cultivated/green land, is informed by the Zionist discourse, which portrays the creation of Israel on the ruins of hundreds of destroyed Palestinian villages as redeeming the land.[36] Such discourse seeks to legitimize and moralize Israel's actions: it depicts Israel as a moral and progressive steward of land efficiency, rather than an immoral settler colonial and apartheid regime.

In line with this discourse of greenwashing and redeeming the land, the construction of the solar plant is considered a favour to Jordan: under the beneficent Abrahamic Accords, barren, "unproductive" land in Jordan will become productive thanks to Israel's environmental development and benevolence. In effect, Prosperity Green moralizes and legitimizes green grabbing and green colonialism as progressive acts that deserve praise rather than condemnation.

ENLT-NewMed

ENLT-NewMed is also portrayed as demonstrating Israeli environmental and moral superiority over its Arab neighbours, including Jordan. After reaching an agreement to develop energy projects with Jordan, Morocco, the UAE, Saudi Arabia, Egypt, Bahrain and Oman, ENLT stated that the project "will bring to light the great experience and expertise of the two Israeli companies in the field of energy".[37] Bringing to light the "experience" and "expertise" of Israel keeps in the dark the experiences of Palestinian and Jawlani struggles against Israeli energy colonialism. Although ENLT-NewMed presents itself as helping to meet the energy needs of seven Arab countries, it too should be understood as an act of energy colonialism, for two main reasons. First, ENLT-NewMed aims to further integrate Israel and place it in a dominant position in the Arab region's economic and energy spheres, thereby creating new dependencies (via energy access and control) that further the normalization agenda and position Israel as an indispensable partner. Second, it will allow ENLT and NewMed, two companies that are deeply involved in Israeli energy projects, to normalize and finance their colonial activities in occupied Palestine and the Jawlan. ENLT operates several renewable energy projects in the Jawlan, with the support of the Israeli government, including Emek Habacha, Ruach Beresheet and Emek Haruchot. ENLT has a 41 percent and a 60 percent stake, respectively, in the first two of these,[38] which are funded by a consortium led by Hapoalim Bank, listed in the United Nations database of businesses and companies that are complicit in the illegal Israeli settlements in the West Bank.[39] ENLT is also involved in renewable energy projects in illegal West Bank settlements. It is developing a 42 MW wind turbine project (in which it holds a 50.15 percent stake) in Yatir Forest, which is located in the Naqab Desert and parts of the West Bank.[40]

The wind farm projects in the Jawlan and in parts of the occupied Palestinian territories are a part of Israel's plans to increase its renewable energy sources. The Jawlanis have protested for years to assert their sovereignty over their land and resources, and they consider these projects another Israeli tool to take over their land.[41] Israel already controls 95 percent of the Jawlan, which it manages for the benefit of around 29,000 illegal Israeli settlers living in 35 settlements in the area.[42] The wind turbines, as a green colonial project, are further disrupting the sustainable relationship between the Jawlani people and their land: since their construction started, the Israeli authorities have restricted Jawlanis' access to their agricultural land. The projects will affect 3,600 dunams (890 acres) of apple, grapes and cherry orchards belonging to Jawlanis. The fight of the Jawlanis against these wind farms is part of decades-long resistance to settler colonial expulsion, plunder of resources and denial of their indigenous sovereignty and identity in relation to the land.[43]

NewMed Energy, which specializes in the extraction of natural gas from the Eastern Mediterranean, is as complicit as ENLT in entrenching Israeli apartheid and settler colonialism. Formerly Delek Drilling, in 2022 it changed its name following its growing business in Arab countries, primarily Jordan, the UAE and Egypt.[44] NewMed Energy has pioneered the Israeli natural gas sector in the Eastern Mediterranean and was involved in most of the Israeli discoveries of gas in the Mediterranean in the last 30 years. One of these, and the most remarkable, was the 2010 discovery of the Leviathan field, the largest natural gas reservoir in the Mediterranean, in which NewMed Energy holds a working interest of 45.3 percent.[45] A year earlier, in 2009, the company, jointly with Chevron, discovered the Tamar natural gas reservoir, also in the Mediterranean.[46] Together, these two gas reservoirs hold an estimated 26 trillion cubic feet of natural gas and have elevated Israel's status in the regional and global energy market, representing a source of geopolitical and economic might in the region and beyond.[47] They are expected to meet Israel's electricity needs for 30 years and to allow it to be a regional exporter of gas (including to the European Union – especially now, in the context of the war in Ukraine). Both Egypt and Jordan (as mentioned earlier) currently import Israeli gas from the Leviathan and Tamar fields.[48]

For years, Israel has denied Lebanon's declaration that a portion of the Leviathan reservoir falls within its Exclusive Economic Zone.[49] Israel has also denied Lebanon's share in Karish, another gas field discovered by NewMed Energy in 2013.[50] In 2022, the two sides reached an unfair US-brokered agreement through which Israel retains full access to the Karish field.[51] Lebanon is only able to develop the Qana field, another disputed gas reservoir in the Mediterranean which could contain nearly 100 billion cubic metres of natural gas, while paying some agreed royalties to Israel. The agreement reflects the asymmetrical power relations between Lebanon on the one hand, and Israel and its staunch supporter, the US, on the other.[52] Meanwhile, and in response to Lebanon's claims, Israel has been intensifying its militarization of the Mediterranean by increasing the presence of its warships there.[53] Even as it fuels disputes over these gas fields and strengthens the position of a highly militarized settler colonial power both regionally and globally, NewMed Energy nevertheless emphasizes its commitment to developing green energy sources.[54]

The main operator and negotiator with Israel regarding its share in Qana is TotalEnergies, a French company which holds a 35 percent stake in the field. TotalEnergies is part of a consortium operating Qana that includes Italian ENI and also the state-owned Qatar Energy, which holds a 30 percent share in the project (Qatar Energy replaced Novatek, a Russian company that was excluded due to the sanctions imposed on Russia following its invasion of Ukraine).[55] Qatar Energy's role in the development of the Qana field was approved by the Israeli government,[56] and makes Qatar complicit in normalization with Israel in the field of energy. This overt normalization by Qatar, which from the 1990s had been engaged in under-the-table normalization, is to the detriment of the Palestinians[57] and reflects a pattern seen in relation to other Arab countries as well: the fact that Egypt, Morocco, Jordan and the UAE are involved in various energy projects (including green ones) with Israel or Israeli companies shows that this normalized relationship is no longer considered scandalous for Arab leaders.

In regard to Egypt, another important point is the fact that the Israeli Leviathan gas bought by the Egyptian government is extracted and transferred via Israel's violent and illegal exercise of control over the Palestinian Exclusive Economic Zone,[58] which is manifested in systematic attacks on Palestinian fishermen by the Israeli navy.[59]

Egypt's relations with Israel concerning natural gas go beyond electrifying Egyptian households. Egypt and Israel, along with Cyprus and Greece, are part of a consortium that aims to supply Europe with gas from the Mediterranean, now as part of EU efforts to end reliance on Russian gas. The consortium aims to build a new pipeline system that will carry gas from Israel and Cyprus to liquefication facilities in Egypt, after which it will be transported to Europe by tankers. The project also includes building a liquification facility on the eastern shore of Cyprus, and constructing "a floating liquefication facility as part of the expansion of the Leviathan field".[60] It is not clear yet if the proposed pipeline and liquefication system will replace the planned construction of the Eastern Mediterranean (EastMed) Pipeline, but it seems that it is being considered as an alternative to EastMed, whose feasibility has been questioned.[61]

No matter what forms the energy projects in the Mediterranean take, two important facts remain. First, the suffering under siege and the traumatic experiences of violence and dehumanization endured by Palestinian fishermen and people in the Gaza Strip cannot be dissociated from the highly militarized gas reservoirs Israel controls in the Mediterranean, and the projects linked to them. Second, the EU is once again showing its hypocrisy: treating Palestinian and Jawlani peoples as less human than Ukrainians by importing Israeli gas as part of efforts to hold Russia accountable for its invasion of Ukraine. As for Egypt and the other normalizing Arab states, by entering into dirty energy deals in the Mediterranean they are now openly taking part in the systematic dehumanization of Palestinians and Syrians at the hands of both Israel and the EU. The dehumanization of the colonized, and the complicity of Arab states in this, are greenwashed by the EU and Israel as they collaborate in what is portrayed as a transition to a greener future and lower-carbon economy. In this respect, portraying fossil gas as a clean source of energy is misleading, to say the least.[62]

ECO-NORMALIZATION: A VIOLENT ONSLAUGHT ON THE PALESTINIANS' RIGHT TO SELF-DETERMINATION

Eco-normalization allows Israel to place itself in both the energy and water sectors regionally and globally, thereby reinforcing its political and diplomatic power in the region and worldwide. With the exacerbating climate and energy crises, countries reliant on Israeli energy and water (as well as technology) may start to see the Palestinian struggle as a matter of less importance than their water and energy security. This allows eco-normalization to reinforce the role of Israeli greenwashing as a money-making machine for Israeli companies while undermining a just agricultural and energy transition in Palestine, inextricably linked to the Palestinian struggle for self-determination.

Mekorot, a major player in Israeli water desalination, has been able to position itself as a leader in water desalination and solutions globally, partly through Israel's greenwashing narrative. For example, Mekorot is responsible for 40 percent of Cyprus's seawater desalination.[63] Its technology and "expertise" generate millions of dollars in revenue from water projects developed across the world, particularly in the global South.[64] This money finances its, and the Israeli government's, practice of water apartheid against the Palestinian people, with Mekorot playing a significant role in building Israeli water apartheid infrastructure, controlling most of the Palestinian water resources in the West Bank and diverting them to illegal Israeli settlements (in addition to the company's role in usurping the Jordan River).[65] The company's infrastructure of wells and bypass pipelines is built in such a way as to ensure that Palestinians living in Area C of the West Bank have no access to water,[66] while, at the same time, it helps the Israeli military to confiscate Palestinian water pipelines and other alternative means to access water in Area C.[67] By practising water apartheid, Mekorot creates a coercive environment that aims to force Palestinians from their land and expand illegal Israeli settlements. Mekorot's active enforcement of Israeli water apartheid policies disrupts the function of the land as a source of subsistence through which Palestinians can sustain their life and identity. It threatens the Palestinian agricultural sector and food sovereignty, which are essential components of a just agricultural transition. For instance, Palestinian farming communities in the Jordan Valley are no longer able to rely on agriculture for their livelihoods due to Israeli restrictions on Palestinian access to water and land.[68]

The same story is seen in the blockaded Gaza Strip where, for decades, Israel has been destroying the agricultural sector. Since 2007, the siege of Gaza has restricted Palestinian farmers' access to their agricultural land and has exacerbated the severe water crisis in the strip.[69]

This entrenching of Israeli energy colonialism and apartheid is also evident in the greenwashing functions of Prosperity Green and ENLT-NewMed. Israel denies the colonized Palestinians (and Jawlanis) sovereignty over their energy resources and perpetuates their captivity to its energy market. Israeli control over Palestinian and Jawlani energy resources is an effective tool of settler colonial dispossession and oppression. At the same time, the Gaza Strip, situated not far from the Leviathan and Tamar gas fields, has been living in darkness for years due to Israel preventing Gazans from having full access to electricity. The electricity crisis in the Gaza Strip is further exacerbated during regular Israeli assaults and massacres.[70] Electricity, water, violence, and a myriad other tools are part of the Israeli settler colonial mechanisms that are used to "manage" and control Palestinians in the designated ghettos. Eco-normalizing and greenwashing energy projects provide Israel with financial aid to consolidate its ghettoization policies in respect of millions of Palestinians in the Gaza Strip and beyond.

The wider Palestinian energy sphere is held captive by Israel. Palestinians inhabiting Area C of the occupied West Bank bear the brunt of Palestinian energy dependence on Israel. Palestinians are denied access to the electricity grid in the area, which has been developed by Israel to serve illegal Israeli settlements. Israel also refuses to issue Palestinians with permits to construct solar panels, which could provide an alternative source of energy. Palestinians are thus forced to build solar panels (often funded by NGOs and the

EU) without Israeli permits, which Israel then confiscates and demolishes.[71] Between 2001 and 2016, Israeli policies and practices in Area C caused Palestinians an estimated loss of 65 million euros, in relation to EU-funded support, including solar installations.[72] The solar energy sector in Area C has been established by Palestinian civil society in order to reinforce the steadfastness of communities in the area, and Israel uses its de-development as a tactic to forcibly displace them.[73]

Even as this is taking place, the Israeli solar energy sector is flourishing, due to expanding illegal settlements and solar farms across the West Bank. In 2016, Israeli revenue from electricity generated from solar farms located in both the West Bank and inside Israel reached 1.6 billion shekels (approximately $445 million). As at 2017, Israel operated four large-scale solar fields in the West Bank. All of these are connected to Israel's national grid, which provides electricity to Israeli households in illegal settlements in the West Bank and within Israel.[74] It is worth pointing out in passing here that the strength of the Israeli solar energy sector sits in stark contrast to the fact that tens of thousands of Palestinian (second-class) citizens of Israel living in 35 "unrecognised" villages in the Naqab Desert have no access to electricity (or water, health care and education), as part of the discriminatory Israeli policies that are designed to force them from their land and replace their villages with Jewish-only settlements and JNF-planted pine trees.[75]

ECO-SUMUD: A VISION FOR A JUST TRANSITION IN PALESTINE

While the word *sumud* (steadfastness) has multiple definitions, in this chapter it is defined as a pattern of everyday practices of resistance to, and adaptation to, the daily difficulties of life under Israeli settler colonial rule.[76] *Sumud* also refers to the Palestinian people's persistence in staying on their land and maintaining their Palestinian identity and culture in resistance to Israeli dispossession and claims that posit Jewish settlers as the only legitimate inhabitants of the land.[77] Eco-*sumud*, a term introduced in this chapter, combines Palestinians' everyday practices of steadfastness with environmentally friendly ways of maintaining a strong attachment to the land of Palestine. It covers the indigenous land-based knowledge, cultural values, tactics and tools Palestinians employ to fight back against the violent Israeli settler colonial disruption of their sustainable relationship with the land. The concept of eco-*sumud* is premised on the understanding that the only sustainable solutions to the ecological and climate crises are those that support the quest of the Palestinian people for a just agricultural and energy transition.

One case study we can use to understand what eco-*sumud* means in concrete and empirical terms is the rain-fed agriculture practised by the Palestinian inhabitants of the village of Dayr Ballut.[78] Dayr Ballut, inhabited by more than 3,000 people, is in the district of Salfit in the central West Bank and has a total area of approximately 11,898 dunams (2,940 acres), of which 5,985 (1,479 acres) is arable land. Only 5.2 percent of the village's land is classified as Area B, under mixed Palestinian and Israeli control, with the rest, 94.8 percent, falling within Area C, which is under full Israeli security and civil control. Dayr Ballut faces myriad settler colonial policies and practices that aim to remove the villagers from their land. These include confiscation of its land to build Israeli bypass roads, illegal settlements and the Apartheid Wall. Israel has established a military checkpoint at the entrance of the village, where it harasses and humiliates Palestinians from Dayr Ballut and nearby villages. The checkpoint also restricts the movement of goods and agricultural produce from Dayr Ballut headed for the West Bank markets.[79]

As in other parts of Area C, Palestinians in Dayr Ballut have less water and electricity than they need, with the Israeli authorities controlling a water well in the village.[80] Despite the ongoing Israeli water apartheid and land grab practices, Palestinians in Dayr Ballut continue to preserve their agricultural land. This has been made possible primarily through adhering to the practice of rain-fed agriculture. Rain-fed agriculture in Palestine (known as *Ba'li* in the vernacular Arabic) is defined as "a suite of planting, tillage, and plant protection strategies that exploit soil moisture for growing crops without irrigation".[81] *Ba'li* agriculture also includes the knowledge and techniques Palestinians use to capture water during the rainy season, such as the use of cisterns and terraces. The captured water is used by farming communities to water crops during the long dry season. *Ba'li* farming has a spiritual dimension, as the word *Ba'li* is derived from *Baal*, the god of fertility and rain worshipped by the Canaanite ancestors of the Palestinian people. Despite the ongoing Israeli control of land, water, borders, movement and markets, *Ba'li* agriculture continues to persist as the main form of farming on most of the agricultural land in the West Bank.

What makes rain-fed agriculture in Dayr Ballut stand out from the rest of the West Bank is the flexibility of farmers in adapting to the changing political, economic, social and environmental circumstances over a century of colonialism. Two key qualities characterize the dynamism of rain-fed farming in Dayr Ballut: the diversification of crops to respond to the challenges facing them, and the centrality of women in farming. Seen through the lens of eco-*sumud*, shifting the cropping system and labour structure in Dayr Ballut is a part of the tactics, techniques and knowledge the villagers apply to sustain their relationship with the land.

After Israel's 1967 capture of the West Bank, the Gaza Strip and East Jerusalem, and the resulting opening up of Israel's labour market to thousands of Palestinian men, the labour structure of Palestinian society was transformed, including in the form of an increased role for women in the agricultural sector.[82] Dayr Ballut saw an even greater role for women in agriculture than other areas of the West Bank, and since 1967 women have been the main cultivators of the Dayr Ballut plain (known as *marj* in the vernacular Arabic) and hilly areas. This was accompanied by a shift in the cropping system, to adapt to different challenges. Until 1948, the plain and hills of Dayr Ballut had cultivated local wheat, barley, sesame, tobacco, sorghum and olive trees. Following the creation of Israel in 1948, new crops, all suitable for *Ba'li* farming, started to be cultivated in the village, influenced by the arrival of refugees who were displaced from the nearby coastal towns and who brought with them different seeds as well as expertise in planting the new crops, which they shared with local villagers. The new crops included garlic, okra, tomatoes, onions, watermelons and cantaloupes, and allowed the villagers to diversify the crops they plant and depend on for subsistence. Then, in the 1970s, women in Dayr Ballut replaced wheat and other small grains with olive trees in the hilly parts of the village. One advantage of this is that olive trees do not require a constant presence by the farmer. Indeed, Palestinian farmers facing Israeli hurdles to access their land, use olive trees as a weapon to maintain their existence on the land. Moreover, olive trees are more resilient and adaptable to climatic changes compared to other crops. By planting olive trees, women were able to reduce the cultivation burden on them in the absence of men and to continue to make profits, especially as demand for olive oil began to increase in light of the shrinking of olive groves due to Israeli settlement expansion.

Palestinian women continue to introduce new crops to the valley in response to the changing circumstances, including by growing the profitable *faqqous* (Armenian cucumber). Every year in May, Palestinians from across the West Bank overcome the Israeli imposed spatial fragmentation and gather in the Dayr Ballut plain to celebrate the *faqqous* harvest festival.

Although Israel floods the Palestinian market with its agricultural produce as part of the dependence of the Palestinian economy on the Israeli one, the crops grown in Dayr Ballut are highly competitive in the Palestinian market. Unlike Israeli agriculture, which depends heavily on synthetic fertilizers, *Ba'li* farming in Dayr Ballut is chemical-free, so the village's crops have a superior taste and quality.

In charge of cultivating and harvesting the land since 1967, the women of Dayr Ballut have been able to protect it from Israeli confiscation and unsustainable Israeli cultivation and land management through their creative diversification of crops that are suited to rain-fed farming. Dayr Ballut thus represents a model of eco-*sumud* that should inform the wider vision of decolonization in Palestine (including in the energy sector). This model of a just transition in Palestine is predicated on five main elements. First, rejection of the systematically constructed inferiority of Palestinians to their colonizers in terms of knowledge and culture, which is a prerequisite for anti-colonial transformation. The women in Dayr Ballut are an exemplar of the Palestinians, who see themselves as true stewards of, and carers for, their ancestral land. In this sense, eco-*sumud* is significant as a counter-narrative to Israeli greenwashing discourse, which, as highlighted above, seeks to legitimize and naturalize Israeli settler colonial dispossession and violence. Second, the relationship with the land and its natural resources should be based on reciprocity and interdependence. Third, land, water and knowledge should be collectively shared, rather than monopolized and commodified as a luxury for the few. Fourth, women are primary rather than secondary actors in the anti-colonial struggle for sovereignty and self-determination. Finally, practising eco-*sumud* is ingrained in a belief in the possibility of defeating Israeli settler colonialism, and in the invincibility of the burning desire of the colonized to determine their own destiny.[83]

CONCLUSION

There is an abiding connection between Israeli greenwashing, which is reinforced through eco-normalization, and the consolidation of apartheid and settler colonialism in Palestine and the Jawlan. As demonstrated above, eco-normalization is socially and environmentally unjust and unsustainable as it obstructs energy democracy and food sovereignty, and any attempt to achieve a just energy and agricultural transition in Palestine and the Jawlan. With increasing Israeli violence and settler colonial expansion in the occupied Palestinian territories, the Palestinian anti-colonial struggle is at a critical juncture. The dark tunnel that is the life of Palestinians under Israeli oppression is getting darker. Yet a glimpse of light can be seen that illuminates the Palestinians' long path to liberation: that light is the increasing resistance of the Palestinian people, who refuse to be isolated, dehumanized and obliterated. The struggle to topple Israel's oppressive regime is also part of the wider struggle for self-determination and emancipation of other peoples across the world. Colonial attempts to further isolate Palestine from the rest of the Arab world through eco-normalization can be thwarted through the collective power of Arabs and other peoples. To this end, social movements, environmental groups, trade unions, student associations and civil society organizations in the Arab region and beyond must intensify their protests against their governments until these governments end their normalization with Israel. Alternative media outlets should challenge mainstream media, which render Palestine invisible and irrelevant to the struggle of Arab (and non-Arab) peoples. Both individuals and institutions, especially in the Arab region, should be more vigilant regarding cultural, academic, social and environmental projects and initiatives: before engaging with them, they should investigate their sources of funding, their participants and their agenda. Environmental movements can also support the Palestinian struggle for self-determination by centring and valuing eco-*sumud* as an indigenous knowledge that can inform solutions to, and strategies to mitigate, the climate crisis. Finally, international grassroots movements should increase their support for boycott, divestment and sanctions against Israel.

NOTES

1. Yoav Galai, "Narratives of Redemption: The International Meaning of Afforestation in the Israeli Negev," *International Political Sociology* 11, No. 3 (2017), pp. 273-91. https://doi.org/10.1093/ips/olx008.

2. The creation of Israel on the ruins of Palestinian cities and villages in 1948 represents the culmination of Zionist efforts that started in the nineteenth century. Zionism is a racist (settler) colonial movement that emerged in Europe in the nineteenth century during a period of European conquest in other parts of the world. The Zionist vision of creating a Jewish state in Palestine by ethnically cleansing indigenous Palestinians was backed by colonial powers like Britain, and then the United States. For more about the Zionist movement, see Edward Said, *The Question of Palestine* (Vintage, New York, 1979).

3. Sara Salazar Hughes, Stepha Velednitsk and Amelia Arden Green, "Greenwashing in Palestine/Israel: Settler Colonialism and Environmental Injustice in the Age of Climate Catastrophe," *Political Geography* 1, No. 6 (May 2022), pp. 1-19, https://doi.org/10.1177/25148486211069898.

4. "The Abraham Accords Declaration," Bureau of Near Eastern Affairs, US Department of State, www.state.gov/the-abraham-accords.

5. *Ibid.*

6. Palestinian BDS National Committee, "The BDS Movement's Anti-Normalization Guidelines Explained," October 30, 2022, https://tinyurl.com/yeykd4bu.

7. *Ibid.*

8. Al-Shabaka: The Palestinian Policy Network, "Environmental Normalization in Palestine with Inès Abdel Razek," Podcast, 24:40, July 28, 2022, accessed September 30, 2022, https://tinyurl.com/54enhcat.

9. Throughout the chapter, the Jawlan, the Arabic name for the Syrian territories occupied by Israel, is used instead of the Golan, the Hebrew name for the area.

10. Israeli Ministry of Energy and Infrastructure, "Jordan, Israel, and the UAE Sign MoU to Advance Project Prosperity, Targeting COP28 for Implementation Plan Development," Press release, November 8, 2022, https://tinyurl.com/7pk9j2da.

11. Bruce Riedel and Natan Sachs, "Israel, Jordan, and the UAE's Energy Deal is Good News," *Brookings* (blog), November 23, 2021, https://tinyurl.com/yc779h5s.

12. Gidon Bromberg, Nada Majdalani and Yana Abu Taleb, *A Green Blue Deal for the Middle East*, Eco-Peace Middle East, 2020, https://tinyurl.com/2y7my29w.

13. Sara Zouiten, "Two Israeli Companies to Launch Renewable Energy Projects in Morocco," *Morocco World News*, August 15, 2022, https://tinyurl.com/de9b8tad.

14. Karen Zraick, "Jordan Is Running Out of Water, a Grim Glimpse of the Future," the *New York Times*, November 9, 2022, https://tinyurl.com/3fmmzm97.

15. Mark Zeitoun and Muna Dajani, "Israel is Hoarding the Jordan River – it's Time to Share it," *The Conversation*, December 19, 2019, https://tinyurl.com/53dad4tk.

16. Sue Surkes, "Israel, Jordan UAE, Sign New MoU on Deal to Swap Solar Energy for Desalinated Water," *Times of Israel*, November 8, 2022, https://tinyurl.com/yc5wpj6u.

17. Sara Salazar Hughes, Stepha Velednitsky and Amelia Arden Green, "Greenwashing in Palestine/Israel: Settler Colonialism and Environmental Injustice in the Age Of Climate Catastrophe," *Environment and Planning E: Nature and Space* 1, (2023), pp. 495-513, https://doi.org/10.1177/25148486211069898.

18. Hadas Gold, "Lakes are Drying Up Everywhere. Israel will Pump Water from the Med as a Solution," *CNN*, August 19, 2022, https://cnn.it/3kBT5AA.

19. Sharon Udasin, "Israel, Jordan, UAE Sign Pivotal Deal to Swap Solar Energy, Desalinated Water," *The Hill*, November 23, 2021, https://tinyurl.com/484vutxs.

20. Zeitoun and Dajani, "Israel is hoarding".

21. "Rehabilitation of the Hula Valley," KKL-JNF, https://tinyurl.com/wz38yf7t.

22. *Ibid.*

23. Zeitoun and Dajani, "Israel is hoarding".

24. The Grassroots Palestinian Anti-Apartheid Wall Campaign, *Israel's Water Company Mekorot Nurturing Water Apartheid in Palestine*, https://tinyurl.com/285fzc3f.

25. Amnesty International, "The Occupation of Water," November 29, 2017, https://tinyurl.com/3yedrnnd.

26. Zeitoun, Dajani, et al., "The Yarmouk Tributary to the Jordan River II: Infrastructure Impeding the Transformation of Equitable Transboundary Water

Arrangements," *Water Alternatives* 12, No. 3 (2019), pp. 1095-1122, https://tinyurl.com/29p2p3am.

27. *Ibid.*

28. Farhana Sultana, "The unbearable Heaviness of Climate Coloniality," *Political Geography*, (March 2022), pp. 1-14, https://doi.org/10.1016/j.polgeo.2022.102638.

29. Israeli Ministry of Finance, "Background – Seawater Desalination in Israel," 2021, https://tinyurl.com/3wmrdane.

30. *Ibid.*

31. Tom Ready, "Jordan's Renewable Sector: Keeping Up the Momentum," *The London School of Economics and Political Science*, November 12, 2020, https://tinyurl.com/ydkahzwj.

32. Suleiman Al-Khalidi, "Jordan gets First Natural Gas Supplies from Israel," *Reuters*, January 1, 2020, https://reut.rs/3kxb1fs.

33. Barak Ravid, "Scoop: Israel, Jordan and UAE to Sign Deal for Huge Solar Farm," *Axios*, November 17, 2021, https://tinyurl.com/ym84amxz.

34. *Ibid.*

35. Udasin, "Israel, Jordan, UAE".

36. The Zionist discourse of land redemption is evident in the narrative constructed around the afforestation project led by the JNF to conceal the remnants of 86 destroyed Palestinian villages after 1948. See : Galai, "Narratives of Redemption."

37. Zouiten, "Two Israeli Companies".

38. "Enlight Renewable Energy," Who Profits, https://tinyurl.com/2p8bxuty.

39. "Named: 112 Companies Linked to Illegal Israeli Settlements by the UN," *Middle East Eye*, February 12, 2020, https://tinyurl.com/yc6u24jy.

40. "Enlight Renewable Energy," Who Profits.

41. Dajani, "Danger, Turbines: A Jawlani Cry against Green Energy Colonialism in the Occupied Syrian Golan Heights," *The London School of Economics and Political Science*, 2020, https://tinyurl.com/5fy396cj.

42. "Illegal settlements," *Al-Marsad*, https://tinyurl.com/47cxt2xk.

43. "Syrians in Occupied Golan Heights Protest over Israel Wind Farm Project," *Middle East Eye*, December 7, 2020, https://tinyurl.com/36ynn4ea.

44. Ricky Ben-David, "Israeli companies Tout Big Plans to Develop Renewable Energy Projects in MENA," *The Times of Israel*, August 17, 2022, https://tinyurl.com/4byh5jvn.

45. *Ibid.*

46. Yuri Zhukov, "Border Disputes and Gas Fields in the Eastern Mediterranean," *Foreign Affairs*, https://fam.ag/3kz5Uvs.

47. *Ibid.*

48. Aidan Lewis and Ari Rabinovitch, "Israel Starts Exporting Natural Gas to Egypt under Landmark Deal," *Reuters*, January 15, 2020, https://reut.rs/402MoZi.

49. Paul Khalifeh, "Karish Gas Field: Are Lebanon and Israel Preparing for War?," *Middle East Eye*, September 2, 2022, https://tinyurl.com/2s4273ny.

50. *Ibid.*

51. "Israel Signs Agreement on Gas Field Shared with Lebanon," *Al Jazeera*, November 15, 2022, https://aje.io/ehf3c4.

52. Danny Zaken, "Sidon-Qana Gas Field could Contain 100 BCM," *Globes*, October 28, 2022, https://tinyurl.com/yey8rzza.
53. Zhukov, "Border Disputes".
54. Ben-David, "Israeli Companies".
55. "Qatar Replaces Novatek in Lebanon's Qana Gas Field Project," *The Maritime Executive*, January 30, 2023, https://tinyurl.com/4uu95p6a.
56. Lazar Berman, "Qatar to Join Consortium Pumping Gas from Offshore Field Straddling Lebanon, Israel," *Times of Israel*, November 23, 2022, https://tinyurl.com/48tx9pu4.
57. Gerald Michael Feierstein, and Yoel Guzansky, "Two Years On, What's the State of the Abraham Accords?," *Middle East Institute*, September 14, 2022, https://tinyurl.com/4n7ehmx4.
58. "Siemens and Chevron: Stop Fueling Apartheid and Climate Disaster," Global Campaigns, BDS Movement, https://tinyurl.com/4phxdfes.
59. Mohammed al-Hajjar, "Three Palestinian Fishermen Wounded by Israeli Navy off Gaza's Coast," *Middle East Eye*, June 16, 2022, https://tinyurl.com/3nkcy6r4.
60. Zaken, "Israel Intensifies Efforts to Increase Gas Exports," *Globes*, February 8, 2023, https://tinyurl.com/4889kmz4.
61. *Ibid.*
62. BDS, "Siemens and Chevron".
63. "Water resource management," What we do, Mekorot, https://www.mekorot-int.com/water-resources/.
64. "Israel's Mekorot Targets World Water Market," *Reuters*, June 7, 2012, https://tinyurl.com/nhwdaswd.
65. PENGON/FOEI, *EU funding for Mekorot: Aiding and Abetting the Israeli Settlement Project*, https://tinyurl.com/4r2fx97r.
66. Under the Oslo Accords signed between Israel and the Palestine Liberation Organization (PLO) in 1993, the West Bank was divided into three areas: A, B and C. Area A, constituting 18 percent of the West Bank, is under the civil and security control of the Palestinian Authority. Area B, comprising 21 percent, is under mixed Israeli and Palestinian control. Area C, which constitutes 61 percent of West Bank land, mainly fertile agricultural land, is under full Israeli control.
67. PENGON/FOEI, *EU funding for Mekorot*.
68. Lina Isma'il and Dr. Muna Dajani, "Agriculture in Palestine," *Heinrich Boll Stiftung*, August 19, 2021, https://tinyurl.com/y4baccwe.
69. *Ibid.*
70. Hadeel Al Gherbawi, "Power Outage adds to "Unbearable" Hardships for Palestinians in Gaza," *Al-Monitor*, August 12, 2022, https://tinyurl.com/2yufrnhb.
71. "COP 26: Destruction of Solar Panels in Area C of the West Bank, an Attempt to Undermine Palestinian Development of Sustainable Energy," *Al-Haq*, November 4, 2021, https://www.alhaq.org/advocacy/19157.html.
72. *Ibid.*
73. *Ibid.*
74. Infographics in "Israeli Solar Fields in the West Bank," Who Profits, January 2017, https://tinyurl.com/mv5t45e7.

75. Amnesty International, "Israel/OPT: Scrap Plans for Forced Transfer of Palestinian Bedouin Village Ras Jrabah in the Negev/Naqab," May 22, 2022, https://tinyurl.com/ymwydy9c.
76. Anna Johansson and Stellan Vinthagen, *Conceptualizing Everyday Resistance: A Transdisciplinary Approach* (Routledge, New York, 2020), pp. 149-52.
77. *Ibid.*
78. Omar Tesdell, Yusra Othman and Saher Alkhoury, "Rainfed Agroecosystem Resilience in the Palestinian West Bank," *Agroecology and Sustainable Food Systems 43*, Vol. 1 (2019), pp. 21-39, https://doi.org/10.1080/21683565.2018.15 37324.
79. Applied Research Center - Jerusalem, *Deir Ballut Town Profile*, (2013), pp. 1-23, https://tinyurl.com/yckrdntt.
80. *Ibid.*
81. Tesdell et al., "Rainfed agroecosystem".
82. Salim Tamari, "Building Other People's Homes: The Palestinian Peasant's Household And Work in Israel," *Journal of Palestine Studies 11*, No. 1 (1981), pp. 31-66.
83. Dayr Ballut is near the village of Al-Zawya, where I was born and grew up within a *fallahi* (peasant) family. Some of the facts given about Dayr Ballut, rain-fed agriculture, the *faqqous* festival and Israeli military checkpoint are my own observations, based on first-hand experience in the village.

4

What Can an Old Mine Tell Us about a Just Energy Transition?

Lessons from Social Mobilization across Mining and Renewable Energy in Morocco

Karen Rignall

When protests over a new solar power plant broke out in southeastern Morocco in 2011, state officials and rural residents alike compared the protests to longstanding conflicts over a nearby cobalt mine.[1] While the officials sought to manage the political dissent in order to protect these mega-projects, the residents had other concerns. They questioned who owned these resources, the land on which the projects were sited and the wealth they created. They wanted jobs. They wanted the economic development such projects were marketed as promoting. Since then, conflicts have intensified over silver, cobalt and phosphate mining *and* renewable energy installations. Residents themselves point to similarities between the economic, ecological and political impacts of these seemingly disparate kinds of projects.[2] Residents are concerned about the material impacts – limited jobs, negligible investment in the local economy and appropriation of scarce water – but they also contest how the familiar political dynamics across these different sectors echo longstanding forms of repression and marginalization. The apparent similarities between mining and solar energy production do more than rehearse power imbalances whereby marginalized rural residents are asked yet again to bear the costs of national development for the benefit of private corporations and state power. Continuities between extraction and renewable energy also raise questions about how to work towards a just transition not only in Morocco, but also in countries around the world that are seeing a surge in renewable energy projects, often in areas with long histories of mining. How to advocate for new forms of energy that do not reproduce the same economic and political inequalities inherent in carbon-fuelled capitalism? Diagnosing what we need to *transition from* is essential to identifying what we need to *transition to*. This diagnosis is about more than a critique. It is also crucial for identifying the collective politics that can produce an equitable transition.

Working towards a just transition requires mapping out how energy production happens in certain places. Beyond high-level national or international policies, what are the bureaucratic and legal procedures that would make these projects come to life for the people and places around them? Such mapping can also document how people mobilize at the local level in ways that are meaningful for them, even when they do not appear to have a lot of power. This political and analytic exercise is particularly important in the Middle East and North Africa (MENA). Discussions of a just transition in the region often turn to democratic governance and how struggles for representation, transparency, redistribution and accountability take precedence in social mobilizations over climate justice or environmental framing.[3] A focus on a just transition as a process that is at least in part worked out in the daily encounters between residents and powerful actors helps to shift the emphasis from democratization as a *prerequisite* for a just transition to democratization as *an important step in that transition*.

A JUST TRANSITION IN MOROCCO

In the autumn of 2021, just as the COP26 was getting under way in Glasgow, a collaborative action research project that included Moroccan partners and the author began its own effort to address the urgent environmental justice demands of the moment. The goal of this effort was to capitalize on nearly a decade of experience of one particular dimension of democratizing the movement for a just transition. The new project aimed to democratize knowledge about extraction and local governance in the southeast of Morocco in an effort to support diverse forms of resident engagement and mobilization. As part of a network of human rights and civil society activists, project partners had previously brokered relations between different groups of residents around the Société Métallurgique d'Imiter (SMI), Imider silver mine and the corporation they were protesting against.[4] This earlier multi-year effort (2012-15) included a corporate social responsibility (CSR) programme across the parent company's southeastern Moroccan mines.[5] While the results of this earlier effort were mixed at best, the activists did learn valuable lessons about how to have the impact they aspire to as brokers, mediators and social justice advocates. They have integrated these lessons in their newly formed Association pour la Promotion de la Médiation au Maroc (APMM) – established in 2017 – and in the action research initiative that they inaugurated in the autumn of 2021.

This action research project asked two kinds of questions crucial to a just transition in Morocco: 1) what are the laws, policies and bureaucratic regulations that govern large-scale extraction projects and renewable energy? and 2) how do power relations around extraction and renewable energy at the local level shape the daily lives of residents? The daily interactions that constitute rural politics are not just parochial or local: they matter at multiple levels, from recognizing the forms of political mobilization that are important to rural residents to identifying points of engagement or resistance that can often change the course of particular projects. A just transition programme for Morocco – and for the region as a whole – needs to widen the lens from an exclusive focus on energy or extraction to also understanding how residents incorporate these projects into their broader political goals. The rural southeast of Morocco is not simply a periphery for the high-income countries (or even for Morocco's urban centres) that are looking to displace the environmental costs of their energy transition onto the global South. It is also a centre for political practice that should be considered as a starting point for what a just transition might look like for the arid lands of North Africa, particularly given the importance of Morocco's rural areas to the country's oppositional politics in the last decade.[6]

This chapter lays out how power mapping and an account of bureaucratic procedures can help to democratize knowledge around extraction and energy production with the objective of supporting local and regional movements for a just transition. This mapping process must be a collaborative, critical project that involves residents of extraction zones regardless of their background, expert knowledge or familiarity with the rights-based language of global social movements. Residents offer unique insights into the politics surrounding extraction, showing the importance of a broad contextual analysis of rural governance for a just transition. The programme laid out here is general enough to be applicable to other contexts and, in fact, grows out of the author's own involvement with just transition work in the central Appalachian coalfields of the US. The chapter proceeds first by describing the contemporary context of conventional extraction and renewable energy in southeastern Morocco. It then outlines a methodology for mapping four continuities between mining and renewable energy production, namely: similar actors and financial interests involved in both sectors; the legal and bureaucratic frameworks that govern both kinds of projects; local revenue systems; and political claims around representation and redistribution.

DIVERSE FORMS OF EXTRACTION
IN SOUTHEASTERN MOROCCO

The impetus for this analysis and the broader action research initiative grows out of the parallels between solar energy and conventional mining identified at the beginning of this essay. The NOORo utility scale solar installation in Ouarzazate was announced in 2010 as a "clean break with the past". The concentrated solar power (CSP) plant was heralded as the flagship project in Morocco's solar plan, which aimed to move the country from nearly complete reliance on imported fossil fuels to producing 52 percent of its power from renewable sources by 2030.[7] Despite international acclaim for this ambitious goal and the government's marked shift towards environmental framing for energy policy, the local and regional dynamics were ambivalent and conflictual from the beginning of NOORo's construction. Activists, residents, and government officials who explicitly compared the protests around the land transfer with labour and environmental mobilizations around the cobalt mine at Bouazzer, less than 200 kilometres away, were deeply aware that the salutary discourse of energy transition notwithstanding, solar energy is embedded in a long history of extraction in the arid southeast of the country.

Extraction in this region goes back centuries. The silver mine at Imider is described by chroniclers in the classical Islamic period. In the project's preliminary research, residents in Zagora Province also noted the possibility that some mines in their region dated from the Almohad period (twelfth century). However, modern-day mines in the region began with the French Protectorate and the aggressive extractivism of speculators, industrialists, and the regional rulers (*caids*) that imposed French rule on southeastern Moroccans.[8] Initial research in the Archives du Maroc reveals the extent to which claims for prospecting and exploitation in the southeast predated both the formal establishment of the Protectorate and the final military victory of the French in Bougafer (current Tinghir Province) in 1933. Mining was bound up with struggles over what kind of power the overlord of the region – the *Pasha* of Marrakech and the *Grand Caid*, Thami El-Glaoui – should hold. Some actors within the protectorate government advocated for enabling his rule while other actors, aligned with European financial interests, advocated for a purely private extraction sector. The former prevailed and the Bouazzer cobalt mine, established in 1928 through an alliance between French capital and El-Glaoui, became one site within the archipelago of mines owned by the company that would eventually be called Managem, the public holding company with majority royal interests. These mines – at Imider, Bouazzer and Bleida – were the site of social mobilizations in the 2010s.

Tracing this genealogy of extraction is an important component of the action research initiative commenced in autumn 2021 because it makes explicit the longstanding ideologies of rule that continue to govern extraction and natural resource management generally. It also reveals the brutal relations of force involved in expropriating these resources on behalf of foreign interests or Moroccan elites. Even seemingly progressive initiatives like those related to solar energy should be contextualized within this history when they draw on the same legal frameworks and relations of power. Residents of the region have never seen the decrees, contracts or other documents that formalized the land transfers and the export of wealth these projects effectuated. These documents live in French archives or in the recently opened Archives du Maroc, far from the people they affect. Democratizing knowledge of extraction means bringing documents home to the original resource owners and bringing to light a history that must be reckoned with in order to understand the contemporary social dynamics of large investment projects. The need for historical reckoning applies no matter the resource in question: cobalt, silver, the water used for commercial watermelon production in the arid Dra' valley or the land used to house the infrastructure for harvesting the sun's energy. This history is also crucial for making sense of the continuities between mining and renewable energy – the focus of the next section, which documents the similar actors and financial interests that are involved in extraction and renewable energy.

SIMILAR ACTORS AND FINANCIAL INTERESTS IN CONVENTIONAL EXTRACTION AND RENEWABLE ENERGY

There is a large amount of scholarship and activist research that tracks how corporations, international financial institutions and global capital flows are linked across extraction and renewable energy.[9] This is an important avenue for understanding how and why transition efforts can deepen rather than challenge relations of dependency rooted in the colonial period. The continuities between fossil fuel commodity chains and those of renewable energies are striking. Geopolitical pressures towards transition in Europe, for example, are not just about meeting decarbonization targets: they also serve financial interests focused on diversifying portfolios and using renewables as a hedge or source of new capital accumulation.[10] Even the surge in renewable energy projects in the MENA region reflects continuities in geopolitical relations between European governments on the one hand, and corporations and MENA fossil fuel producers on the other. The availability of capital and expertise in energy infrastructure, and the goal of diversifying revenue sources mean that many fossil fuel producers are simultaneously leaders in renewable energy. One of the clearest expressions of this positioning is the Desertec Initiative, an effort to link the entire southern Mediterranean rim – and its deserts – to the European grid. Although formally defunct, Desertec's underlying logic informs Morocco's renewable energy policy and other regional initiatives, such as "green hydrogen".[11]

Tracking these relations also sheds light on the technical and economic decisions shaping renewable energy policies and projects in Morocco and beyond. Utility-scale renewable energy, for example, is touted as a way of achieving economies of scale and utilizing existing infrastructure to channel renewable energy to the grid. These were the justifications for the Moroccan solar plan's technical choice in favour of concentrated solar power (CSP), a relatively new utility-scale technology with untested financial prospects, when it was selected for Ouarzazate.[12] But choosing CSP over photovoltaic technologies *and* over decentralized renewable energy is as much about centralizing economic and political power as it is about economies of scale. Community solar or small-scale energy generation that avoids the grid forecloses the opportunities for capital accumulation presented by Morocco's mega-project approach to renewable energy development. Such capital accumulation can occur regardless of whether the projects themselves are profitable once in operation – existing projects are not, and the Moroccan state still must subsidize the energy produced in these new projects to make the power competitive with fossil fuel-generated electricity.[13] Rather, the multiple contracts for the construction, and to a lesser extent the operation, of the plants creates various opportunities for profit (or, more precisely, rents). Many companies winning these contracts are subsidiaries of fossil fuel companies or, at a minimum, are financed by excess capital in the oil-producing Gulf (especially Saudi Arabia, home to ACWA power, which won the contract for the inaugural NOORo installation in Ouarzazate). These companies represent an effort to diversify out of fossil fuels and build on the strong geopolitical ties which the Moroccan government forged in large part through oil.

Understanding how these geopolitical calculations play out in people's daily lives is important. Documenting how these commodity chains work can democratize knowledge about state–corporate alliances for extractive capital accumulation. But documentation is not enough: how do we translate these complex relations for residents so that they can make the links between their local realities and global processes? At the local level, global commodity chains might not appear as relevant as the way state and private sectors obscure their roles and blur the lines of authority between them. The *makhzen* (the institutions of government associated with the king and unelected institutions of government) and the *sulta* ("authorities" – especially the Ministry of Interior and the security services) are often the frontline authorities responding to mobilizations and securing stability for Managem, the company that is formally private and listed on the Casablanca stock exchange but which, as noted above, began as a company of El-Glaoui, the regional *caid* in the Protectorate period, and subsequently passed into the royal holdings. When residents talk of projects as coming from "our commander", referring to the king by his longstanding title of Commander of the Faithful, the distinction between the *makhzen* and the private company held by the king becomes difficult to discern.

However, it is not certain that exposing these global commodity chains is the most effective way to support just transition efforts in Morocco's rural southeast. For both the privately owned mines at Imider and Bouazzer and the parastatal solar energy installation at Ouarzazate, popular demands over the past decade have focused on employment, rural investment and transparency about what resources (especially water) are being used to the detriment of local populations. These claims have been similar across sites and resources – the triggers for the eight-year-long sit-in near the Imider silver mine were water expropriation and the lack of employment at the mine, while ongoing concerns at the solar energy installation at Ouarzazate are water and the paucity of jobs for local residents.[14] Although official estimates of water consumption at Ouarzazate are between 2.5 and 3 million cubic metres a year, actual consumption appears to be substantially higher, even by official admission. This is due to high water requirements for washing the solar reflectors in the desert environment, and to possible inefficiencies in the steam turbine technology used in the Ouarzazate CSP installation. In preliminary research in Midelt, the site of the next installation in the Moroccan solar plan, local government officials commented that the plant under construction included newer, less water-intensive technologies, with the goal of producing 300 MW more than the Ouarzazate plant with one-sixth of the water requirement. The political troubles the current director of the Moroccan Agency for Sustainable Energy (MASEN) has encountered in the past year have been attributed by some in the southeast to geopolitical concerns relating to the country's fraught ties with Germany, but also with the slow pace of the solar plan, as well as the economic inefficiency and resource intensity of the Ouarzazate installation.

A focus on one particular project or commodity chain, however, can obscure the similarities across renewable energy and conventional extraction. A place-based approach to a just transition broadens our focus to cover the range of resources and strategies for asserting the control necessary for such large projects, regardless of what is being extracted. In Morocco, these strategies centre on controlling collectively owned land, perhaps the most hot-button issue in the rural parts of the country (and some urban areas) for the past two decades. This broader approach contextualizes the extracted resources within other resource politics, especially land and water. Beyond extraction companies and renewable energy contractors, actors include export agricultural investors vying for water, and members of ethnic collectivities or other social groups with historical grievances. This is not desk research; it requires extended engagement with residents and the different ways they mobilize ties to state actors or other authorities. Entering into research and activist alliances with diverse groups in the regions surrounding these mega-projects is one way to build an understanding of these complex local politics. This approach eschews talk of how projects impact "the community", rejecting the idea that there is such a thing as one community, and instead actively seeks differing perspectives and positionalities.

LAND CONFLICT AND RESOURCE POLITICS IN THE LEGAL AND BUREAUCRATIC CONTEXT

Mapping extractivism across conventional extraction and renewables becomes more concrete for rural residents when the focus shifts to the laws and bureaucratic procedures used to implement a particular project in their region. To be sure, the histories of the *projets structurants* (mega-projects) in the southeast of Morocco diverge significantly: from the cobalt mine of Bouazzer, inaugurated in 1928 before the French had even secured military control over the entire region, to the future-oriented, globalized discourse of renewal that frames the solar plan. However, mining and renewable energy are concentrated in the same regions and deploy the same laws and bureaucratic procedures for securing the resources necessary for extraction. Beyond tangible resources, such as land, metals or minerals, these resources include public investment in infrastructure – the roads necessary to transport materials and the mined resource, for example – and the use of state power to control popular dissent. The history of mining from the colonial period to the present reveals a striking continuity in the way wealth is extracted from the poorest areas of the country, with minimal reinvestment in social and economic infrastructure.

Histories of modern Morocco critique the colonial bifurcation of the country into a "useful" centre that received resources and "development", and a "useless" periphery that was neglected. These terms represent a particularly explicit description of the extractive capitalism common to all colonial contexts. However, such a neat binary does not fully describe how the colonial and independent states *did* invest in the rural margins of Morocco. Infrastructure and other economic investment in the rural periphery extracted resources and labour for the benefit of populations elsewhere. The southeast of Morocco was – and is – most definitely "utile": the question is, useful for what and for whom? Understanding how histories of extraction intersect with land governance, agricultural policy and state power in the southeast shows how similar strategies are used to secure state or corporate access to all sorts of resources.

Identifying continuities in the rules and procedures governing both mining and renewable energy projects is important because it makes it possible to document the full range of mechanisms used for expropriation. Some of these mechanisms are buried in the complicated language of regulations and administrative procedure, out of sight of local residents. At the same time, documenting bureaucratic frameworks can help to identify openings for making claims that expand the political tools at the disposal of residents living with extraction. The contemporary legal and bureaucratic framework for mining in Morocco has been influenced by the global expansion in metals and mineral mining, especially efforts to apply new technologies to make older operations viable again, and the rush to secure strategic sources for the rare earth metals so essential for the technology sector and renewable energy production.[15] These novel ways of valuing extraction are evident in Morocco's new Mining Code of 2015. The code elaborates a detailed legal framework to encourage more investment in extracting metals and minerals – beyond the dominant phosphate sector – on the premise that a burdensome regulatory environment has suppressed full development of extractives.[16] Conventional extraction is, therefore, not a "legacy" sector, an outdated antecedent to renewable energy that will fade away as part of the transition away from fossil fuels. On the contrary, extraction becomes even more important to support the increased need for the metals and minerals that are key to renewable energy production. In addition, the huge complexes associated with the solar plan require standard building materials and carbon-intensive inputs, such as expanded paved roads and high voltage transmission infrastructure.

Efforts to grow the mining sector and renewable energy in Morocco parallel the country's agricultural development strategy over the past decade (the *Plan Maroc Vert*).[17] The guiding philosophy of that plan was to map each agroecological zone in the country to find new ways to promote export agriculture for the benefit of commercial interests over small farmers.[18] In the southeast, the *Plan Maroc Vert* has spurred the growth of large date, apple and watermelon agribusiness farms that, along with mines and renewable energy installations, compete for water and land. This competition squeezes residents with a limited capacity to defend their land rights or secure their own livelihoods. Researchers documenting the legal and bureaucratic procedures at the root of extractivist policies, therefore, need to look beyond the recent flurry of legislation and investment facilities in emergent sectors like solar energy or rare earth metals. They also need to account for agricultural policies, new and old, as well as the archaic and ambiguous legal frameworks governing land and water. Here, the broad discretion given to state authorities by colonial policies designed to facilitate expropriation have contemporary benefits for powerful actors. State authorities use that discretion to quickly and quietly secure land and other resources.[19]

Thus, despite the distinctive nature of each form of extraction, the shared bureaucratic frameworks governing land, water and natural resource extraction draw all these resources into the same political dynamics. But just as legal frameworks for land and water have been used to dispossess local residents, activists also wonder if there are any entry points citizens can use to contest how extraction projects are implemented on the ground. This requires sustained organizing, and a democratization of knowledge about how to use legal codes for oppositional politics. State or corporate appropriations of sub-surface rights are difficult to contest. As in most countries around the world (the US is a notable exception, with some complexities regarding First Nations), the Moroccan state claims sovereignty and ownership over the sub-surface. While the French colonial state also claimed public domain over water in Morocco, water rights today are complex and subject to layers of positive, formally Islamic, and customary law. Social mobilizations, such as the occupation at the Imider silver mine, have highlighted water depletion and contamination, but sovereignty claims over natural resources and extracted wealth have not figured prominently in Moroccan social movements.

Collectively owned land has also been the site of contestation for extraction projects. The solar installation in Ouarzazate, for example, used colonial laws to expropriate communal land in order to acquire its 3,000 hectare parcel.[20] This is the legal framework that is used to govern property transfer in all of Morocco's collectively owned lands. But, like the *Plan Maroc Vert* and the Mining Code of 2015, a new collective land law was passed in 2019 to facilitate private investment and expropriation of land deemed underutilized for the purposes of national development.[21] There have been discussions for decades about the intractable problems associated with collectively owned land in Morocco. It is difficult to define who has rights in that land, and proponents of privatization say that collective ownership precludes investment. These issues were used to justify the 2019 law, which ostensibly rationalizes collective land management. However, preliminary research reveals a widespread fear that it will only accelerate the transfer of land for large-scale investment projects and the imposition of market logics on land that was never for sale.

Documenting which bodies of law and administrative procedure govern a given project requires a deep dive into diverse areas of law, some directly related to the extracted resource and others spanning diverse resources, local government administration, taxation, and budgeting. These are dry and difficult areas for even the most experienced scholars or activists to penetrate without legal or fiscal expertise. They do not capture the popular imagination in the same way social mobilizations do. And it may seem like a questionable exercise to develop popular education materials on seemingly arcane corners of law. However, experience in extractives organizing in Latin America and elsewhere indicates that these legal and bureaucratic mechanisms can provide openings for popular resistance or civic engagement.[22] The surge in civil society organizing in Morocco around decentralization, transparency and the rule of law, especially among a handful of observers in rural areas and smaller regional capitals, shows similar potential. Democratizing knowledge of legal and bureaucratic frameworks is important in and of itself, but it can also be a tool for making claims for restitution or accountability, even if these openings are small and change occurs over a long time.

WHY TAXES MATTER, OR AN ARGUMENT FOR REPARATIONS

A similar analysis applies to documenting the procedures and practices for allocating revenue from extractives and renewable energy. Public debates about the costs and benefits of both the Ouarzazate solar power plant and the southeastern mines have focused on how project operations affect residents in three key ways: environmental impact, employment and other direct impacts under the control of the company or contractor. This has tended to limit the discussion of costs and benefits to the projects' direct operations and the CSR programmes for each site, which are similar across extractives and renewables. In the case of NOORo, the Moroccan Solar Energy Agency (Masen) initially responded to unrest with ad hoc measures before transitioning to a formal community development programme anchored by AgriSud, a French NGO responsible for coordinating agricultural development initiatives in the commune surrounding the plant. Over time, however, as Masen rebranded from the Moroccan Agency for Solar Energy to the Moroccan Agency for Sustainable Energy (the change took effect in 2016 and shifted its focus to financing and technology transfer), Masen progressively distanced itself from direct involvement in CSR programmes. In Ouarzazate, ACWA Power, the Saudi-led contractor, is now charged with community relations and CSR initiatives. The political controversy swirling around Masen in 2021, ostensibly related to slow progress in implementing the solar plan, financial issues related to the COVID-19 pandemic, and operational inefficiencies, has also muted the visibility of CSR in the solar plan. For Managem, unrest around the mines in Bouazzer and Imider – and, as preliminary research is showing, other mines in the company's portfolio – emerged around the same time as that relating to the solar installation in Ouarzazate. A two-pronged CSR programme was one of a series of responses from the company, and the Ministry of Interior was charged with securing the sites in the interests of general social order. The first, a *Programme d'urgence* (2012-13) aimed to reduce tensions with residents at Imider, although it also represented a company effort to redirect attention away from the highly visible occupation at Mount Alban. A broader *Plan stratégique* (2013-16) involved all Managem's southeastern Moroccan mines and included a needs assessment process with each commune. This process produced a list of projects for which the commune governments needed to provide match funding, usually from the ministries associated with the intervention (especially education and public health). Discussions with participants in the coordinating committee and commune governments revealed a mixed experience, but also a sense that they learnt much from the process of engaging with the company and government authorities.

However, all these initiatives were by definition voluntaristic, limited interventions, as is typical with CSR programmes. They involved no structural or systematic dialogue about how extraction fits into long term rural development or relations between residents, the state and the private sector. The quotidian operations of local government and revenue systems may represent a more fruitful site for making claims about returning wealth, sustaining investment and involving residents in resource allocation. Here, the gaps between policy and practice are important, as is the changing landscape of decentralization reforms in Morocco's historically centralized fiscal regime. In the initial stages of the "advanced project of regionalization", as the process that King Mohamed VI began after assuming power in 1999 is called, revenues from taxes paid on extraction went entirely to regional governments, not local communes or provinces. The Mining Code of 2015 shifted this allocation and currently 50 percent of tax revenues from mining production goes to the regions and 50 percent to the communes. Information about this change is uneven, as there is widespread confusion among residents and some commune officials about what provisions of the new Mining Code have taken effect, and when. This new allocation regime, however, does raise questions and offer new possibilities for assessing how different projects relate to local economic development planning, service provision, and broader discussions about how much wealth is extracted from some of the poorest communes in the country.

In many sacrifice zones, such as the Appalachian coalfields in the US, the concessionary property tax and revenue regimes offered to extraction companies in one sector create a path dependence whereby new forms of extraction follow upon previous ones because they can take advantage of revenue systems designed for other resources.[23] The long-term result is minimal investment in infrastructure or diversification because of diminishing tax bases or a lack of capacity or will among local officials, and even some activists, to demand redistributive measures that funnel accountable and transparent resources to extraction zones. It can seem unimaginable to demand reparations for decades, even centuries, of extraction and dispossession conducted for the benefit of others.

Project research into how this path dependence might work in the Moroccan context is only just beginning, but documenting revenue systems for residents around these projects is an important step in democratizing knowledge about the relationship between wealth extracted and returned in the form of government revenues or investment. Initially, this has been only a descriptive exercise in Morocco – documenting production levels over time, taxes paid and revenues allocated to the communes where projects are located – but a burgeoning academic literature on the effect of resource dependency on economic growth, government transparency and other measures of well-being indicates further avenues for documenting how extraction affects residents and regional political economies.[24] Empirically describing the wealth effects and economic impact of extraction does not in and of itself offer a structural account of the historical dispossession associated with extractivism, but it can provide an additional tool for organizing and claims-making. Such applied research builds on civil society activists' strategies for participating in and contesting local politics to hold the state to its own promises regarding the rule of law and the devolution of fiscal responsibility to local communes. While the master's tools can never be used to dismantle the master's house,[25] understanding and using these administrative frameworks can widen the space for popular participation and claims-making around extraction and energy projects.

The engagement of fiscal experts and legal scholars may be necessary to make sense of these regulations for activists and researchers, but popular education strategies are key to translating them for the broader public. In addition to documenting the formal procedures of revenue allocation, this involves accounting for direct and in-kind expenses associated with the extraction or investment projects that are shouldered by local governments, wealth generated and exported, and wealth returned in the form of tax revenue, employment and other multiplier effects (positive or negative). This is a highly political exercise that involves identifying how to account for externalities or ecosystem services that attempt to quantify values that are inherently unquantifiable for many, including the historical stewards of these resources. Variations of this cost-benefit accounting in other places and for other resources also reveal that the analysis looks quite different when conducted at a national scale or at the local and regional scales that are at the heart of the approach advocated here.[26] Mines that may represent a relatively small part of the country's overall economy may have a transformative impact on regional and local socio-ecologies and relations of power. Consideration of this transformative impact is often dismissed as local or parochial, the unreasonable demands of uninformed local populations who should be willing to shoulder the inevitable cost of a necessary transition. A just transition depends not simply on acknowledging these demands or better distributing the benefits of renewable energy, but also on providing reparations for previous waves of dispossession and disinvestment. A just transition also depends on rethinking why and how these zones are being asked yet again to shoulder the burden of provisioning wealthy consumers elsewhere.

SOCIAL MOBILIZATION AND SHARED POLITICAL CLAIMS ACROSS EXTRACTION AND RENEWABLE ENERGY

For advocates of a just transition, one of the first indications that renewable energy might be repeating the historical inequities of mining has been the similarities in social mobilizations across both sectors. In the Moroccan southeast, participants and officials explicitly made these comparisons in the protests around NOORo and the mines in the region. A structural analysis of these similarities needs to go beyond the mere observation that rural peoples have always been marginalized and will continue to be dispossessed by dominant approaches to renewable energy production. That is an important observation, to be sure, but it does not answer the question of why histories of dispossession are repeating themselves. Nor does it allow for the agency of rural peoples and account for their often ambivalent relationship to extraction or energy production.

There are many possible ways in which both types of extraction (renewable energy and mining) might deepen inequality. The action research project in southeastern Morocco that commenced in autumn 2021 focuses on 1) the processes that naturalize and perpetuate a regional political economy dependent on exporting wealth with minimal reinvestment; and 2) dominant discourses that argue that marginalized residents must "sacrifice" their resources or well-being for national development or a low-carbon energy transition. At the same time, a grounded analysis does not assume that extraction is the only – or even the most important – driver for local or regional politics. Both kinds of projects are drawn into a complex mosaic of political claims that extend beyond mining or energy. Rural politics, like politics anywhere, are multidimensional, and people factor these projects into different, often competing, aspirations and priorities. Documenting these diverse claims clarifies how and why rural residents mobilize in the way they do, or why they respond with other forms of political expression besides overt mobilization.

Recognizing the broader context for extraction politics acknowledges the sometimes overwhelming nature of state and corporate power but does not assume a predetermined outcome to the extraction encounter. Residents not only exert agency in their ability to resist or respond, but they also nego-tiate or use the presence of extensive projects to craft their own political projects. This approach also recognizes the possibility of internal dissen-sion or differences among state and corporate actors and takes their own moral universes seriously.[27] Few people living in and around extraction in Morocco describe purely "good" and "bad" actors or institutions, reflect-ing the competing imperatives and moral complexities associated with these projects. These complexities produce multiple agencies among residents and workers whose critiques or goals may not mesh easily – if at all – with social movement framing.[28]

Popular responses to extraction are also diverse, running the gamut from organized resistance movements to a social fracturing that produces violent conflict.[29] In southeastern Morocco, the occupation at Imider captured the imagination of many Moroccans and international observers through the savvy and at the same time culturally grounded combination of cus-tomary idioms and globalized discourses of resistance. The protests at the Bouazzer cobalt mine or NOORo plant were similarly intelligible to social movement activists and observers. However, these were only a few among many responses, some of which were less visible to those not versed in the practice of rural politics in the Moroccan southeast.

Embedding extraction in broader claims around land, resource control and political representation brings multiple forms of political practice into view, especially in areas where social movements or overt resistance are not prominent.[30] Even failures – of extraction projects or social mobilizations – can produce politics, enabling residents to build alliances or expertise and feeding into their diverse political projects.[31] So, for example, while the occupation camp at Imider might have been dismantled in 2019, the nearly decade-long effort can hardly be termed a failure. It was one of several forms of political expression that changed how local and communal politics unfold around the mine, as witnessed by a changing of the guard towards younger elected officials in the past two communal elections. Residents may work with timescales and aspirations that differ from those of climate justice movements. Just as extraction can enact a slow violence, so too can the envi-ronmentalism of the poor or other political responses unfold over extended timeframes.[32]

This approach also avoids an *a priori* judgement about how residents should respond to extraction. They may balance critique with desires for development, the jobs that a large-scale project might bring, and an emo-tional connection with the people and identities associated with extraction.[33]

Preliminary research on resource conflicts in rural Morocco indicates that mining and renewable energy deepen inequalities, but that people can use these conflicts to imagine and experiment with a different kind of politics or approach to rural governance. This new imaginary can be considered an emergent politics of the commons, which includes non-movement forms of political agency.[34] In other words, advocates for a just transition need to listen to people's diverse goals and acknowledge their preferred form of action rather than use a predetermined frame of analysis that privileges organized social movements.

Even resistance may not conform to dominant environmental discourses as some groups rely on customary or seemingly apolitical practices to articulate their political claims.[35] This non-movement social mobilization may represent an effective, culturally resonant set of approaches to engaging with extraction that recognizes people's complex relationships to large-scale projects – few want to reject them out of hand but rather try to reimagine what they do, how they operate and whom they benefit. The project team's history of research and activism in the southeast indicates that discourses of environmental justice do not resonate with many residents in the region. The power of the analysis the team is engaged in lies in the way it engages diverse forms of political agency to render the push for a just transition less abstract – seeing it as a concrete, emplaced encounter that does not need to look like other environmental justice movements for it to promote a just transition. At the same time, this approach is not a replacement for, or argument against, formal social movements. Rather, it represents an expansive and critical recognition of the necessity of diverse forms of political practice.

CONCLUSION

Given the seemingly overwhelming power held by the state, corporations and international finance institutions, the notion of working with procedure and law at the local or regional levels to influence the extraction encounter may appear naïve. On its own, this approach will not achieve a just energy or economic transition for Moroccans – or people anywhere – who are dispossessed through successive and seemingly unrelenting waves of extractivist policies. It is, however, a crucial step for engaging residents who live with the complex reality of extraction as both a source of dispossession *and* development. Democratizing knowledge about extraction as a mode of governance that spans both mining and renewable energy is one way to recognize the people who live with extraction projects as equal partners in social movements, whether or not they adopt the frames of resistance or climate justice. For researchers and activists alike, honouring different modes of political practice means committing to critical engagement with the discursive frames residents themselves adopt. Understanding the history and social dynamics of a place beyond the extraction encounter de-centres extraction as the only political force shaping people's lives. Their agency unfolds alongside the imbalanced power relations that set companies and state agencies apart and above those dynamics.

However, developing place-based just transition strategies is not simply about extended engagement with local residents on their own terms. Such an approach also provides the grounds for solidarity with other place-based movements and strategies, where successful models for engagement can be used and adapted for new contexts and mutual reinforcement. When viewed in this light, the trenchant critiques, incremental strategies and long-term visions of activists and residents in the Moroccan southeast are as essential a part of just transition efforts in Morocco and North Africa as any in the climate justice movement.

NOTES

1. Karen Rignall, "Solar Power, State Power, and the Politics of Energy Transition in pre-Saharan Morocco," *Environment and Planning A*, Vol. 48, pp. 540-57, 2016, https://doi.org/10.1177/0308518X15619176.

2. Koenraad Bogaert, "Imider vs. COP22: Understanding Climate Justice from Morocco's Peripheries," *Jadaliyya*, November 21, 2016, www.jadaliyya.com/Details/33760; Soraya El Kahlaoui and Koenraad Bogaert, "Politiser le regard sur les marges: Le cas du mouvement "sur la voie 96" d'Imider," *L'Année du Maghreb* 21, pp. 181-91, https://doi.org/10.4000/anneemaghreb.5555; Hamza Hamouchene, "The Ouarzazate Solar Plant in Morocco: Triumphal 'Green' Capitalism and the Privatization of Nature," *Jadaliyya*, March 23, 2016, https://tinyurl.com/55ne5bfr; Atman Aoui, Moulay Ahmed el Amrani and Karen Rignall, "Global Aspirations and Local Realities of Solar Energy in Morocco," *Middle East Research and Information Project*, June 10, 2020, https://tinyurl.com/23ye2vj8.

3. Jeannie Sowers, "Environmental Activism in the Middle East and North Africa," in H. Verhoeven (ed.) *Environmental Politics in the Middle East*, pp. 27-53, November 2018, https://doi.org/10.1093/oso/9780190916688.003.0002.

4. Mohammed Benidir, "Brokerage, Compensation and Reproduction of the Discharge: Community Reparation and Development of Mining Areas in South-Eastern Morocco," *International Development Policy*, 13 (1), 2021, accessed November 2021, https://doi.org/10.4000/poldev.4476.

5. The parent company, Managem, is publicly traded on the Casablanca stock exchange and is in turn a subsidiary of the royal holding, Al Mada.

6. Koenraad Bogaert, "The Revolt of Small Towns: The Meaning of Morocco's History and the Geography of Social Protests," *Review of African Political Economy* 42(143), pp. 124-40, September 24, 2014, https://doi.org/10.1080/030 56244.2014.918536.

7. Amrani and Rignall, "Global aspirations".

8. Ouail Bouimezgane, "Développement des zones minières et le mouvement des habitants : Cas du sud/est, " (PhD dissertation. Ibn Zohr University: Agadir, 2016); Mohamed Oubenal, "Les Transformations Socio-Économiques Dans Le Sous." *Regards Croisés Sur Les Sociétés Amazighes*, 2021.

9. Mathieu Blondeel, Michael J. Bradshaw et al., "The Geopolitics of Energy System Transformation: A Review," *Geography Compass* 15, No. 7 (2021), https://doi. org/10.1111/gec3.12580; Matt T. Huber and James McCarthy, "Beyond the Subterranean Energy Regime? Fuel, Land Use and the Production of Space," *Transactions of the Institute of British Geographers* 42(2017), pp. 655-668.

10. Luigi Carafa, Gianleo Frisari and Georgeta Vidican, "Electricity Transition in the Middle East and North Africa: A De-Risking Governance Approach," *Journal of Cleaner Production*, Vol. 128 (2016), pp. 34-47, https://doi.org/10.1016/j. jclepro.2015.07.012.

11. Roberto Cantoni and Rignall, "Kingdom of the Sun: A Critical, Multiscalar Analysis of Morocco's Solar Energy Strategy," *Energy Research and Social Science*, Vol. 51 (2019), pp. 20-31; https://doi.org/10.1016/j.erss.2018.12.012; Hamza Hamouchene, "Green hydrogen: The New Scramble for North Africa," *Al Jazeera*, November 20, 2021, accessed December 1, 2021, https://tinyurl. com/nwjvpn7j.

12. Cantoni and Rignall, "Kingdom of the sun".

13. Lucile Daumas, "Le secteur de l'énergie renouvelable au Maroc concentration aux mains du secteur privé," Committee for the Abolition of Illegitimate Debt (CADTM), October 8, 2016, accessed December 1, 2021, www.cadtm. org/Le-secteur-de-l-energie; Gonzalo Escribano, "The Geopolitics of Renewable and Electricity Cooperation between Morocco and Spain," *Mediterranean Politics*, No. 24, (2019), pp. 674-81.

14. Amrani and Rignall, "Global aspirations".

15. Gitanjali Poonia, "How the Rise of Copper Reveals Clean Energy's Dark Side," *The Guardian*, November 9, 2021, accessed November 10, 2021, https://tinyurl. com/562635am.

16. Abdessamad El Atillah, Mustapha Souhassou and Zine El Abidine El Morjani, "Le cadre législatif de l'exploration et la recherche minière au Maroc entre le Dahir de 1951 et la loi 33 -13," *International Review of Economics, Management and Law Research* 1, No. 1(2018), accessed November 15, 2021, https://revues. imist.ma/index.php/IREMLR/article/view/12679.

17. Nicolas Faysse, "The Rationale of the Green Morocco Plan: Missing Links between Goals and Implementation," *The Journal of North African Studies*, Vol. 20, September 2015, https://doi.org/10.1080/13629387.2015.1053112.

18. Azzedine Akesbi, Najib Akesbi, Khadija Askour et al., "Le Plan Maroc Vert : Une analyse critique," *Questions d'Èconomie Marocaine* (Rabat: Presses Univér-sitaires du Maroc, 2011), pp. 9-48.

19. Karen Rignall, *An Elusive Common: Land, Politics, And Agrarian Rurality in a Moroccan Oasis* (Cornell University Press, Ithaca, New York, 2021), https://doi.org/10.7591/cornell/9781501756122.001.0001.

20. Rignall, "Solar Power, State Power, and the Politics of Energy Transition in pre-Saharan Morocco," *Environment and Planning A: Economy and Space*, Vol. 48, No. 3, 2016, pp. 540-57. https://doi.org/10.1177/0308518X15619176.

21. Amrani and Rignall, "Global aspirations."; David Balgley and Rignall, "Land Tenure in Morocco: Colonial Legacies, Contemporary Struggles," in Horman Chitonge and Ross Harvey (eds.), *Land Tenure and Reform in Africa: Addressing Challenges and Complexities* (Springer Nature, Switzerland AG, 2022).

22. Henry Veltmeyer and James Petras (eds.), *The New Extractivism: A post-Neoliberal Development Model or Imperialism of the Twenty-First Century?* (Zed Books, London, 2014).

23. Rignall, K, L. Shade, C. Starr, and L. Tarus, "The Role of Land in a Just Transi-tion," in S. Scott and K. Engle (eds.) *A Just Transition in Appalachia* (University Press of Kentucky, Lexington – in press).

24. For an introduction, see Jeffrey D. Sachs and Andrew M. Warner, "The Curse of Natural Resources," *European Economic Review* 45, No. 4 (2001), pp. 893-906; Frederick van der Ploeg, "Natural Resources: Curse or Blessing?" *Journal of Economic Literature*, Vol. 49, No. 2 (2011), pp. 366-420, DOI: 10.1257/jel. 49.2.366.

25. Audre Lorde, "The Master's Tools Will Never Dismantle the Master's House," Catalyst Project, https://tinyurl.com/kzxymcss.

26. Stratford Douglas and Anne Walker, "Coal Mining and the Resource Curse in The Eastern United States," *Journal of Regional Science*, Vol. 57, No. 4, (2017), pp. 568-90, https://doi.org/10.1111/jors.12310.

27. Mette M. High and Jessica M. Smith, "Introduction: The Ethical Constitu-tion of Energy Dilemmas," *Journal of the Royal Anthropological Institute*, Vol. 25, No. S1: Special Issue: Energy and Ethics? (2019), pp. 9-28, https://doi.org/10.1111/1467-9655.13012; Fabiana Li, "In Defense Of Water: Modern Mining, Grassroots Movements, and Corporate Strategies in Peru," *Journal of Latin American and Caribbean Anthropology* 21, Vol. 1, (2016), pp. 109-129, https://doi.org/10.1111/1467-9655.13012.

28. Jessica Smith Rolston, "Specters Of Syndromes and Everyday Lives of Energy Workers in Wyoming," in Sarah Strauss, Stephanie Rupp, and Thomas Love (eds.), *Cultures of Energy: Anthropological Perspectives on Power*, (Left Coast Press, San Francisco, 2013), pp. 584-92.

29. Jerry K. Jacka, "The Anthropology of Mining: The Social And Environmen-tal Impacts of Resource Extraction in the Mineral Age," *Annual Review of Anthropology*, No. 47 (2018), pp. 61-77; Zilliox, S., and Smith, (2018) "Colora-do's Fracking Debates: Citizen Science, Conflict and Collaboration', *Science as Culture* 27 (2), pp. 221-241.

30. John Gaventa, "Power and Powerlessness in an Appalachian Valley – Revisited," *Journal of Peasant Studies*, Vol. 46, No.3 (2019), pp. 440-56, https://doi.org/10.1 080/03066150.2019.1584192.

31. Dana E. Powell, *Landscapes of Power: Politics of Energy in the Navajo Nation*, (Duke University Press, Durham, 2017).

32. Joan Martinez-Alier, *The Environmentalism of the Poor: A Study of Ecological Conflicts and Valuation* (Edward Elgar, Cheltenham, UK, 2002); Rob Nixon, *Slow Violence and the Environmentalism of the Poor*, (Harvard University Press, Cambridge, MA, 2011).

33. Shannon Elizabeth Bell and Richard York, "Community Economic Identity: The Coal Industry and Ideology Construction in West Virginia," *Rural Sociology* 75, No. 1 (2010), pp. 113-43, https://doi.org/10.1111/j.1549-0831.2009.00004.x; Colin Filer and Martha Macintyre, "Grass Roots and Deep Holes: Community Responses to Mining in Melanesia," *The Contemporary Pacific* 18, No. 2 (2006), pp. 215-31.

34. Asef Bayat, *Life as Politics: How Ordinary People Change the Middle East* (Stanford University Press, Stanford, 2013); Rignall, *An Elusive Common*.

35. Rignall, *An Elusive Common*.

5

Towards a Just Agricultural Transition in North Africa

Saker El Nour

The bleak reality of global climate change becomes clearer with each new report issued by the Intergovernmental Panel on Climate Change.[1] North Africa is extremely vulnerable to climatic and environmental crises, which are a daily occurrence in the lives of the millions of people living in the arid, semi-arid and desert areas of the region. Over the last few decades, drought rates and temperatures have risen continuously, leading to increasing desertification. The region also suffers from severe water scarcity,[2] land degradation and livestock depletion.[3] The accelerated environmental crises directly and indirectly affect agriculture, including grazing, and fishing activities. They also intensify poverty and erode food sovereignty.[4] Approximately 52 percent of the total population in North Africa live in rural areas[5] and this population, which includes small-scale farmers and farm workers, is among the poorest and most impacted by the stark effects of agroecological crises.

North Africa's perilous situation in regard to climate change stands in contrast to the fact that the region accounts for a very small percentage of global greenhouse gas emissions. In 2021, the African continent as a whole produced around 3.8% of the world's carbon dioxide emissions, while the continent's average per capita emissions remained the lowest in the world, at around 0.94 metric tonnes per year.[6] In North Africa, Egypt was responsible for 0.68% of global emissions, Algeria 0.46%, Tunisia 0.08% and Morocco 0.2%.[7] A recent study shows the global unevenness of greenhouse gas emissions: while the global North's rates stand at 90 percent, the global South produces only 10 percent.[8] However, countries in the global South bear the brunt of the crises brought on by climate change, and are in dire need of a just transition – to help mitigate the harmful impacts of environmental change and to adapt to their long term consequences.

Agriculture is both negatively impacted by climate change and a significant contributor to it. Due to the dominance of global capitalist food systems and industrial agricultural production, land use and forest management accounted for a total of 23 percent of greenhouse gas emissions between 2007 and 2016.[9] North African countries are no exception to this pattern, dominated as they are by a high-emissions corporate food regime.[10] Against this background, it is vital to assess the possibilities for, and obstacles to, a just transition in the North African agricultural sector.

Table 5.1 Selected economic, social and demographic indicators shaping agriculture in North Africa[11]

Indicator	Algeria	Egypt	Tunisia	Morocco
Share of agriculture in GDP (2020)	14.2%	11.5%	11.7%	12.2%
Percentage of the labour force active in the agricultural sector (2020)	10%	21%	14%	33%
Agri-food trade balance (in $1 million): a comparison between Europe and the world (2017)	World −9,063 Europe −2,815	World −8,750 Europe −1,070	World −797 Europe 95	World 242 Europe 1,907
Agricultural arable land in 2018 (million hectares)	7.5	2.9	2.6	7.5
Percentage of irrigated land out of total agricultural land	3.2% (2017)	100%	3.9% (2013)	4.6% (2011)
Rural population in 2020 (in millions)	11.5	58.6	3.6	13.5
Percentage of the rural population out of the total population (2020)	26%	57%	30%	36%

The agricultural sector in North Africa has experienced significant transformation in the last few decades. As Table 5.1 shows, the share of agriculture in GDP is low. Yet despite the declining share of agriculture in GDP, the agricultural sector remains a primary source of employment, particularly in Egypt and Morocco. Likewise, the percentage of the population that lives and works in rural areas remains high, despite ongoing urbanization. In recent decades, North Africa has also witnessed a sharp increase in rural poverty, malnutrition and social inequalities.[12] Finally, with the exception of Morocco and Tunisia, North Africa also has a negative trade balance with Europe.

Fighting hunger and coping with the impacts of climate change on agriculture and rural populations necessitate an economic, social and environmental transition. What such a transition should look like, how it takes place, and who should carry it out has been the subject of much debate.

Box 5.1   A just transition versus just a transition

The term "just transition"[13] refers to a set of principles, processes and practices that create a shift away from an extractive economy towards a globally equal, low-carbon economy.[14] The concept of a just transition first appeared in debates between the environmental movement and the labour movement in North America. It then developed in the 1990s as a concept linked to workers' needs for decent employment and green jobs, and was adopted by the International Labour Organization (ILO), as highlighted in the Paris Climate Agreement.[15] More recently, the concept of a just transition has become more comprehensive, bringing together socioeconomic and environmental dimensions, both at the level of the nation state and globally. The term also creates space for engaging with questions of gender, class and varied forms of anti-colonialism in relation to the transition towards a low-carbon alternative to the *status quo*.[16] This broader approach towards a just transition enables discussions about a far-reaching social and economic restructuring that addresses the sector-specific and context-specific roots of inequality.

In this context, this article looks at the challenges, components and characteristics of a just transition within the agriculture sector in North Africa. As in many other countries,[17] the last few years have seen local and traditional knowledge of food systems, and ecological and regenerative agriculture, put forward as solutions to the dominant agri-food system and ecological crises in North Africa. However, these new dynamics have not been sufficiently studied: there is no overview of these developments or the practices and networks upholding them. This chapter fills this gap by evaluating and comparing agricultural policy transformations and the possibilities of a just transition in the agriculture sectors in Algeria, Egypt, Morocco and Tunisia.[18] The chapter is divided into three sections. The first section analyses agricultural policies and the trajectory of agricultural development in the region. The second section explores questions of environmental and climate debt, as well as the effects of uneven environmental changes on natural resources and opportunities for development. The third section presents and discusses ecological and regenerative agriculture, local initiatives and networks of actors who are building a just transformation of agriculture in North Africa.

AGRICULTURAL POLICY TRANSFORMATIONS IN NORTH AFRICA

This section analyses the shifts in access to resources and agricultural policies that took place in North Africa in the post-colonial era, in order to better understand the transformation of the agricultural economy and the dominant development model in the region over time.

Access to land and water in the post-colonial era

Discussions about the agrarian question were prominent during anti-colonial struggles and in the aftermath of national liberation projects.[19] After the colonial era ended, countries pursued multiple pathways in regard to managing their agricultural resources and the colonial heritage within the sector.[20] Algeria, Egypt, Tunisia and Morocco implemented a variety of agrarian reform models in the period 1950-70, which produced crucial shifts in agricultural policies and the state of rural societies across these countries.

Following Algerian independence in 1962, the National Liberal Front (FLN) adopted agrarian reforms that amounted to an agricultural revolution. It promoted rural development by facilitating the access of small-scale and landless farmers to land and by providing them with social and technical support.[21] Additionally, 250,000 hectares were redistributed to war veterans who were grouped into 250 productive peasant cooperatives. The lands previously held by colonists were distributed to over 2,200 farms, the majority of which were large farms with an average of 1,000 hectares, for a total area of 2.5 million hectares.[22] During the 1970s, uncultivated lands were nationalized while large land holdings were restricted.[23]

In Morocco, agricultural modernization became a central pillar of the country's development path after independence in 1956. In 1962, for example, the National Institute of Agricultural Research was established with the aim of modernizing the agricultural sector. Under pressure from the *Union Marocaine des* Travailleurs (UMT), the *Union Nationale des Forces Populaires* (UNFP), the *Parti du Progrès et du Socialisme* (PPS) and the *Istiqlal* Party, the government passed agrarian reform laws in 1963 to recover the lands of colonists. These were implemented in two phases, ending in 1973. Expropriation of previously colonial land was significant, amounting to one million hectares of arable land[24]: the monarchy redistributed the lands formerly in the hands of French colonists to rural elites as a means of securing power and buying loyalty towards the *Makhzen*.[25] In 1969, the

Agricultural Investment Charter was approved, and in 1972 a law was passed which granted farmers agricultural lands from state-owned private property. A law on peasant cooperatives, giving them access to modernized plots in former collective lands, was also enacted. The state also invested in building dams and undertook large-scale irrigation projects, with the aim of developing a new, loyal class of middle-income farmers. Nevertheless, the system of land control remained in the hands of the state. Indeed, it served as a tool to purchase local elites' loyalty and to reduce conflict.[26]

In Tunisia, three years following independence, Law 48 of 7 May 1959 enabled the state to take possession of neglected and unused collective agricultural properties, covering an area of approximately 500,000 hectares. In the same period, local notables, merchants, the self-employed and powerful members of the ruling Constitution Party were able to buy some of the colonial lands.[27] Then, on 12 May 1964, a law was passed that nationalized 300,000 hectares of colonial lands. Thus, by the end of the 1960s the Tunisian state owned 800,000 hectares of agricultural land: approximately 10 percent of the total area of agricultural land in the country.[28] These lands helped initiate the short-lived experiment of peasant cooperatives in Tunisia, which disintegrated in 1969, just eight years after it was launched. After this, Tunisia began to shift towards a more market-based, neoliberal approach. In a move that benefited local leaders and powerful individuals, Tunisia privatized collective lands through the Law of 14 January 1974.[29]

In Egypt, agrarian reform was a central policy during the first era of the July 1952 regime, in the early post-colonial period. Between 1952 and 1970, 343,000 hectares (12.5 percent of agricultural land) were redistributed to 343,000 families, consisting of 1.7 million individuals – almost nine percent of the rural population.[30] As a result of the Nasser regime's agrarian policies, villages saw significant changes in their class composition: while the larger, more influential landlords lost much of their lands, there was an increase in the area owned by small- and medium-scale farmers, and there was improved rent security for tenants. Also, there was a minor improvement in the situation of landless farmers and agricultural workers.[31] The "green revolution" instituted by post-colonial governments relied on agricultural mechanization, chemical fertilizers, pesticides and hybrid seed varieties to increase agricultural production.

Ultimately, North African agriculture development models in the two decades following independence focused on modernizing the agricultural sector and preserving large farms, whether through state administration or through highly centralized and controlled cooperatives. To various degrees, North African countries adopted progressive, state capitalist and "green revolution" policies. This was achieved through a combination of strategies, such as providing technical and material support to farmers, supporting production inputs, inaugurating large irrigation projects, boosting and disseminating modern agricultural knowledge and guidance, establishing research centres and agricultural schools, and setting up agricultural cooperatives. In this era, the state in these countries utilized discourses of modernization reliant on mechanization, commercial and export agriculture, and the marginalization of small-scale local knowledge. In fact, despite the emphasis on food self-sufficiency, the export of cash crops continued to follow the same pattern that had been dominant in the colonial era, especially for commodities such as citrus, vines, vegetables, cotton and olives.[32]

The impact of neoliberalism on agriculture and natural resources

The turn towards neoliberalism in North Africa began in the 1980s. Under pressure from international financial institutions, namely the International Monetary Fund (IMF) and the World Bank, countries in the region began to liberalize foreign trade, devalue local currencies and allow an increased dominance of the market, through both the continued privatization of public companies and the gradual erosion of public services. Priority was accorded to reducing public debt, social spending and employment rates in the public sector.[33]

As a result of neoliberal transformations, North African countries saw a major change in water and land management. The state withdrew from the management of natural resources, allowing the private sector to take over. This led to an increase in the penetration of private investment companies in the agricultural sector, with the private sector acquiring more resources, particularly in vast desert areas, through access to groundwater and land that the state made available to major agricultural investors.[34]

In Algeria, the era of state farms came to an end in 1980s, with the latter being divided into small farms of ten to 70 hectares. In 1987, these lands were progressively moved into the hands of agricultural investors. Accompanying this change was a gradual shift towards market forces,[35] notably with the long-term liberalization of agricultural production inputs, leading to an increase in the price of fertilizers, pesticides and farming equipment. This in turn led to an increase in the prices of agricultural products as a whole. Following the 1994 agreement between Algeria and the IMF, state support for agricultural inputs was completely removed.

In Morocco, the neoliberal transformation in the agricultural sector intensified in 2003. This was exemplified by the privatization of two public companies that had managed the bulk of the lands recovered from colonists: the *Société de développement agricole* (Sodea) and the *Société de gestion des terres agricoles* (Sogeta). With this move, the ownership of 90 percent of former colonial lands was transferred to private investors, the state's major administrative notables, the army and the security apparatuses.[36]

In Tunisia, neoliberal policies were implemented before the initiation of the Structural Adjustment Programme (SAP) under the World Bank in 1986. The state geared agricultural production towards export and high value-added crops, facilitating private sector access to land and putting an end to state commercialization of agricultural products.[37] These policies were coupled with the state's progressive withdrawal from traditional agricultural sectors.[38]

Since 1979, Egypt has pursued a policy of economic openness. State-owned farms were dismantled, agrarian reform laws were amended, and the Agricultural Cooperative Union was dissolved. Also, the state applied a set of measures to reduce subsidies to farmers in the Nile Valley and Delta, such as removing pesticide and fertilizer subsidies, and allowing the private sector to control agricultural production inputs.[39] Further, the ownership limit imposed on agricultural companies was abolished, enabling investors to own more reclaimed lands. In 1992, Law 96 was passed, regulating rental relations between landlords and tenants. This law put an end to rental security, triggering a sustained wave of protests in the Egyptian countryside.[40]

In North Africa as a whole, during this period, states focused on expanding their hold over desert agriculture for the export market while accelerating the commodification of state lands, making them available to agricultural investors.[41] Since the 1990s, policies of agricultural development in the desert have been regarded as a solution to the food provision and production crisis in North Africa.[42] International financial institutions supported policies of agricultural expansion in the desert based on a capital- and technology-intensive model of production of mostly export crops, with associated degradation of water and land resources.[43]

As a result of these neoliberal transformations, food self-sufficiency policies were terminated in favour of more market-based food security policies. The latter meant that food came to be sourced through market mechanisms, often irrespective of provenance – whether this be global commodity markets, domestic production or even food aid. Accordingly, major shifts occurred in diets, leaving North African countries exposed to a sharp increase in nutritional diseases and food dependency. Algeria and Egypt became among the biggest importers of wheat globally.

Following 40 years of neoliberalism, the key features of the current dominant agri-food system in North Africa can be summarized as follows:

- The removal of subsidies for small peasant farmers and the gradual withdrawal of the state from all forms of technical and material support for agricultural production. This includes the state abandoning its role in centrally controlling agricultural operations and practices, such as fertilization and the types of seeds and pesticides used. This withdrawal has given the private sector unfettered access to food staples and import channels. The state also entirely surrendered to the forces of the market its role in determining the prices of agricultural inputs and outputs, ceasing agricultural input and credit subsidies.

- The promotion of a model of industrial agriculture based on large-scale farms. This was achieved by reclaiming desert spaces and enabling agricultural investors to access large areas of land. Thus, colonial structures were repurposed and reproduced through a system in which land is now in the ownership of the few; these dynamics are particularly visible in the cases of Morocco and Egypt.

- The adoption of a policy of primarily export-driven agriculture through financial incentives, the provision of chillers in airports, etc. Most importantly, North African states form part of a system of international trade that serves to bolster the interests of the global North at the expense of local populations in the global South.

- The dominance of a globalized, consumerist diet with a high level of cheap carbohydrates, leading to an increase in the incidence of food-related diseases, high levels of obesity and malnutrition. Additionally, there has been a replacement of food self-sufficiency policies with market-based food security policies.

The current situation: a marginalized peasantry and an extractive capitalist mode of agriculture

The decline of the welfare state in the post-colonial, neoliberal era saw the emergence and reproduction of a localized dualism that had existed in the colonial era: the existence of two agricultural sectors – one characterized by private, large-scale farms in receipt of state support, the other based on small-scale farmers in the plains, valleys and oases, dependent on rain-fed agriculture and characterized by under-development and marginalization.

In North Africa, agriculture is a major sector of employment for women, accounting for 55 percent of women's employment, in comparison to only 23 percent for men.[44] With the migration of men and women (whether it be for economic reasons, or as a result of wars and conflict), the number of seasonal migrant workers is continuing to increase. In Egypt, for instance, according to the 2010 agricultural census,[45] the total number of women workers in the agricultural sector amounted to five million in that year, 40 percent of whom undertake unpaid labour for their own families. Further, the growth of capitalist forms of agriculture has amplified the feminization of agricultural work, along with the dependence on girls, who can be as young as eight years old, who work in very poor and exploitative conditions.[46] The nature of agricultural work is problematic on many fronts, starting with the working conditions and health and safety issues (see the next paragraph), and extending to the local and global division of labour and its relationship with women's empowerment and development. The working conditions of women farm workers are especially important in light of the COVID-19-related health crisis,[47] as well as fears of a new food crisis, which would exacerbate already existing tensions in the region. For instance, the 2021 Food and Agriculture Organization (FAO) Food Price Index (FPI) shows a worldwide large increase in the prices of meat, dairy, cereals, vegetable oils and sugar between November 2020 and November 2021.[48]

Agriculture is one of the most dangerous production sectors in the world. According to estimates of the ILO, approximately 170,000 agricultural workers are killed every year. Workers in agriculture are at least twice as likely to die at work as workers in other sectors. Millions of agricultural workers are exposed to serious work injuries in accidents linked to agricultural equipment or poisoning with pesticides and other chemicals.[49] Indeed, due to the under-reporting of deaths, injuries and work-related diseases in the sector, it can be assumed that the real picture of health and safety for agricultural workers is likely to be worse than official accounts.

Relationships of unequal exchange in the global system underpin the agricultural crisis in North Africa. Countries in the region are subjected to unequal exchange with the global North, particularly the European Union (EU), through a variety of trade agreements that enable the EU to benefit from North African agricultural products at preferential rates. These agreements not only facilitate the exploitation of the region's resources, but they also maintain and further entrench the difference in wages in the agricultural sector in the South compared to the North, and the extraction of surplus value for the benefit of European consumers.[50] As the biggest

trading partner of North African countries, much of the region's production is geared towards export to the EU market. The EU, therefore, directly
impacts development policies and the dominant trade and agriculture plans
in the region. Under the slogan of "trade for development",[51] the EU, in partnership with local elites, pushes North African countries to sign free trade
agreements, which, in turn, aggravate the structural crisis.[52]

As dependency theorists argue, while colonialism may have gone, the
development model of the colonial era has remained dominant in different ways, perpetuating the disparities between the global North and
South. Under neoliberalism, former colonizers played a key role in integrating peripheral economies into the global economy and trade system,
and creating patterns of dependency.[53] Meeting the needs of the European
market necessitates monocropping, large farms and catering to the prefer-
ences of European citizens – for example, in the way in which olive oil is
prepared, or in the cultivation of specific varieties of dates, strawberries,
flowers and citruses.

In sum, these agricultural policies and practices have created another
form of dualism. On the one hand, industrial agriculture degrades land and
water. Based on the intensification of capital and energy, capitalist agriculture further pushes agricultural workers – men and women – into precarity.
It also exacerbates inequalities and centralizes land ownership. This is clearly
the case in desert agriculture, where large areas are allocated to big investors
while small-scale farmers are restricted to limited spaces.[54] On the other
hand, the absence of subsidies for peasant farming has led to the impoverishment of small farmers and the degradation of natural resources in oases and
rural areas. Further, the legacy of the "green revolution", with its intensive
use of fertilizers, pesticides and hybrid seeds, has culminated in the neglect
of intergenerational local agricultural and ecological systems. As a result,
natural resources, such as land and water, have deteriorated, the biodiversity
of seeds has declined, and the balance between humans and the environment has been disrupted, causing what is referred to as a "metabolic rift".[55]

JUST TRANSITION: FACING AN UNEQUAL
ECOLOGICAL EXCHANGE

As previously argued, the concept of "unequal exchange" advanced by proponents of the dependency theory focuses on the movement of labour
power and capital. However, despite its importance in providing valuable
conceptual insights, this concept fails to provide an in-depth insight into
the mechanisms of a just transition. Understanding the possibilities of a just

transition requires looking at the process of unequal *ecological* exchange, a concept which is more comprehensive than the former. To achieve this, it is key to investigate four clusters of resources: 1) the raw materials and energy used to produce goods and services; 2) the land required to directly or indirectly produce those goods; 3) the services consumed in order to produce those goods; and 4) labour in supply chains. Such unequal socioeconomic and environmental flows prevent countries in the global South from achieving development on their own terms.[56]

Box 5.2   From unequal ecological exchange to climate debt

The concept of unequal ecological exchange emerged and developed within academic debates, while the concept of ecological debt materialized within the environmental justice movement.[57] As a term, climate debt was introduced during the 1992 Earth Summit in Brazil, with the aim of highlighting the continuity of historical and colonial forms of exploitation of resources in the global South. Above all, ecological debt is an economic concept that is shaped by two struggles relating to distribution. The first one is unequal ecological exchange that can be summarized as the cumulative product of unequal trade-centred environmental exchange, while the second is the climate debt that can be summarized as a historical but persistent unequal distribution of global carbon sinks to the benefit of advanced capitalist countries.

Social and environmental movements in the global South have faced difficulties with the first aspect of the concept of ecological debt. They have, therefore, focused on calculating and estimating climate debt. This was first done in 1999, through the Committee for the Abolition of Illegitimate Debt (CADTM). The 2010 World People's Conference on Climate Change and the Rights of Mother Earth in Cochabamba, Bolivia, also adopted the concept of climate debt. In the proceedings of that conference, climate debt is defined as the total of "emissions debt" and "adaptation debt". The former refers to the cost of historical and current excessive emissions per person in the global North, which deprive countries of the South of their fair share of air. The latter points out the exorbitant costs incurred by countries of the global South in adapting to the significant damages and risks of greenhouse gas emissions and climate change, despite their limited contribution to the environmental crisis. Climate debt is, therefore, seen as part of a broader debt to mother Earth.[58] In the Cochabamba conference proceedings, developed countries were called upon to take a set of measures which can be summarized as follows: 1) decolonizing the atmosphere by reducing greenhouse gas emissions; 2) remunerating countries of the global South for losing development opportunities due to life under a colonized airspace; 3) taking responsibility for climate change-based migration; and 4) tackling debts related to climate change mitigation and adaptation, and handling the damage of the excessive emissions of the global North.[59]

In North Africa, historically unequal ecological exchange is intertwined with relationships of exchange with European countries. Here, unequal exchange affects the allocation of water, land, climatic resources, energy and labour power, all of which are geared towards food production for European markets. North African countries bear the environmental costs, as their local ecosystems are destroyed and their natural resources depleted. They also bear the economic costs by generating surplus value through international trade with European countries. This, in turn, has far-reaching consequences for the sustainability of resources, energy and land in North Africa, as well as for the ability to develop frameworks for food sovereignty and to achieve a just transition locally. Unequal environmental exchange perpetuates an imperialist way of life in the capitalist core countries, while severely restricting the chances of a just transition in the South. What is presented as an environmentally and socially just transition for Europe is not necessarily the case for the peripheries attached to the continent in the southern Mediterranean and West Africa.

Discussions about just transition focusing only on the capitalist core in the global North, whether in relation to the crisis of the Western mode of production and consumption, or indeed the introduction of technological, ecological modernity as a solution to the crisis, completely overlook the situation of countries in the South, as well as the possibilities for, and hindrances to, achieving a just transition in those contexts. Here, a critique of global North-centric just transition is essential: while such a transition is portrayed as global, it broadly disregards questions of ecological and climate debt in relation to countries of the global South.[60] As studies of Moroccan women workers on farms in the south of Spain have shown,[61] unequal exchange and climate debt should be at the heart of debates about a just transition in North Africa. The export of vegetables, fruit and cheap labour to Europe is a by-product of the destruction of nature.

There have been many estimates of the scale of climate debt. For example, at the Copenhagen Summit in 2009, a study by the International Institute for Environment and Development estimated the cost of climate change to developing countries at up £6.5 trillion over the following two decades.[62] Likewise, another study by the African Development Bank in 2011 demonstrated that the costs of adaptation in Africa range from $20 to $30 billion per annum over the following 20 years.[63] Submitted to the Secretariat of the United Nations Framework Convention on Climate Change (UNFCCC) after the Paris Climate Summit, these reports highlight the plans of North

African countries (among others) to reduce emissions and adapt to climate change, and the expected costs of such changes. For example:

- Tunisia stated that in order to adapt to climate change and achieve a 41 percent reduction in emissions by 2030, in comparison to the 2010 level of emissions, the state needs international funding, capacity-building, and technology transfer, the total cost of which would be $20 billion.[64]
- Morocco estimated the cost of reducing greenhouse gas emissions by 42 percent at $50 billion.[65]
- Egypt identified the need for $73 billion to alleviate the impacts of climate change, without setting specific quantitative goals for reducing emissions.[66]
- Algeria reiterated its commitment to reducing greenhouse gas emissions by 22 percent by 2030. These plans were put forward without a specification of the value of this commitment or the efforts of climate change adaptation. Such a change, however, requires external support in terms of funding, technology development and capacity-building.[67]

Although these figures are dramatically higher than the development support North Africa receives, they portray but a small aspect of the economic burdens of climate change, and the global responsibility for bearing its consequences.

AGROECOLOGICAL AND REGENERATIVE AGRICULTURE AS VEHICLES FOR A JUST TRANSITION IN NORTH AFRICA

North African countries are, to varying degrees, integrated into the contemporary global food system, which is dominated by transnational corporations, international trade and export-led agriculture. As previously argued, these patterns of global unevenness have led to the rapid degradation of natural environments and resources, and to the marginalization of small-scale farmers and peasants, and the local communities in which they are embedded.

The region, therefore, needs to rewrite its agricultural, environmental, food and energy policies. It is necessary for alternatives to be locally centred and to be able to flourish autonomously, independent of European interests. This necessitates a bottom-up, rather than top-down approach, one that is informed by the daily practices and struggles of agricultural workers, local activists and actors in the region. It is evident that some peasant practices and ideas spreading in the region intersect with the principles of regenerative ecological agriculture – also known as agroecology (see Box 3). These form the building blocks of an ecological transition in the agricultural sector. The adoption of these practices is driven by a number of factors, including peasants' need to cope with climate change, and the high prices of pesticides and chemical fertilizers. There has also been a renewed interest among sections of both the rural and urban population in re-invigorating traditional agricultural technologies and using innovative ways to confront water scarcity, soil degradation and rising temperatures. Grounded in concrete realities, these practices delineate a possible starting point for building a bottom-up just transition project. A just transition must empower the local population and redefine development as development that is based on participation, and the preservation and renewal of resources.

Box 5.3 Agroecology as a science, a practice and a social movement

Agroecology can be defined as a science, a practice and a social movement.[68] The main aim of agroecology is to transcend the dominant agricultural paradigm and to develop agro-ecosystems that have minimal dependence on external inputs through practices that work with natural cycles and which centre farmer autonomy and agency in decision-making and the production of knowledge.[69] Regenerative agriculture is a branch of agroecology which represents a more reparative farming system. Regenerative agriculture and agroecology directly address the challenges of climate change as they focus on soil health, biomass, biodiversity and soil carbon sequestration.[70] Regenerative agriculture and agroecology are guided by some of the following principles:[71] 1) the interdependence of all parts of the agrarian system, including the farmer and the family; 2) the importance of ecosystem balance; and 3) the need to multiply ecological interactions and the workings of natural cycles in order to reduce the need for chemicals and other industrial inputs.[72] Thus, agroecology and regenerative agriculture enable farmers to meet their food needs through sustainable production methods while also revitalizing natural and agricultural environments. [73]

Practices of agroecology, regenerative agriculture and food sovereignty

Table 5.2 shows a selection of eco-regenerative agricultural practices identified through studies of local and indigenous knowledge in relation to water preservation in North Africa, as well as the few studies dealing with ecological and regenerative agriculture in the Maghreb, namely in Tunisia, Morocco and Algeria.[74] These have been complemented by the results of my own fieldwork in the countrysides of Egypt, Tunisia and Morocco between 2008 and 2019, as well as interviews with scholars and activists in the North African Network for Food Sovereignty.

Table 5.2 Selected practices of eco-regenerative agriculture in North Africa[76]

Category	Practices
Soil management, soil improvement and carbon sequestration	No-till farming Crop rotation (alternating cereals with leguminous crops) Diversity of crop compositions in farms Unprocessed organic fertilizers Processed organic fertilizers (compost) Liquid organic fertilizers (compost tea) Organic worm-based fertilizers (vermicompost) Liquid worm-based organic fertilizers (vermicompost tea)
Water resource management	Khattaras, Foggaras, cisterns (Al-majel) in Morocco, Algeria and Tunisia, respectively Bridges (Tunisia) Growing country-specific varieties Night irrigation (Egypt) Crop condensation North African oases three levels farming system[77]
Energy saving	Manual labour Use of animals Flow irrigation Night irrigation Solar irrigation
Environmental landscape management and wildlife control	Ecological traps Manual collection of grass Multiplying varieties and not planting the same crops in the same plot of land
Sustainable agricultural production	Terrace cultivation (mountainous regions of Morocco and Algeria) Oases systems Mixed agro-pastoral systems
Seed sovereignty	Seed self-production Municipal/domestic seed usage

These practices are linked to an increase in soil biomass,[75] a high level of organic matter, the enhancement of biodiversity and an increase in effective ecological/biophysical interactions within the agricultural system. Additionally, these practices renew and preserve the agricultural landscape, maintain and provide water resources, improve the livelihoods of agricultural workers, and provide safe, healthy and culturally appropriate food for local populations.

The aim here is not to give a complete overview, but rather a snapshot of practices related to ecological and regenerative agriculture in the contexts studied. Despite increasing experimentation with agroecological practices – often with the support of grassroots initiatives and organizations – fully integrated eco-farms remain very rare in North Africa. More commonly, peasants mix ecological farming practices with capitalist farming practices, examples being the use of both chemical and organic fertilizers, or resorting to both ecological and non-environmental modes of crop irrigation.

The emergence of these practices can be explained in part by the strategies small-scale farmers develop to bypass difficult environmental and economic conditions. For instance, small-scale farmers in Egypt are more inclined to switch to the use of animal waste and organic fertilizers when faced with the exorbitant prices of chemical fertilizers and pesticides. Similarly, they favour local seeds and rely on seed saving and sharing practices to circumvent the high prices of imported seeds. Likewise, in the Maghreb, small-scale farmers and peasants use local knowledge and technology in relation to environmental water preservation in the face of increasing water scarcity. While these practices do not necessarily stem from a radical environmental vision for agriculture, they can nevertheless be transformative. They serve as attempts to improve the livelihoods of impoverished farmers, helping them to continue their farming work in the face of capitalist exploitation. In this case, practices of agroecology and regenerative agriculture can be depicted as a kind of agroecology of the poor, as they are a product of poor people's focus on their own livelihoods.

Local actors and networks

There are a number of civil society organizations and government research institutions that support the transition towards ecological agriculture at different scales. This section highlights some of these initiatives.

Some of the institutions, associations, organizations and networks mentioned in Table 5.3 below play multiple roles in promoting agroecology and regenerative agriculture, through, for example, providing training tools for agroecological practices, producing research and reports, and facilitating networking between actors. In North Africa, farmers' cooperatives occupy a key position in supporting ecological farming practices,[78] particularly when understood in the context of the Maghreb-specific concept of cooperatives (ta'adoudya), which encompasses notions of solidarity, cooperation and sisterhood. These local forms of joint action, solidarity and alliance building are crucial: they help further integrate agoecological systems through knowledge dissemination and the extension of practical help in the form of training courses in soil maintenance and renewal, the provision of organic fertilizers and the propagation of native seeds.

Table 5.3   Examples of initiatives supporting eco-regenerative agriculture in North Africa[79]

Organizations	Geographical area of work
The North African Network for Food Sovereignty	North Africa
Alexandria Research Centre for Adaptation to Climate Change (ARCA)	A government institution in Egypt
Organic Agriculture Association	Egypt
Fayoum Agro Organic Development Association (FAODA)	Fayoum, Egypt
The Integral Development Action of Minia	Province of Minia, south of Egypt
Egyptian Association for Sustainable Agriculture	Province of Asyut, south of Egypt
Arid Regions Institute	A government institution in Tunisia
Observatory of Food Sovereignty and the Environment (OSAE)	Tunisia
Shapes and Oasis Colours Association (AFCO)	Chenini Oasis, south of Tunisia
Torba Association	Algeria
Pedagogical Ecological Farm	Zéralda region, Algeria
Network of Agroecological Initiatives in Morocco (RIAM)	Morocco
Worm-breeding groups – producing worm-based organic fertilizers	Egypt, Tunisia, Morocco, Algeria
Agricultural cooperatives	Egypt, Tunisia, Morocco, Algeria
Peasant/agricultural trade unions	Egypt, Tunisia, Morocco, Algeria
Food baskets linking consumers and producers (linking farmers to consumers in cities)	Egypt, Tunisia, Morocco, Algeria
Local agricultural markets	Egypt, Tunisia, Morocco, Algeria
Agricultural women workers' trade unions	Tunisia, Morocco

These mutually beneficial partnerships are necessary to widen and popularize agroecological experiences. Raising issues around workers' health and the use of chemical fertilizers, agricultural workers' unions push for organic methods of pest control, while associations facilitate the building of participatory relationships through direct selling, unionization, and mutual aid in a way that transcends the narrow confines of the market and private, individual interests.

As previously argued, despite the growing emphasis placed on the importance of local forms of regenerative agriculture and agroecology in confronting climate change, these practices remain largely marginalized in North Africa, at the level of both agricultural development policies and climate change mitigation policies. Indeed, these practices are primarily implemented at an individual level (farms) or at a local scale (community) with the support of civil society organizations and some research institutions. These dynamics do not allow for major changes to take place in agricultural policies, and they do not help rebuild food sovereignty on the basis of regenerative ecological agriculture. This problem is compounded by the dominance of industrial agricultural science and technology in the curriculums of agricultural colleges. For instance, in Egypt, pesticide, fertilizer and seed companies fund academic conferences in colleges of agriculture, while the curriculum promotes genetic engineering and the biotechnological revolution as solutions to the global food crisis.[80]

Despite these limitations, observations from the field demonstrate growing bottom-up pressure to build food sovereignty while supporting regenerative ecological agriculture in the region. It is on this basis that it is possible to set in motion a just transition of the agricultural sector in North Africa.

CONCLUSION

This chapter has shed light on the opportunities for, and challenges to, a just agricultural transformation in North Africa. Mainly export-led and intensive in its use of energy and capital, industrial agriculture remains the dominant framework for agricultural policies in the region. These policies are incapable of confronting climate change and the environmental crisis in the region. In fact, they add to it. Further, they are unable to achieve food sovereignty in North Africa, and actively contribute to the marginalization and impoverishment of agricultural workers and rural populations. This chapter has highlighted some of the dynamics within rural communities and their efforts to innovate and regenerate through local knowledge, with the aim of counteracting the degradation of natural resources and peasants' livelihoods. Additionally, the chapter has shown the pluriverses of agroecology and regenerative farming practices. However, these practices remain interwoven with capitalist farming methods. This can be mainly attributed to the absence of organized and sustained public policy support for an agroecological transition.

North Africa needs to rewrite its agricultural, environmental, food and energy policies. At the heart of any serious just transition programme should be the goal of achieving autonomy, ending dependency, reducing poverty, and mitigating the effects of climate change and environmental degradation. Building such a programme requires a more radical and local participatory approach, in order to regenerate and preserve local natural resources. This move offers a road to liberation from dependency; it requires building novel and locally-rooted knowledge systems and skills that support ecological and regenerative agriculture. The green revolution of the post-independence state would not have been possible without state intervention and support. State support consisted not only of providing production inputs, irrigation projects and mechanization, but also of providing agricultural extension services and establishing extension farms and research centres and institutes. Therefore, ecological and regenerative agriculture in North Africa is in need of a locally-oriented just transition plan. However, this will not be achieved without pressure from below, informed by the needs and aspirations of small-scale farmers, peasants and farm workers, who remain indispensable in a just transition in the region and beyond.

NOTES

1. Richard P. Allan, "Summary for Policymakers," *Climate Change 2021: The Physical Science Basis. Contribution of Working Group I to the Sixth Assessment Report of the Intergovernmental Panel on Climate Change*, (Cambridge University Press, 2021), pp. 3-32, doi:10.1017/9781009157896.001.

2. Imed Drine, *Climate Variability and Agricultural Productivity in MENA region*, *Working Paper No. 2011/96*, UNU-WIDER, 2011, https://tinyurl.com/msdtf27v.

3. Quentin Wodon and Nicholas Burger et al., *Climate Change, Migration, and Adaptation in the MENA Region*, The World Bank, 2014, accessed August 5, 2021, https://mpra.ub.uni-muenchen.de/56927.

4. Jeannie Sowers, Avner Vengosh and Erika Weinthal, "Climate Change, Water Resources, and the Politics Of Adaptation in the Middle East and North Africa," *Climatic Change*, 104, No. 3 (2011), pp. 599-627, https://doi.org/10.1007/s10584-010-9835-4.

5. Guy Jobbins and Giles Henley, "Food in an Uncertain Future: The Impacts of Climate Change on Food Security and Nutrition in the Middle East and North Africa," World Food Programme and Overseas Development Institute, 2015, https://tinyurl.com/39eatbcp.

6. Statistica.Com, "World energy carbon dioxide emissions by region", 2021, https://www.statista.com/statistics/205966/world-carbon-dioxide-emissions-by-region/.

7. Crippa, M., et al., CO_2 emissions of all world countries – JRC/IEA/PBL 2022 Report, EUR 31182 EN, Publications Office of the European Union, Luxembourg, 2022, doi:10.2760/730164, JRC130363.

8. Jason Hickel, "Quantifying National Responsibility for Climate Breakdown: An Equality-Based Attribution Approach for Carbon Dioxide Emissions in Excess of the Planetary Boundary," *The Lancet Planetary Health* 4, No. 9 (2020), pp. 399-404, https://doi.org/10.1016/S2542-5196(20)30196-0.

9. Pete Smith et al., "How Much Land-Based Greenhouse Gas Mitigation can be Achieved Without Compromising Food Security and Environmental Goals?," *Global Change Biology* 19, No. 8 (2013), pp. 2285-302, https://doi.org/10.1111/gcb.12160.

10. Harriet Friedmann, "Discussion: Moving Food Regimes Forward: Reflections on Symposium Essays," *Agriculture and Human Values* 26, No. 4 (2009), pp. 335-44, https://doi.org/10.1007/s10460-009-9225-6.

11. Sources: World Bank data 2021; Omar Bessaoud, Jean Paul Pellissier et al., "Rapport de synthèse sur l'agriculture en Algérie," CIHEAM-IAMM Montpelier, 2019, https://hal.science/hal-02137632.

12. Habib Ayeb and Ray Bush, *Food Insecurity and Revolution in the Middle East and North Africa: Agrarian questions in Egypt and Tunisia,* (Anthem Press, 2019).

13. Transnational Institute et al., *Just Transition: How Environmental Justice Organizations and Trade Unions are Coming Together for Social and Environmental Transformation,* February 11, 2020 https://www.tni.org/en/justtransition.

14. "Just Transition," What We Do, Climate Justice Alliance, accessed September 14, 2021, https://climatejusticealliance.org/just-transition.

15. Harald Winkler, "Towards A Theory of Just Transition: A Neo-Gramscian Understanding of How to Shift Development Pathways to Zero Poverty and Zero Carbon," *Energy Research & Social Science* 70, (2020):101789, https://doi.org/10.1016/j.erss.2020.101789.

16. Damian White, "Just Transitions/Design for Transitions: Preliminary Notes on a Design Politics for a Green New Deal," *Capitalism Nature Socialism* 31, No. 2 (2020), pp. 20-39, https://doi.org/10.1080/10455752.2019.1583762.

17. Claire Lamine et al., "The Place Of Agroecology in the New Dynamics Within the Agricultural World in Brazil and in France," Conference paper, XIII. *World Congress of Rural Sociology*, (Lisbon, July 29 to August 4, 2012).

18. The analysis is limited to these countries and excludes Mauritania and Libya because I was unable to access sufficient data on agricultural transformations in those countries.

19. Fathi Abdel-Fatah, *alnaasiriat watajribat althawrat min 'aelaa almasalat alziraeia* [Nasserism and the Experience of the Revolution from Above: The Agricultural Question], (Cairo: Dar Al-Fikr Publishers, 1987).
20. In 1956, Tunisia was liberated from French colonialism and Morocco from French and Spanish occupation, while in the same year Egypt saw the withdrawal of the British from the Suez Canal. As for Algeria, it achieved independence in 1962.
21. Hamid Aït Amara, "La terre et ses enjeux en Algérie," *Revue des mondes musulmans et de la Méditerranée* 65, no. 1(1992), pp. 186-96, https://doi.org/10.3406/remmm.1992.1564; Hichem Amichi, Gilles Bazin et al., "Enjeux de la recomposition des exploitations agricoles collectives des grands périmètres irrigués en Algérie: le cas du Bas-Cheliff," *Cahiers Agricultures* 20, No. 1-2 (2011), pp. 150-56, https://doi.org/10.1684/agr.2010.0459.
22. Omar Bessaoud, "alfilahat fi aljazayir: min althawrat alziraeiat 'iilaa al'iislahat alliybralia (1963-2002) [Agriculture in Algeria: From Agricultural Revolutions to Liberal Reforms (1963-2002)]," *Insaniyat* 22, (2003), pp. 9-38, https://doi.org/10.4000/insaniyat.7027.
23. Slimane Bedrani, "Algérie: une nouvelle politique envers la paysannerie?," *Revue des mondes musulmans et de la Méditerranée* 45, No.1 (1987), pp. 55-66, https://doi.org/10.3406/remmm.1987.2170.
24. *Ibid.*
25. The *Makhzen* is a term unique to Morocco: it denotes the ruling elite that revolves around the king. It consists of the monarchy, notables, landowners, tribal leaders and sheikhs, senior military personnel, security directors and chiefs, and other members of the state bureaucracy.
26. Fatah Ameur, Hichem Amichi and Crystele Leauthaud, "Agroecology in North African Irrigated Plains? Mapping Promising Practices and Characterizing Farmers' Underlying Logics," *Regional Environmental Change* 20, No. 4 (2020), pp. 1-17, https://doi.org/10.1007/s10113-020-01719-1.
27. Ayeb, *De la construction de la dépendance alimentaire en Tunisie*, Thimar and Observatoire de la Souveraineté Alimentaire, 2019, accessed July 30, 2021, https://tinyurl.com/35mx2ezj.
28. Mathilde Fautras, *Paysans Dans la Révolution : Un défi Tunisien*, (Karthala, 2021), p. 494.
29. *Ibid.*
30. Mahmoud Abdel-Fadil, *altahawulat aliaqtisadiat aliajtimaeiat fi alriyf almisrii 1952-1970 : dirasat fi tatawur almas'alat alziraeiat fi misr* [Economic and Social Transformations in the Egyptian Countryside 1952-1970: A Study in the Development of the Agricultural Question in Egypt], (The Egyptian General Book Organization, Cairo, 1978).
31. Abdel-Fatah, *Nasserism.*
32. Bessaoud, "Agriculture in Algeria."
33. Mohammed Said Saadi, "siasat altaqashuf ladaa sunduq alnaqd alduwlii wa'athariha ealaa alhimayat alaijtimaeia [International Monetary Fund Austerity Policies and their Impact on Social Protection]," Arab NGO Network for Development, 2014, accessed September 15, 2021, https://tinyurl.com/2p8nx9t6.

34. Kuper Marcel, Hammani Ali et al., "Que faire avec les eaux souterraines en Afrique du nord ?," In Pesche Denis et al. (eds.), *Une nouvelle ruralité émergente: Regards croisés sur les transformations rurales africaines*. (CIRAD-NEPAD, Montpellier, 2016), pp. 64-65.

35. Amichi et al., "Enjeux de la recomposition".

36. Omar Aziki and al., "difaean ean alsiyadat alghidhayiyat bialmaghrib dirasat maydaniat hawl alsiyasat alfalahiat wanahb almawarid [For Food Sovereignty in Morocco : A Field Study of Agricultural Policy and Resource Robbery]," ATTAC Morocco, 2019, https://tinyurl.com/y558dhd4.

37. Fautras, *Paysans Dans la Révolution*.

38. Bessaoud, "Agriculture in Algeria."

39. Mohammed Mandour and Gamal Siam et al, *al'ard walfalah fi masra: dirasat fi athar tahrir alziraeat almisria* [The land and the farmer in Egypt: A Study of the Effects of Liberalizing Egyptian Agriculture], (Centre for Agricultural Economic Studies and El-Mahrousa Center, Cairo, 1995), p. 28.

40. Saker El Nour, (2017), *al'ard walfalah walmustathmira: fi almas'alat alziraeiat walfalahiat fi misr* [The Land, the Farmer and the Investor: On the Agricultural and Farming Question in Egypt], (Dar Al-Maraya, Cairo, 2017), p. 192.

41. Nada Arafat and Saker El Nour, "How Egypt's Water Feeds the Gulf," *Mada Masr*, 2019, accessed September 15, 2021, https://tinyurl.com/4dnyrt75; Najib Akesbi, Noureddine El Aoufi, and Driss Benatya, *Agriculture marocaine à l'épreuve de la libéralisation*, (Economie Critique, Rabat, 2008); Hafiza Tatar, "Transformations foncières et évolution des paysages agraires en Algérie," *Méditerranée* 120 (2013), pp. 37-46, https://doi.org/10.4000/mediterranee.6660.

42. Daoudi Ali et Lejars Caroline, "De l'agriculture oasienne à l'agriculture saharienne dans la région des Zibans en Algérie. Acteurs du dynamisme et facteurs d'incertitude," *New Medit* 15, No. 2 (2016), pp. 45-52, https://agritrop.cirad.fr/580861; Ali Daoudi, Jean-Philippe Colin et al., "Mise en valeur agricole et accès à la propriété foncière en steppe et au Sahara (Algérie)," *Les Cahiers du Pôle Foncier* 13(2015), p. 34; David Sims and Mitchell Timothy, *Egypt's Desert Dreams: Development or Disaster?* (The American University Press, Cairo, 2016), https://doi.org/10.5743/cairo/9789774166686.001.0001.

43. Khaled Laoubi and Masahiro Yamao, "The Challenge of Agriculture in Algeria: Are Policies Effective?" *Bulletin of Agricultural and Fisheries Economics* 12, No. 1 (2012), pp. 65-73.

44. Stefan Kühn, *World Employment Social Outlook Trends* 2019 , International Labour Office, (ILO, Geneva, 2019).

45. Ministry of Agriculture and Land Reclamation of Egypt, *Agricultural Census 2009/2010*, 2010.

46. Zhour Bouzidi, Saker El Nour and Wided Moumen, "Le travail des f ail des femmes dans le secteur agricole: Entr emmes dans le secteur agricole: Entre précarité et écarité et empowerment-Cas de trois régions en Egypte, au Maroc et en Tunisie," Gender and Work in the MENA Region Working Paper No. 22 (Population Council, Cairo, 2011).

47. Ali Amouzai and Sylvia Kay, *Towards a Just Recovery from the COVID-19 Crisis: The Urgent Struggle for Food Sovereignty in North Africa*, Transnational Institue, 2021, https://tinyurl.com/4pzjj59f.

48. "FAO Food Price Index," World Food Situation, Food and Agriculture Organization of the United Nations (FAO), accessed December 4, 2021, https://fao.org/worldfoodsituation/foodpricesindex.

49. "Agriculture," Hazardous Work, International Labor Organization, accessed October 8, 2021, https://tinyurl.com/4b72cefp.

50. Ana Uzelac, Incoherent at Heart The EU's economic and migration policies towards North Africa, Briefing Paper, (OXFAM: 2020), http://doi.org/10.21201/2020.6805.

51. Emmy Månsson, Discourses of Ecologically Unequal Exchange Processes of "Othering" in the European Union's Framing of Trade, (Lund University, 2020), *LUP Student Papers*, http://lup.lub.lu.se/student-papers/record/9011422.

52. Hamza Hamouchene and Layla Riahi, *Deep and Comprehensive Dependency: How a Trade Agreement with the EU Could Devastate the Tunisian Economy*, Transnational Institute, 2020, https://tinyurl.com/228fpphs.

53. Juan Infante-Amate and Fridolin Krausmann,. "Trade, Ecologically Unequal Exchange and Colonial Legacy: The Case of France and its Former Colonies (1962-2015)," *Ecological Economics*, Vol. 156 (2019): 98-109.

54. Saker El Nour, "Grabbing from Below: A Study of Land Reclamation in Egypt," *Review of African Political Economy* 46, No. 162 (2020), pp. 549-66.

55. John Bellamy Foster, *The Return of Nature: Socialism and Ecology*, (Monthly Review Press, 2020); Ulrich Brand and Markus Wissen, *The Imperial Mode of Living: Everyday Life and the Ecological Crisis of Capitalism*, (Verso, 2021).

56. Brand and Wissen, *The Imperial Mode of Living.*

57. Joan Martínez-Alier, *The Environmentalism of the Poor: A Study of Ecological Conflicts and Valuation*, (Edward Elgar Publishing, 2002), p. 328.

58. Rikard Warlenius, "Calculating Climate Debt. A Proposal," in *12th Biennial Conference of the International Society for Ecological Economics*, 2012.

59. "Peoples Agreement," World People's Conference on Climate Change and the Rights of Mother Earth, Cochabamba, April 22, 2010, accessed October 22, 2021, https://tinyurl.com/57zd5w9z.

60. Climate debt is the outcome of damages related to the export of raw materials and other products from the global South at prices that do not include compensation for damages. Climate debt also arises from the global North exploiting environmental services and goods in countries of the South without return and without any acknowledgment of the rights of those societies. Lastly, climate debt also arises from the uneven global distribution of waste, of which the global South is the recipient. This causes significant environmental destruction in the South. See: Gert Goeminne and Erik Paredis, "The Concept of Ecological Debt: Some Steps Towards an Enriched Sustainability Paradigm," *Environment, Development and Sustainability* 12, No. 5 (2010), pp. 691-712.

61. Emmanuelle Hellio, "Importer des femmes pour exporter des fraises (Huelva)," *Études rurales* 182(2008), pp. 185-200, https://doi.org/10.4000/etudes rurales.8867.

62. Andrew Simms, *Ecological Debt-Balancing the Environmental Budget and Compensating Developing Countries*, IIED, 2001.

63. Vivid Economics, *The cost of adaptation to climate change in Africa*, (African Development Bank, 2011), https://tinyurl.com/bdd9k2kd.

64. "NDC Registry," United Nations Framework Convention on Climate Change, Tunisia, https://unfccc.int/NDCREG.

65. "NDC Registry," Morocco.

66. "NDC Registry," Egypt.

67. "NDC Registry," Algeria.

68. Wezel et al., "Agroecology as a science."

69. Miguel A Altieri, "Agroecology: The Science of Natural Resource Management for Poor Farmers in Marginal Environments," *Agriculture, Ecosystems & Environment* 93, No. 1-3 (2002), pp. 1-24, https://doi.org/10.1016/S0167-8809(02)00085-3.

70. Soil, specifically its biomass, is considered a natural carbon deposit when managed sustainably. This leads to mitigating the effects of climate change by sequestering carbon in the soil and mitigating emissions of warming gases into the atmosphere. See: FAO, Soils Help to Combat and Adapt to Climate Change by Playing a Key Role in the Carbon Cycle, 2015, accessed September 17, 2021, https://www.fao.org/3/i4737a/i4737a.pdf.

71. Charles A. Francis, Richard R. Harwood and James F. Parr, "The Potential for Regenerative Agriculture in the Developing World," *American Journal of Alternative Agriculture* 1, No. 2 (1986), pp. 65-74, http://doi.org/0.1017/S0889189300000904.

72. Nyong, A., Adesina, F. and Elasha, B.O., "The Value of Indigenous Knowledge in Climate Change Mitigation and Adaptation Strategies in the African Sahel," *Mitigation and Adaptation Strategies for Global Change* 12, No. 5 (2007), pp. 787-97; Carl Folke, "Traditional Knowledge in Social-Ecological Systems," *Ecology and Society* 9, No. 3 (2004), p. 7, http://www.ecologyandsociety.org/vol9/iss3/art7; Jan Salick and Nancy Ross, "Traditional Peoples and Climate Change," *Global Environmental Change* 19, No. 2 (2009), pp. 137-316, https://doi.org/10.1007/s11027-007-9099-0; Donna Green and Gleb Raygorodetsky, "Indigenous Knowledge of a Changing Climate," *Climatic Change* 100, No. 2 (2010), p. 239; Mohamed Berriane et al., "Les savoirs locaux peuvent-ils inspirer des solutions adaptatives dans les arrière-pays du Maroc ?," *Collection Dialogue des deux rives*, (Fondation Roi Abdulaziz, Casablanca, 2017), pp. 87-109.

73. V. Ernesto Méndez, Christopher M. Bacon and Roseann Cohen, "Agroecology as a Transdisciplinary, Participatory, and Action-Oriented Approach," *Agroecology and Sustainable Food Systems* 37, No. 1 (2013), pp. 3-18, DOI: 10.1080/10440046.2012.736926.

74. Ameur et al., "Agroecology in North African."; Koladé Akakpo et al., "Challenging Agroecology through the Characterization of Farming Practices' Diversity in Mediterranean Irrigated Areas," *European Journal of Agronomy* 128 (2021), p. 126284.

75. Soil biomass consists of all the living organisms in soil that have arisen from their decomposition to make up the mass of soil organic matter. Soil biomass

includes living components such as microorganisms, worms, roots and stems of plants below the soil surface, and residual biomass, such as organic matter from decomposing plants and animals. See: Rui G. Morgado et al., "Chapter 3 – Changes in Soil Ecosystem Structure and Functions Due to Soil Contamination," *Soil Pollution*, (Academic Press, 2018), pp. 59-87.

76. Author's fieldwork in Egypt and Tunisia (2018-2010); Ameur et al., 2020; Hamamouche, M. F., Kuper, M., Amichi, H., Lejars, C., & Ghodbani, T. "New reading of Saharan agricultural transformation: Continuities of ancient oases and their extensions (Algeria)", *World Development*, 107, (2018), pp. 210-223. Faiz, Mohammed El, and Thierry Ruf. "An introduction to the Khettara in Morocco: two contrasting cases." *Water and sustainability in arid regions: Bridging the gap between physical and social sciences* (2010), pp. 151-163. Ayeb, H., & Saad, R. *Gender, Poverty and Agro-Biodiversity Conservationin Rural Egypt and Tunisia.* (American University in Cairo Press, 2013). Remini, Boualem, Bachir Achour, and Jean Albergel. "Timimoun's foggara (Algeria): an heritage in danger." *Arabian Journal of Geosciences*, Volume 4, Issue 3-4 (2011), pp. 495-506.

77. Mohamed Taher Sraïri and Khaoula Bentahar, "Work Organization and its Economic Efficiency in Oasis Crop-Livestock Farms," *2nd International Symposium on Work in Agriculture: Thinking the Future of Work in Agriculture*, March 29-April 1, 2021, Clermont-Ferrand, 2021, accessed December 5, 2021, https://tinyurl.com/yszex64p.

78. Ameur et al., "Agroecology in North African".

79. Compiled by the author based on interviews with research participants conducted in 2021.

80. Saker, *The Land*.

6

The Electricity Crisis in Sudan: Between Quick Fixes and Opportunities for a Sustainable Energy Transition

Razaz H. Basheir and Mohamed Salah Abdelrahman

THE CRISIS

Over the last few years,[1] the electricity sector in Sudan has been in a state of crisis: 60 percent of the Sudanese population have been living without electricity, while millions of people currently suffer from hours of continuous power cuts, as the available electricity capacity covers a mere 60 percent of the demand.[2] Frequent tariff increases, reaching 13,000 percent for some social groups, have also exacerbated the crisis.

Several factors linked to supply and demand have caused this deteriorating situation. Investigating the context for this requires identifying the key political transformations, involving both civilian and military governments, that have occurred throughout Sudan's modern history, as well as reviewing the energy policies of the colonial era and discussing subsequent long-term changes in the energy sector. In this chapter, we also uncover the extent of the environmental vulnerability of energy production and consumption in Sudan, in addition to its link to a sustainable energy transition. Finally, this chapter seeks to interrogate the role the energy sector might play in meeting the political demands of the glorious December 2018 Sudanese revolution: freedom, peace and justice.

SUPPLY

Sudan's two main sources of power generation are hydroelectric and thermal, each producing approximately 50 percent of the current capacity of 3.5 gigawatts.[3] According to 2018 estimates, only 32 percent of the Sudanese people, mostly living in urban centres, are connected to the national grid. This uneven geographical distribution excludes the five federal states of Darfur and the region of South Kordofan, whose metropolitan areas are supplied by local networks that operate for an average of six hours per day.[4] These are the same sites that experience frequent conflict – a situation that has been shaped largely by historical developmental inequality.

Colonialism and developmental inequality

Since the colonial era, Sudan has been marked by severe developmental disparity correlated with the country's significant cultural diversity. That diversity prompted the separation of northern Sudan and the southern regions, including some areas of Darfur, Kordofan and the Blue Nile,[5] under the law of "closed districts". Until 1946, that law cut off large areas and diverse population groups from the socioeconomic development of the rest of the country. In 1955, just prior to the declaration of independence, war erupted between northern and southern Sudan. The country's wide developmental disparity and the northern monopoly of power were among the key factors behind the conflict.

The colonial policies did not stop after independence. Conflict escalated from one where government participation had been limited, to taking on an explicitly religious character following the first military coup in 1958.[6] It reached a pinnacle under the military "salvation government",[7] which governed from 1989 to 2019. During this period, the country's military rulers introduced what is locally known as the "Hamdi Triangle", comprising areas within Dunqulah, Al-Abyad and Sennar, the core of its developmental policies. These policies resulted in the concentration of development in a geographically limited area which, on the one hand, is culturally homogenous and, on the other, facilitated the formation of an Arab-Islamic alliance that was intended to provide the core of a homogeneous state capable of operating effectively even after geographical divisions. Thus, these policies provided an overt ideological cover for the state on the basis of developmental marginalization, while cultural and religious exclusion divided Sudan.[8] During this period, energy was one of the most important services provided by the state, and was thus inherently political, in that it served to perpetuate the power of the country's military rulers.

Sudan's hydrological dam projects are another colonial legacy that continues to heavily impact the energy sector. A quick glance at the history of dam construction shows that it was linked to Anglo-Egyptian colonialism in Sudan. For example, studies of the Second Cataract of the Nile began in

1897, that is, before British colonists entered Khartoum. This demonstrates the central position of control over the Nile waters in the colonial power's strategy. From the initial phase of colonialism, preparatory studies of the waterfall were undertaken, and in 1904 a detailed strategy was developed. These studies were carried out in accordance with the colonial priorities at the time, which focused on storing water for the benefit of expanding agriculture in Egypt and then in Sudan, in order to provide agricultural products for the colonizer at a low cost.

Sir William Garstin, a renowned scientist who studied the hydrology of the Nile and who had a long history of working in India and Egypt, was commissioned to research water storage options on the Nile. Garstin was the first person to conceptualize the construction of dams on Lake Albert, located in Uganda and extending into the Democratic Republic of the Congo, and the Jonglei Canal in South Sudan. Garstin pointed to the need to build a dam on Lake Tana in Ethiopia, as well as a dam on the Atbarah River in Sudan, to regulate the flow of the Nile's water. He also highlighted the possibility of benefiting from the lands between the Blue and White Nile through the Gezira Scheme in Sudan. In order to achieve this, Garstin proposed building the Sennar Reservoir in Sudan. Garstin's report was published in 1904; it contained various proposals, including initiating projects in Egypt.[9] It is thus clear that storing water for the benefit of Egypt was one of the motives of the British colonial project in Sudan.

The colonial project of agricultural expansion in Egypt relied on preserving water and protecting Egypt from floods, particularly following the floods of 1945-46.[10] This was exemplified in the 1946 report on the future maintenance of the Nile, which contained a detailed discussion of the question of the storage of the Nile waters through reservoirs. Moreover, a 1953 report entitled "Control of the Nile Waters", and a 1954 report by H. A. Morris, who was the Sudanese government's adviser on irrigation,[11] briefly alluded to potential and expected energy production from the Nile. Similarly, the official documents of the Sudanese Dams Implementation Unit make clear that the colonial enterprise had, since the 1940s, drawn up plans to preserve water for the benefit of Egypt, and that the strategy had changed from building the Merowe Dam in Sudan to establishing the High Dam as a means of securing its presence within Egyptian territory.[12] The shift in the aim behind constructing dams away from preserving water for the benefit of Egypt to only energy production began to clearly crystallize after the World Bank report of 1983, which detailed the possibilities of utilizing the proposed dams for energy production.[13] This is what the salvation government relied on in its studies thereafter, when it refocused all of its projects solely on energy production.

The overall vision for maximum exploitation of the Nile's waters was developed according to the colonizer's priorities at the time, which were to store water in order to ensure agricultural expansion in Egypt, after the failure of all previously adopted measures.[14] Nevertheless, the projects that have more recently been proposed by different national governments, specifically those related to dams under Omar al-Bashir's rule, do not differ greatly from the colonial vision that was put forward in Garstin's report.

The main transformation in these plans that took place in the second half of the twentieth century was a change in the declared primary aim of exploitation of the Nile's waters from water storage for the purposes of agricultural expansion (to benefit Egypt) to dam construction to produce energy, and thereby achieve development objectives. This camouflaged old colonial projects under a mask of development which carries promises of energy production. However, these promises were to remain unfulfilled, both in theory and in the face of reality and practical experience.

In its first stage, the period of Al-Bashir's rule was associated with isolationism and economic blockade, which drastically reduced the possibilities for expanding services. Soon after, in the early 2000s, oil was discovered in Sudan and, at the same time, a political settlement was signed with the largest movements and political parties. Together, these developments provided an economic spinoff which was reflected in the provision of various services, notably in the energy sector. Nevertheless, this period was not without political challenges, including in regard to managing the post-settlement transitional phase. At the time, the government's priority was the survival of Al-Bashir's rule, and by extension, the system of political Islam in Sudan. This was reflected in the government's energy-related policies, which played a political role.

The strategy of survival was evident in Al-Bashir government's attempt at the beginning of the twenty-first century to mobilize community networks for political purposes. This initiative, which saw one of the biggest political projects ever witnessed in Sudan, sought to transform energy production and distribution operations in the centre of the country, i.e., the Hamdi Triangle. By expanding the electricity supply network serving the residential sector in this area, the government aimed to garner political support. As part of this initiative, the Merowe Dam was marketed as a saviour that would guide Sudan away from darkness towards light and development. Al-Bashir highlighted this in his speech inaugurating the dam: "The Merowe Dam is the project of the century, the project of the beginning of the end of poverty, and the project of the great launch of the Greater Sudanese state."[15]

Al-Bashir's government presented the Merowe Dam as a major development project. Indeed, Al-Bashir himself attempted to market the project at the opening ceremony as a response to the 2009 International Criminal Court's memorandum. He declared in the same dam inauguration speech: "They will issue their decision tomorrow, and after that they will issue a second and third decision, and the people will not pay attention to them. They will be preoccupied with decisions and we will continue to develop."[16] At the time, the slogan of Al-Bashir's supporters was "the dam is the response".[17] However, the fog of developmental discourse was soon cleared away as the reality of growing electricity cuts and the rising costs of electricity itself became clear.

As the Sudanese government focused its construction and marketing operations on pro-government companies, construction costs rose, a result of increasing corruption and nepotism, and a lack of oversight. This resulted in an exorbitant debt of $3 billion related to the construction of the Merowe Dam, even as the dam's ability to generate power actually decreased compared to the initial promises. During its opening, it was proclaimed that the dam would produce 1,250 MW; however, its actual capacity dropped to less than 600 MW.[18]

The lack of transparency surrounding the dam project was a crucial factor in increasing its environmental costs. The government assigned the engineering aspects of the work to Lahmeyer International, a German company which had been implicated in corruption cases related to water projects in the Lesotho Highlands in southern Africa,[19] as a consequence of which the World Bank had discontinued dealings with it for seven years. In Sudan, Lahmeyer received funds from parties that had rarely considered transparency a priority. The company continued operating as an engineering consultant for other dam projects, even expanding its work under the period of the Salvation government's rule.[20] The construction of the Merowe Dam under Lahmeyer's guidance involved violations related to environmental studies – such studies were not approved for the dam until 2007. A report on the environmental situation in Sudan, which was issued after the protracted armed conflict between 1983 and 2005, made clear that the government did not adhere to its own legal standards when approving studies of the environmental impact.[21] The report described how studies that were presented to the relevant Sudanese authorities had not been approved as they lacked basic components relating to integrity. This put pressure on the financiers to stop the flow of funding to the related projects. As a result, the government removed the minister and all departments involved in approving environmental impact reports. These reports were approved only about a week after the appointment of new departments replacing those that were removed. This demonstrates that the Merowe Dam was considered to be of extreme importance. Furthermore, it demonstrates the lack of attention to the environmental and social costs of building dams, such as increased evaporation rates. Indeed, the Merowe Dam's[22] evaporation rates are as high as approximately 1.5 billion cubic metres of water. This is in addition to the general increase in the number of artificial lakes in Sudan, which clearly impacts the production rates of staple crops and orchards in areas north of the dam. The dam also contributed to the displacement of tens of thousands of people, and the loss of their means of livelihood.[23]

A few years later, in 2013, it was announced that an operation to increase the height of the Roseires Dam had been completed. This dam is located in Blue Nile state, approximately 550 km southeast of Khartoum. After that, in 2017, the Upper Atbara and Setit Dams, in the states of Kassala and El-Gadarif, about 460 km east of Khartoum, were completed. Theoretically, these two dams produce 280 and 320 megawatts, respectively. Various projects in Sudan have been built using loans from Gulf and China. However, many specialists have questioned the usefulness of Chinese and Gulf financing for hydro-energy projects. In regard to China, it is argued that it provides Sudan with loans in return for Chinese government-owned companies being commissioned to construct dams in the country. As for the Gulf countries, it is argued that they provide loans in exchange for fertile land, as a means of addressing their own food security issues.[24]

Financing based on loans is one of the major problems facing energy production projects, especially dams. Instead of mobilizing the resources of host countries by funding through progressive taxation, the creation of public shareholder companies, and the provision of opportunities for the affected population to contribute to projects and solutions that guarantee broad participation and benefits, these projects are financed by loans that not only reduce national sovereignty in relation to strategic projects, but also increase the debt burden.

Projects such as the Merowe Dam, the heightening of the Roseires Reservoir, and the construction of the Upper Atbara and Setit dams, offer clear examples of these loan-related dynamics. Chinese companies obtained construction contracts for these projects, while Saudi Arabia acquired more than one million feddan (i.e., 420 thousand hectares) of Sudanese land for a period of 99 years. Saudi Arabia's land acquisition equals the total area of the new Upper Atbara project, which is located on fertile lands that Saudi Arabia wishes to exploit as part of a project to provide food security for itself.[25] The residents of this area were forcibly displaced from their lands, receiving unfair compensation: those who owned less than ten agricultural feddans (i.e., 42 thousand square metres) were compensated with a residential plot of 300 square metres, and those who owned more than ten feddans were compensated with two residential plots with a total area of 600 square metres.[26] Thus, in addition to approximately 700,000 citizens being forcibly displaced from their homes, the people in this area lost their agricultural lands, and shepherds lost the natural grazing paths utilized by more than seven million heads of livestock.[27]

The energy return from these hydro projects is low in comparison to their exorbitant economic, social and environmental costs. These projects have exacerbated development inequality as they have involved a large section of the population losing its traditional means of livelihood. At the same time, the areas closest to these dams, such as the localities of El Buhaira and El ʿAzaza near the Roseires Reservoir, and most of the villages on the banks of the Atbara River, have neither electricity nor regular access to water. These hydropower projects thus create zones of sacrifice for the benefit for "development" and capitalist accumulation in other spaces. This helps reproduce developmental disparity, deepens historical inequality and further increases conflict in various degrees and forms.

Hasty solutions

In addition to denying more than 60 percent of the Sudanese people access to the national grid, the relatively large increases in annual consumption levels (ten percent) worsened the national supply gap. As a result, the energy sector was under pressure to provide more electrical capacity. These pressures were addressed through the construction of new thermal power plants, which are heavily reliant on imported fuels: more than 1,500 thermal megawatts were added between 2008 and 2019. In 2017, the amount spent on fuel was estimated at $1.3 billion, with the government's support for the sector reaching 15 percent of state expenditure.[28] These plants cause significant emissions, equalling about 6.25 million tonnes of carbon dioxide.

The speed and relatively low initial cost of these new thermal plants, and their contribution to national electricity capacity, has tended to obscure the significant operational challenges facing the country, which lost more than

75 percent of its oil reserves and their associated profits following the secession of South Sudan in 2011. This event rendered Sudan largely dependent on imported fuel, as well as exposing it to unstable exchange rates and accelerating inflation rates. In addition to increasing electricity prices, the expansion of thermal production did not take into consideration the negative impact of this form of production, which causes a significant increase in the emission of greenhouse gases.

Despite the presence of these thermal projects, hydro-generation options have remained at the heart of future plans for electricity supply in Sudan. The salvation regime[29] repeatedly expressed its intention to construct a group of large dams on the Nile River north of Khartoum, in the areas of Dal, Kajbar and Al Sheraik, located on the second, third and fifth cataracts in the north of Sudan. They would have a total combined operating capacity of 990 MW. These dams would be in addition to various other projects in Daqash, Mukrat, Sheri and Sablouka.[30] However, as a result of widespread rejection by the local populations in these areas, these projects face challenges. In the view of these populations, these projects will not be useful and will flood most of the residential, agricultural and archaeological spaces from north Khartoum to Old Halfa. The local populations are also challenging these projects on the basis of their high costs compared to the small returns they will provide. The endeavour of marketing hydro-energy projects as a solution collides with the reality of the construction of Ethiopia's Renaissance Dam, which will change the nature of the Nile and help stabilize water flow throughout the year. Such a development will not only render these projects technically useless, but also impose challenges that make their implementation unrealistic.

In sum, all of these factors render the electricity sector plan, which aims to generate 80 percent of the electricity supply by 2031, unrealistically ambitious.[31] In addition to the gap in the available capacity, the significant cost of extending transmission and distribution networks means that a large segment of the population are currently left in the dark. This is due to the fact that electricity supply lines are concentrated in the centre and north of the country, which is the historical centre of economic and political power in Sudan.

There are some common features between Sudan and some sub-Saharan African countries in terms of a decline in electrification rates and population density, and the fact that these countries have not succeeded in creating thriving markets and industries from energy alternatives. However, unlike those countries, Sudan's situation is in part the result of the international isolation imposed on Al-Bashir's regime due to US sanctions. For example, compared to Tanzania,[32] which has 109 insulated solar plants, with a total capacity of 158 MW, Sudan has only one plant, with a capacity of no more than five MW. The first solar initiative was launched in Sudan in 2014, involving piloting of the solar home systems model whereby individual homes are provided with solar systems in instalments, in cooperation with local banks. The initiative first targeted 100 users with a capacity of 100 watts per user (the aim is to reach a total capacity of 110 MW by 2031).[33] According to the latest report on this initiative, the number of homes benefiting from the service reached 1,500 in 2018.[34]

It was only in 2020 that Sudan's first solar plant was established, in El Fasher, the capital of North Darfur State and one of the most important cities in the Darfur region.[35] This plant has a capacity of five MW. A twin plant in the city of El Daein in Darfur encountered various obstacles that have so far prevented its completion, including issues related to funding, delays in receiving materials and equipment, and some cases of equipment theft. The two plants were financed by the Sudanese Hydro Generation and Renewable Energy Company (SHG&REC) and are implemented by a local private company called Top Gear.[36]

While the crisis currently affecting the electricity sector can be traced back to the Al-Bashir era and its corruption, the post-revolution transitional government of 2018 directly and indirectly aggravated it. The neoliberal doctrine of the World Bank and the International Monetary Fund dictated all macroeconomic reforms implemented by the transitional government. Abdalla Hamdok, the former prime minister who has previously worked in the United Nations, did not attempt to resist this neoliberal tide, even arguing that these reforms were prerequisites for debt relief and for obtaining new loans and subsidies.[37] Furthermore, floating the Sudanese pound and lifting subsidies on basic commodities led to the official value of the pound plummeting against the dollar. By early 2023, it stood at 570 pounds to the dollar, in comparison to 55 pounds to the dollar in January 2021. The floating of the pound also led to a massive increase in fuel prices, from 100 pounds per gallon to 2,500 pounds per gallon.[38]

The direct ways in which the transitional government destabilized the electricity supply are also rooted in these neoliberal economic reforms, which targeted the energy sector. The implementation of these reforms, which came at a critical time, was subject to a great degree of coordination and planning. The sector's failure in this matter rendered users the victims of unjust reform recommendations and their poor implementation. The situation then worsened after the *coup d'état* of 25 October 2021, in response to which all foreign aid was suspended. As a result, the state treasury was under increasing pressure and has accelerated the implementation of the package of flawed reforms.

CONSUMPTION

One of the most important features of the oil years in Sudan, from 1999 to 2011, was the change in the lifestyle of the country's urban middle class. The Greater Khartoum area, which hosts 20 percent of Sudan's population (about nine million people)[39] and houses the country's main industries, services and transactions, consumes 60 percent of the country's electricity, which goes also to residential areas. 60 percent of this consumption goes to the residential sector. Here, the architecture of residential buildings is relevant. Traditional architecture, based on earth materials and incorporating spacious, well-ventilated courtyards that are suitable for the desert climate of Sudan, have frequently been replaced by concrete jungles of poorly ventilated vertical buildings inspired by the architecture of Dubai and other major cities in the Gulf region. These architectural transformations in the city have led a large proportion of citizens to rely entirely on air conditioning units, with the attendant high rate of electricity consumption. As a result, electricity demand rates in the summer are twice as high as in the winter. However, unlike Dubai, strict laws and institutions regulating buildings and the manufacture and import of electrical appliances are completely absent. Average household electricity consumption in Greater Khartoum is 308 kWh per month, which is almost six times the average for sub-Saharan Africa.[40] It is considered that this offers a possibility for improving the state of the energy sector.

On the one hand, these high rates of urban electricity consumption put pressure on the electricity sector. On the other, they exert political pressure on the government to secure a more stable supply. Indeed, it is this pressure that has pushed the sector towards seeking quick-fix emergency solutions, such as the significant increase in thermal capacity, which the five-year plan drawn up in 2018 under Al-Bashir intends to produce. The 2018 plan proposed increasing the available capacity by an additional 8.7 gigawatts, 60 percent of which would be from thermal plants. In its report of mid-2019, the World Bank reviewed this plan, while simultaneously evaluating the current situation of the electricity sector and presenting recommendations for its recovery. This report now acts as a reference point for reforming the electricity sector. Below, we discuss the most important arguments and axes presented in the report.

THE WORLD BANK REPORT

The main reforms proposed in the World Bank report can be divided into three elements: first, the lifting of tariff subsidies; second, the energy mix to ensure future capacity; and third, the private sector's involvement.

The lifting of subsidies on electricity tariffs

According to the World Bank report, electricity tariffs in Sudan are the lowest in sub-Saharan Africa, regardless of the income levels of the countries compared. The report shows that electricity tariffs represent between one and three percent of the average monthly income of families – compatible with the recommendation that electricity tariffs should not exceed five percent of the average monthly income of families. However, this recommendation does not take into account the fact that in a country in which more than 65 percent of the population is employed in the informal sector, many families do not have a fixed monthly income.[41] The World Bank report also adds that the bulk of the electricity tariff subsidy provided by the government is directed to the wrong social groups, as a greater percentage of electricity is consumed by classes higher up the income pyramid. In other words, the largest proportion of government subsidies actually goes to the rich. The report concludes that these subsidy rates are very generous and it recommends gradually reducing them over five years to reduce the country's fiscal deficit.

Between January 2021 and January 2022, electricity tariffs were adjusted three times, at exponential rates. For instance, compared to the pre-2021 tariff, the Lifeline Tariff, which is the least expensive type and which targets the lower income groups, has been reduced from 200 to 100 kWh. At the same time, the kWh cost has increased at a rate exceeding 3,000 percent: from 0.15 pounds to five pounds.[42] In addition, the commercial and agricultural tariffs have increased by 13,000 percent and 5,000 percent respectively, among other increases.[43] The outcomes of these severe increases have been reflected in hikes in the prices of all manufactured products and commodities, which in turn has only worsened the suffering of a population whose resources are consumed by inflation.

While we can agree on the necessity for electricity tariff reforms in light of the current conditions of the energy sector, as well as the overall economic situation that Sudan has inherited from Al-Bashir's corrupt regime, the details of these reforms remain a matter of debate. It is unfair to assume that different social groups have the same ability to absorb great increases in electricity tariffs, and the prices of goods and services accompanying them. It would have been possible to achieve a better reform formula by balancing the rates of increase for the different groups that make up the class pyramid. Examples of such nuanced measures include maintaining the same tariff for the social groups that consume the least electricity (an average of 177 kWh per month), and setting the tariff for higher consumption groups (about 600 kWh per month, for example) at its real cost. Not only would this increase the sector's revenues, it would also stimulate the one percent that consumes more than a quarter of the residential sector's supply to rationalize their electricity use.[44]

Despite the significant increases in tariffs, so far the lifting of subsidies has been only partial, which means that there are more increases on the way in order to meet the World Bank's full recommended rise. Such increases may exacerbate the popular anger that has engulfed citizens following the coup measures, especially in light of the current situation in which tariff increases do not translate into supply stability. Additionally, the latest wave of increases in 2022 sparked strong opposition within the ranks of smallholders, who saw the tariffs they pay increase by 2,000 percent.[45]

In early 2022, smallholders in northern Sudan, with the support of other civil forces, protested against the increase in agricultural tariffs and demanded a reduction. They expressed these demands by constructing barricades on the national road that links northern Sudan with Egypt and that facilitates the movement of significant quantities of goods and people. This protest culminated in what was later known as the "North Barricades". This action was successful: the protestors seized suspicious commercial goods (gold and raw materials) smuggled into Egypt. They also expanded their protest from one point along the route to 14 further points. In addition to the demand to reduce electricity tariffs, the region's residents added a set of historical demands, related to the suffering they have endured as a result of hydropower generation projects since independence. The North Barricades continued for more than four months and succeeded in reducing the electricity tariffs for local smallholders from 21 to nine pounds.[46]

In sum, without controlling Sudan's macro-economic indicators, inflation will devour all additional profits gained from reducing subsidies, leaving the electricity sector with a double bill for fuel from thermal stations, current and future. Given the inflation rates, which exceeded 260 percent in March 2022,[47] and the very low purchasing power of the population, more than half of whom are below the poverty line,[48] the liberalization of prices in this context can only lead to further deviation from the sector's goals for electrification by 2031. Indeed, successful electrification experiences in countries such as Ghana and South Korea[49] have only come as a result of government efforts to design a fee and tariff package that is appropriate for citizens with limited and irregular incomes.

The energy mix for future capacity

As a result of the lack of full coordination between the various relevant authorities during Al-Bashir's era, it is difficult to compare the plan developed in 2018 for Sudan's future energy capacity and the country's Nationally Determined Contributions that were agreed upon at the COP21 in Paris and updated in 2021.[50] The 2018 plan aims to construct new plants with a capacity of 3,000 MW before 2031, with a mixture of wind and solar energy, both on and off the grid. Meanwhile, the government plan recommends an additional 60 percent capacity from thermal plants, a proposition which the World Bank report slightly revised, reducing the thermal generation rate to 50 percent and increasing renewable energies – excluding hydropower generation – to 800 MW for both wind and solar energy. This equates to ten percent of the future additional capacity of each of the two sources. Also, the report adopts criteria of the least-cost plan in order to reach the energy mix

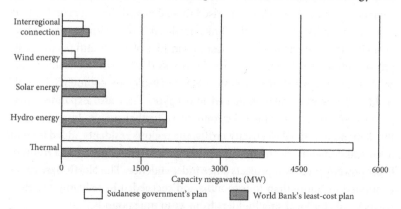

Figure 6.1 A comparison between the five-year plan of the Government of Sudan and the one proposed by the World Bank.

for future capacity. This approach takes into account only the financial costs of projects, ignoring their social and environmental costs.

In light of this discussion, one question cannot be overlooked: why do future capacity plans not primarily focus on renewable energies? This question is particularly important for two reasons. First, daily solar radiation rates are extremely high in Sudan, and wind speeds exceed 7 m/s in several locations, which makes it an ideal environment for producing wind energy.[51] Second, there has been a discernible decline in the prices of renewable energies over the last few years, with prices reaching a level of competitiveness comparable with those of conventional energy sources.

Hyper-centralization is another feature that remains dominant in the planning mindset in Sudan. The experiences of countries in the region and other countries of the global South offer many examples of alternative and decentralized models being successfully utilized as appropriate solutions for electrification problems, challenges with attempts to transition to clean energies and the huge financing obstacles that hinder centralized projects. For example, Kenya successfully increased its electrification rates from 32 percent in 2013 to 75 percent in 2018. It achieved this by resorting to a mixture of technical and institutional solutions, including conventional on-grid, solar home systems and mini/micro-grids.[52] Moreover, the cost of micro-grids is steadily decreasing, and the capacities of what is currently known as the technologically flexible third generation micro-grids offer real prospects, such as the possibility of connecting groups of micro-grids together or connecting them to the national grid in the future.[53]

Although the parameters of the future energy capacity plan emphasize the crucial role of micro-grids in the project of universal electrification in sparsely populated areas of Sudan, there are in fact no details about their role in the short term. Further, the World Bank report follows the model of government policies, which are characterized by directing limited financial resources to strengthen consumer supply within the grid. This strategy came about as an urgent response to the urban supply crisis but it was at the expense of electrification projects for off-grid users, despite the high social costs of depriving more than 60 percent of the population of their right to access electricity, which should be seen as a basic service.

Involvement of the private sector

In the concoction of liberalization that has been described over and over again globally, the magic ingredient is the opportunity for the private sector to participate in order to attract funding and expertise. In 2010, the National Electricity Authority, which monopolizes all electricity supply operations, was divided into five companies, as per their technical functions, such as thermal and hydropower generation, transmission and distribution, and so on. This reform provided the private sector with the opportunity to enter the generation sector, as independent power producers (IPPs). However, with the exception of isolated plants whose construction and operation were privatized, accounting for no more than three percent of the total capacity, the private sector did not engage in any way in subsequent generation projects. The World Bank report attributes this lack of involvement to the absence of a comprehensive framework detailing the partnership process and its regulations, in addition to the low (subsidized) electricity tariffs, which are deemed unattractive to investors.

The framework that the World Bank report proposes pushes the current sector reforms towards becoming more attractive to investors, rather than increasing electric capacity and restructuring public companies so they can efficiently lead future projects. Not only does the profitability (il)logic at the core of the electricity sector reforms threaten the latter's durability, these reforms may be difficult to attain in the first place, which will render the sector prey, once again, to quick-fix emergency solutions.

The guidelines and best practices in the power generation industry,[54] which guide the world's energy investors, contrast sharply with the current state of the sector in Sudan. Among the guidelines and best practices are a range of determinants considered important for a successful energy sector, such as environments that are conducive to investment, clear and consistent policy frameworks, practices of competitive bidding, stable fuel supplies, and so on. Such conditions are difficult to imagine in the near future in Sudan, especially in light of the political and economic instability currently affecting the country. For instance, Turkish independent power producers, which own some isolated plants in Sudan, currently have suffered difficulties in securing fuel supplies, as well as experiencing delays in receiving government payments. In response to these problems, they have cut off supplies to some Sudanese cities for days at a time.[55]

Perhaps the most reasonable and just scenario, especially for the poorer classes, lies in extricating the private sector from attempts to solve the electricity crisis in Sudan. Indeed, this scenario would involve directing those very development loans directly to public companies in the sector, as well as departing from free market conditions and determinants. Most of these free market conditions have been shown to be in stark contrast to electrification projects that incorporate a social dimension, such as the Lifeline Tariff, cross-subsidy,[56] fair import taxes, and localization requirements for industries and labour. In a general sense, the need for provision of financing – which is used to justify the involvement of the private sector – has become a means of privatizing aid. In turn, this serves as a neocolonial tool that increases the dependence of the global South. Furthermore, it does not help to achieve sustainable solutions to distinctive problems facing societies there. Rather, privatized aid becomes a channel through which the budgets of development support are diverted to private companies. Power Africa, for instance, is the largest energy project initiative in sub-Saharan Africa, with total funding commitments of $54 billion. This project is being used by the US government to increase the profits accrued by the private sector in the US. In 2016, it directed nearly 90 percent of its commitments – amounting to $7 billion – to private US banks and financial institutions in order to implement or finance energy projects in sub-Saharan Africa.[57]

Therefore, the argument that local energy companies lack the operational competence required to lead such projects is refutable. During the oil years in Sudan, the economic recovery associated with this era provided the state treasury with significant surpluses. Despite the corruption and nepotism present at that time, the National Electricity Authority did not lack the technical and administrative expertise to carry out successful and large-scale reforms. These included the shift to a pre-payment system,[58] which raised collection rates to 93 percent – among the highest in the region.[59] Also, generous amounts were spent on rehabilitating and training cadres within and outside Sudan, which put the country in a good position: high financial benefits preserved the electricity sector from the phenomenon of brain drain that characterizes most other sectors in the country. In the last few years, projects have also been undertaken to localize the manufacture of equipment, such as transformer assembly plants, and the manufacture and programming of pre-payment meters.

CONCLUSION

It can be argued that it would be better to take even the smallest of steps in the right direction, one which is sustainable and socially just, rather than attempting to take big leaps to solve the problems of the few (the rich and the upper classes), including by making choices that have significant environmental and social ramifications, often for the poor and marginalized classes. In addition to the problem of access to the grid, from which the rural population, nomads, and the precarious urban classes in Sudan have continuously suffered, the extent of frequent power outages in urban areas has only worsened in recent years. While this crisis situation motivates the search for emergency and less costly solutions, it can also be an opportunity to rethink conventional ways of generating and managing the electrical supply.

Since independence the complete disregard for the livelihoods of local communities and their problems has been a steady feature of large energy projects in Sudan, such as hydropower dams. This neglect has been manifested in forced displacement and the destruction of traditional livelihoods. Additionally, this trajectory of increasing thermal plants in the country, driven by their low initial cost and their relative ease of construction, severely harms the environment. Furthermore, the dependence on imported fuels has been proven to inflate the operating costs of these plants, especially in light of the continuous deterioration of the local currency and the rise in global fuel prices.

The current roadmap for resolving the electricity crisis relies on the full and unconditional adoption of neoliberal reforms dictated by international financial institutions. The latter claim to support the Sudanese people in their aspirations for a democratic transition. Nevertheless, the outline of these future reforms is no different from the sector's previous strategies, or even as regards the hydro-energy projects implemented in the colonial era. Among the key features of these reforms is the gradual lifting of subsidies on electricity tariffs, attracting private investors and estimating future capacities using the criteria of the least-cost plan. These three themes have been discussed throughout this chapter, as well as their social, economic and environmental consequences, especially for the most vulnerable social groups in Sudanese society.

It is important to note that this chapter does not claim to provide a comprehensive answer to the thorny questions which the current electricity crisis in Sudan poses. Rather, it has attempted to point out basic criteria and priorities which must be taken into account when considering solutions to this crisis, and the reason why current approaches to the crisis are flawed. When it comes to the question of supply, there is the problem of prioritizing the private sector and centralized generation plants with energy sources that are not environmentally friendly, for purely financial reasons. When it comes to consumption, citizens outside the grid's range and in marginalized areas are given secondary importance, while the high tariffs are disproportionate to the conditions of the poorest social groups.

Moreover, it is worth emphasizing that there is no single solution that can meet the needs of all of Sudan's social groups. The current complex situation calls for studying and understanding the needs and contexts of different social groups, which will dictate different technical solutions and institutional structures. In turn, such a process will require different forms of financing that can enable the mobilization of financial resources, ranging from self-financing and public and private capital to low – or zero – interest rate development aid. Sudan can also join other countries in pushing the climate reparation agenda[60] in order to obtain financial resources that do not carry the burden of additional debt. Indeed, this would enable Sudan to adopt energy generation technologies which can help reduce emissions and dependence on biomass as a source of energy that threatens forests and vegetation. This move would also enable access to appropriate solutions based on centralized, decentralized or stand-alone supply systems and a more socially just budget allocation.

As with many problems in countries in the global South, the current electricity supply crisis in Sudan requires an urgent, sustainable and feasible solution. Such a solution will need to involve a great degree of integrated planning between various state apparatuses, as well as rearranging priorities to accord with the requirements of a socioeconomic development which is just, sustainable and suitable for local contexts. A green and just energy transition in Sudan must take into consideration the importance of formulating policies independent of the imaginaries of the old colonial legacy – a legacy which is based on huge infrastructure and political symbols and icons that serve the elites. A just transition will also need to eschew the neocolonial facades and their promises of financing, whose effects are only reflected in the stock prices of transnational corporations.

NOTES

1. This chapter has been drafted in 2022, way before the start of the civil war in Sudan in April 2023. Therefore, it doesn't unfortunately take into account the new developments that have occurred since.
2. World Bank, *From Subsidy to Sustainability: Diagnostic Review of Sudan's Electricity Sector*, June 30, 2019, accessed August 5, 2020, https://tinyurl.com/486279jw.
3. *Ibid.*
4. Ministry of Water Resources and Electricity (MoWRE), *Long-and Medium-Term Power System Plans*, Lahmeyer International GmbH, 2012.
5. South Sudan seceded and declared independence in 2011.
6. Abdallah Bola, "shajarat nisab alghul fi mushkil "alhuiat althaqafiati" wahuquq aliansan fi alsuwdan [The Ghoul Genealogy Tree in The Problem of "Cultural

Identity" and Human Rights in Sudan]," *The Sudanese Society for Research in Arts, Fine Arts and Humanities*, 2005, https://tinyurl.com/3jt2kh4p.

7. "Salvation" is the term used to refer to the Sudanese Islamist Movement, which came to power through a military coup led by Omar al-Bashir. As the coup leaders claimed at the time that their motive to seize power was to save the country from high prices and corruption, they constantly referred to themselves using this term.

8. Sudanese Online, "alnasu alkamil liwaraqat eabdalrahim hamdi .... w khatat taqsim alsuwdan [The full text of Abderrahim Hamdi's paper... and plans for the division of Sudan]," 2005, https://tinyurl.com/yckv6frr.

9. Sir William Garstin, "Report Upon the Basin of the Upper Nile: With Proposals for the Improvement of the River;" William Willcocks, "The Nile in 1904," National Printing Department of Egypt, 1904.

10. Mahmoud Saleh Othman Saleh, *alwathayiq albiritaniat ean alsuwdan 1940-1956* [British Documents on Sudan 1940-1956], (Riad el-Rayyes Books, 2011).

11. Morris, H. A. "Development of the Main Nile for the Benefit of Egypt and Sudan," 1954.

12. Dafaallah Ahmed Abdullah, "mashrue sadi marwi: alqisat alkamilat min alfikrat 'iilaa al'iinjaz [Merowe Dam Project: The Complete Story from Conception to Completion];" Dams Implementation Unit, Ministry of Water Resources, Irrigation and Electricity, 2016.

13. World Bank, *Sudan: Issues and options in the energy sector*, 1983, https://tinyurl.com/bdf42pay.

14. Ali Khalifa Askouri, *khazaan alhamaadabi: namudhaj al'iislam alsiyasii lil'iifqar wanahb almawarid* [The Hamdab Dam: Political Islam's Model of Impoverishment and Plundering of Resources], (AJSP Printing Services, 2014).

15. Al-Bashir's speech at the opening of the Merowe Dam in 2008.

16. *Ibid.*

17. In other words, the dam was a response to the arrest warrant for Al-Bashir issued by the International Criminal Court.

18. Mohammed Jalal Ahmad Hashim, risalat kajbar: min 'ajl alsuwdan la min 'ajl qaryati. qadaya alsudud bishamal alsuwdan [Kajbar's Message: For the Sake of Sudan, Not for the Sake of a Village. The Question of Dams in Northern Sudan], (*Shafuq lil-Nashr wa al-Intaj al-I'lami*, 2016).

19. Probe International, *World Bank Sanctions Lahmeyer International for Corrupt Activities in Bank-Financed Project*, 2006, https://tinyurl.com/ypfdfexs.

20. The expansion of Lahmeyer was linked to the director of the Dams Implementation Unit, Motazz Moussa, who later became the prime minister of Sudan (2018-19). This was documented through field interviews conducted by the authors in 2019 and a discussion group in 2021.

21. United Nations Environment Programme, *Sudan: Post-Conflict Environmental Assessment*, 2007, https://postconflict.unep.ch/publications/UNEP_Sudan.pdf.

22. Muhammad Salah Abdelrahman, *sir aldhahab: altaklifat aliajtimaeia walbiyiya liltaedin* [The Price of Gold: The Environmental and Social Cost of Mining] (Maktabat jazirat alward, 2018).

23. *Ibid.*

24. Harry Verhoeven, "Big is Beautiful: Megadams, African Water Security, and China's Role in the New Global Political Economy," *Global Water Forum*, October 16, 2012, https://tinyurl.com/3bdt7fpz.
25. Sudan's Transparency Initiative, "Saddiy 'aeali nahr 'atbarah wa stit: fasad wa tashrid [The Dams of Upper Atbara and Setit: Corruption and Displacement]," *Al Rakoba*, 2017, https://tinyurl.com/cmhtwuup.
26. Ibid.
27. Ibid.
28. World Bank, *From Subsidy to Sustainability*.
29. Souad Lakhdar, "wazir alkahraba' yatamasak bi'iiqamat sudud kajabar wa dal wa alsharik [The Minister of Electricity Adheres to the Construction of the Kajbar, Dal and Al-Shareek Dams]," *Al Rakouba*, 2016, https://tinyurl.com/n3f39y7z.
30. Wilson, E.D.F. and Scott, *Pre-Feasibility and Feasibility Studies of DAL Hydropower Project*, (Unpublished), 2010; Hashim, "Kajbar's message."
31. MoWRE, *Long- and Medium-Term*.
32. Lily Odarno, Estomih Sawe et al., "Accelerating Mini-Grid Deployment in Sub-Saharan Africa: Lessons from Tanzania," *Tanzania Traditional Energy Development Organization (TaTEDO), World Resources Institute*, October 4, 2017, accessed March 1, 2021, https://tinyurl.com/3ukfds6x.
33. MoWRE, *Rural Electrification with Solar Home Systems Project* (Presentation), accessed May 5, 2022, www.unescwa.org/sites/default/files/event/materials/2p4.pdf.
34. World Bank, *From Subsidy to Sustainability*.
35. Sudan News Agency, "shamal darfur tatasalam rasmiana 'awal mahatat liltaaqat alshamsiat fi alsuwdan [North Darfur Officially Receives the First Solar Power Plant in Sudan]," 2020, https://suna-news.net/read?id=694938.
36. The Sudanese company of hydrogeneration and renewable energy (SHG&RE Co. Ltd), "Solar Energy Shines in the West of the Country," Facebook, video, 8:00, www.facebook.com/watch/?v=404676296987592.
37. World Bank, *Reforms, Arrears Clearance Pave the Way For Sudan's Full Reengagement with the World Bank Group*, March 29, 2021, https://tinyurl.com/45k35w4j.
38. "Sudan Gasoline Prices," *Global Petrol Prices*, 2022, accessed July 2022, www.globalpetrolprices.com/Sudan/gasoline_prices.
39. Enrico Ille and Griet Steel, "Khartoum: City Scoping Study" *African Cities Research Consortium*, June 2021, https://tinyurl.com/3e2atevu.
40. World Bank, *Sudan Energy Transition and Access Project*, Project Information Document No. PIDC30301, September 22, 2020, https://tinyurl.com/2p8b38zn.
41. The Challenge Fund for Youth Employment, *Sudan Country Scoping*, February 2021, https://tinyurl.com/49r29srd.
42. Anadolu Agency, *Sudan Will Quadruple the Electricity Tariff*, 2021.
43. *Open Sudan*, "tafasil al'asear aljadidat lilkahraba' fi alsuwdan bidayatan min 24 yanayir 2022 [Details of the New Electricity Prices in Sudan starting from 24 January 2022]," January 25, 2022, https://opensudan.net/archives/22037.
44. World Bank, *From Subsidy to Sustainability*.
45. *Open Sudan*, "Details of the New Electricity Prices".

46. Beam Reports, "(ters alshamal).. kayf tashakalat harakat almuqawamah wama matalib almuhtajiyna? [Northern Barricades: How was the Resistance Movement Formed and What are the Protesters' Demands?]," February 12, 2022, https://tinyurl.com/432j4ubk.

47. Food Security Analysis, *WFP Market Monitor – Sudan*, World Food Program, March 2022, https://docs.wfp.org/api/documents/WFP-0000138508/download/.

48. "Sudan Income Poverty," *Trading Economics*, 2022, https://tinyurl.com/yn42pd4r.

49. Philipp A. Trotter and Sabah Abdullah, "Re-focusing Foreign Involvement in Sub-Saharan Africa's Power Sector on Sustainable Development," *Energy for Sustainable Development* 44, (June 2018):139-146, doi:10.1016/j.esd.2018.03.003.

50. The Council of Ministers and the High Council for Environment and Natural Resources, *Sudan's Updated First NDC,Interim Submission*, May 21, 2121, https://tinyurl.com/mr2y4v2r.

51. "Wind power: How Windy Does It Have To Be?" *Renewable Energy Technology, Renewable First*, https://tinyurl.com/3e48nv9p.

52. Collen Zalengera, Long Seng To et al., "Decentralization: the Key to Accelerating Access to Distributed Energy Services in Sub-Saharan Africa?" *Journal of Environmental Studies and Sciences* 10:270-289, 2020, https://doi.org/10.1007/s13412-020-00608-7.

53. ESMAP, *Executive Summary*. In *Mini Grids for Half a Billion People: Market Outlook and Handbook for Decision Makers*, World Bank, September 27, 2022, https://tinyurl.com/mr86tj5z.

54. Katharine Nawaal Gratwick and Anton Eberhard, "An analysis of Independent Power Projects in Africa: Understanding Development and Investment Outcomes," *Development Policy Review* 26, No. 3 (2008), pp. 309-338, https://tinyurl.com/3tmbscw8.

55. Shabakat 3ayin, "'iitfa' kamil lilkahraba' fi thalath mudun rayiysiat bidarfur [A Complete Shutdown of Electricity in Three Major Cities in Darfur]," January 4, 2020, https://tinyurl.com/2bwreupu.

56. Cross subsidization is the process of charging higher prices to one group of consumers in order to support a price reduction in favour of another group.

57. Trotter and Abdullah, "Re-Focusing Foreign Involvement".

58. Prepaid meters are tools through which the cost of electricity is paid before it is used by purchasing a specific number of electrical units and feeding them into the meter. This type of meter ensures that consumers pay the amounts required of them without having to be chased by the collectors of electricity companies. This improves the companies' collection rates for the amounts that are due.

59. World Bank, *From Subsidy to Sustainability*.

60. Climate reparations refers to the provision by countries that have historically contributed to the climate crisis, through activities causing harmful emissions, of support to countries that did not contribute to this crisis, and which are considered the most vulnerable in the face of climate change.

PART II

Neoliberal Adjustments, Privatization of Energy and the Role of International Financial Institutions

7

International Finance and the Commodification of Electricity in Egypt

Mohamed Gad

Between 2008 and 2014, Egyptians experienced an unprecedented increase in power outages, with dire consequences that ranged from the suspension of hospital services, to the disruption of various lines of industrial production, to the breakdown of ventilation systems during the country's blisteringly hot summer months. These outages revealed the severe shortfalls in the funds available to modernize and expand the national electricity generation network, with the state relying primarily on its own financing or borrowing from development partners during this period. At the same time, the state maintained a monopoly over the processes of electricity production and distribution, while the private sector occupied a limited position in energy production via the build-operate-transfer (BOT) system.

In response to the power outage problems, the state initiated an urgent plan in 2014 for the maintenance and expansion of the electricity production network. This was followed by successive projects to establish extensive power stations for the production of electricity from gas, involving a number of companies, such as the German company Siemens. At the same time, the state enacted legislative amendments allowing for the expansion of the role of the private sector in the renewable energy sector, which contributed to the return of dependable electricity to Egypt.

The expansion of the electricity production network that took place after 2014 was made possible due to the state's openness to various forms of international commercial finance, notably Engineering, Procurement, Construction and Finance (EPC-F), through which the companies contracted to implement various projects have obtained international loans to finance the power stations themselves. Meanwhile, the public deficit did not increase.

In fact, it was reduced by several means, the most prominent of which was ending the state subsidy system for electricity tariffs.

With the liberalization of electricity prices, the private sector became more involved in the energy sphere, both as a funder of infrastructure projects and as an energy producer. Indeed, the state encouraged such involvement through enacting legislative amendments. In sum, the state abandoned its monopoly over the electricity production process, opened up to new forms of infrastructure financing and moved away from subsidizing electricity prices for a wide range of income groups.

This chapter analyses the political economy of the liberalization of electricity prices in Egypt, highlighting the main policy shifts in regard to electricity pricing. It does so by focusing on the role of international finance, including both commercial loans from international banks and financing from international institutions, such as the International Monetary Fund (IMF) and the World Bank. The chapter demonstrates how the energy crisis in 2014 paved the way for a replacement of public by private financing in the energy sector, a process which saw the institutions of national energy production in Egypt shift from public service providers to private company-like bodies in competition with private sector firms.

This chapter also challenges the World Bank's narrative concerning this transformation that occurred in Egypt. The bank claims that the main objective of liberalizing electricity prices was to cut subsidies to the rich and redirect resources towards the neediest social groups. According to the bank, liberalization was a "socioeconomic" reform. In reality, price liberalization led to the influx of international finance, which was presented as a necessity due to the power outage crisis. While the social component of the alleged reform process is acknowledged, the chapter argues that its role was not central. Indeed, the reform has imposed several downward pressures on living standards, so that in various respects liberalization has achieved the opposite of what was promised in regard to social reform.

AFTER DECADES OF PREVARICATION, LIBERALIZATION

Egypt's electricity production dates back to the nineteenth century.[1] During the 1940s, Egypt issued a law regulating private sector investments in public utilities.[2] The law regulated the duration of any concessions granted to companies to provide public services and the prices to be charged to the public for those services, ensuring that provision remained profitable. A specialized department was established to regulate and administer concessions in the electricity sector in 1948.[3]

The private sector remained a player in electricity production in Egypt until the wave of nationalization that took place in the 1960s.[4] The nationalization processes reflected an ambitious state vision to expand coverage of the electricity network. In 1964, a public body was established to assume responsibility for planning the extension of electricity supply across the country.[5] Seven years later, a specialized department was created to provide electricity to the Egyptian countryside.[6]

From the 1960s to the late 1980s, the "Arab socialist" economic model included the state's involvement in electricity production and the provision of essential services. Electricity was made available to the population at prices in line with prevailing wage levels. In this model, the state played a crucial role in determining the availability of jobs within the private sector.[7]

However, the worsening of the external debt crisis through the latter half of the 1980s prompted the Egyptian state to embark upon a comprehensive programme to dismantle the "Arab socialism" model – a move which was also a condition for a loan from the IMF received in the 1990s.[8] The essence of the programme was to withdraw the state from the economy in order to allow the private sector to expand, and to end price-distorting policies such as controls over interest rates and both currency and commodity exchange, as well as opening up the country to external financing in lieu of Central Bank borrowing (i.e., printing money).

Adjusting the electricity price tariff was one of the propositions advanced in the new economic programme of the 1990s, with the aim being for the price of energy to match the actual cost of energy by 1995. However, this was not achieved.[9] The retreat from liberalization of the electricity sector was one of several instances in the 1990s where the state withdrew from its commitments to the IMF measures due to fear that it would cause popular discontent. Decades of "Arab socialism" had contributed to the politicization of economic life; relinquishing this political responsibility demanded a gradualist approach on the part of the state.

In contrast to the resolute efforts of the state to withdraw from its social role during this period – clearly embodied in the acceleration of privatization within the state industrial sector in the 1990s – there was a great deal of prevarication over the liberalization of both fuel and energy prices. On this basis, it is clear why the 1990s and 2000s saw many legislative amendments that aimed to liberalize the electricity sector. However, despite these amendments, the sector continued to operate largely as it had over the previous decades. Thus, the World Bank describes the dynamics of liberalization during this period as involving more a change in appearance than in substance.[10] One such apparent change was the transfer of responsibility for electricity distribution from the Ministry of Electricity to the newly created Ministry of Public Enterprise Sector in 1993. The latter was established to manage the greater part of both the public industrial and services sectors and had the ability to dispose of their assets, including those related to electricity. Although this transfer of responsibility between the two ministries was seen as a step towards the privatization of public entities under control of the new ministry, this did not occur.

In 2000, the public electricity production and distribution companies were merged with the Egyptian Electricity Transmission Company (EETC) and turned into a holding company. This new structure allowed these companies to offer a percentage of their shares as an Initial Public Offering (IPO). However, the IPO did not materialize.

In 1996, an important legislative amendment was made to the law regulating the Egyptian Electricity Body,[11] which allowed for the proliferation of a novel form of exploitation under the BOT system, albeit with provisions for the return of the assets concerned to the government after an agreed period. This was the most serious attempt to liberalize the electricity sector during this period. The amendment was broadly consistent with the fiscal policies objectives of the late 1990s, which focused on reducing the budget deficit. However, the partial nature of liberalization did not foster the continuation of the BOT system. While the 1996 amendment enabled the private sector to step into energy production, the Ministry of Electricity maintained a monopoly over the purchase of energy from companies. Given the ministry's commitment to keep prices consistent for the end user, investing with the state remained risky. Any crisis could widen the gap between the price at which the private sector sold electricity to the state and the price at which it was sold to the end user. In such a scenario, the state might fail to meet its commitments to the private sector.

The probability of this risk increased in 2003 when Egypt was forced into devaluing its local currency following two major crises: the East Asian financial crisis and terrorist events that damaged the tourism industry.[12] Since the state's contracts with private electricity producers required adjustments to prices in line with fluctuations in exchange rates, this could lead to exorbitant costs. Hence, the state reneged on its plans to expand the usage rights of independent producers.[13] Only three independent producers were granted these rights, accounting for approximately 10 percent of total electricity production in Egypt.[14]

In this period, the state further expanded the electricity infrastructure, relying on financing from state-owned banks and development partner countries. This was accompanied by limited tariff increases applied to end users, which the World Bank deemed insufficient to cover the real cost of electricity production in view of inflation.[15]

As the first decade of the 2000s drew to a close, the contradictions within the electricity subsidy model had reached a critical point. The public treasury was no longer able to finance the expansion and adequate maintenance of the electricity production and distribution networks. This was due to two factors: first, the rapid growth of demand, reinforced by the expansion of the service to almost the entire territory of Egypt and by energy-intensive investments attracted by cheap energy pricing, and, second, the strict limitation of public spending by the deficit-reduction policy, and the implementation of non-progressive tax policies as a means of attracting investment, which limited the collection of public revenues.[16]

As for transmission losses in Egypt, they were relatively high at up to 15 percent of the total energy produced, according to the Ministry of Electricity. This was due to the poor quality of the distribution networks and electricity theft. In addition, power plant efficiency and availability were five to eight percent below the accepted average.[17]

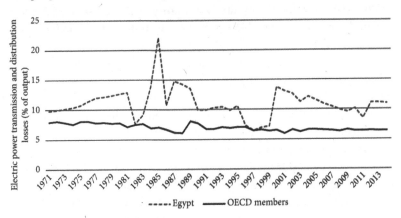

Figure 7.1 A comparison between Egypt and OECD countries in terms of electric power transmission and distribution losses (in percentage of output) - World Bank.

Moreover, power stations operated according to the "normal cycle" system, one major shortcoming of which is the high level of fuel consumption involved: only 40 percent of the fuel used produces electricity, the remainder is wasted.[18] The continued dependence on traditional fuels, such as natural gas and mazut,[19] means that any shortage or increase in the price of petroleum products causes a crisis in the electricity production.

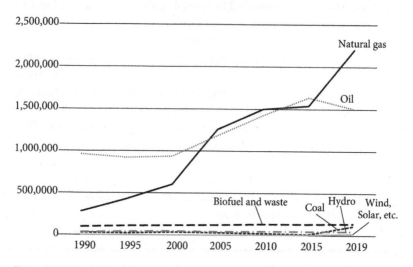

Figure 7.2 Egypt's total energy supply by source in Tera Joule (TJ) – International Energy Agency (IEA).

With the entry into the twenty-first century, such a crisis arose due to a drastic decrease in the availability of petroleum products, for various reasons. In fact, during the second decade of the 2000s, Egypt's position shifted from being a net exporter to being a net importer of petroleum products. This reliance on imported fuel inputs to generate electricity worsened the production cost crisis.

The decline in the production of petroleum products was not simply the outcome of natural resource limitations. Energy subsidies for the end consumer – including fuel for cars, energy for industry, and electricity for consumers – alongside the shortage of financial resources, also contributed to widening the gap between the price of petroleum products purchased from oil extraction entities, the majority of which were and still are foreign companies,[20] and the price of the final energy product. Consequently, the state's cumulative failure to pay debts owed to extraction companies[21] resulted in decreasing private investments in Egypt.[22]

After the revolution of January 2011 and the instability of the regimes that followed, the Ministry of Petroleum withdrew from new exploration

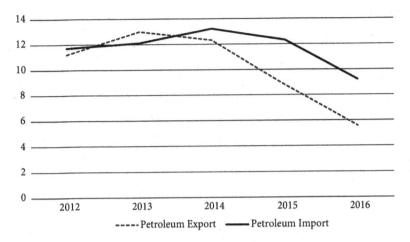

Figure 7.3 Petroleum exports and imports (US $ billions).

agreements, a decision also fuelled by the latent anger of Egyptian Gas companies over the 20-year period of fixed purchase prices for foreign partners' shares.[23] The weak investment in the energy sector and the decline in petroleum resources led to a higher dependence on imports, resulting in increased costs of petroleum inputs for electricity production, which further strained the subsidy system. The general political instability and the deterioration of foreign exchange reserves after the 2011 revolution led the government to persist with the policy of power cuts as a cost-cutting measure. These outages took a heavy toll in terms of public discontent among households – the major electricity consumers – and economic losses in both the production and service sectors.[24] These developments paved the way for the accelerated liberalization of the electricity sector and the further integration of the private sector into the energy production system, whether as a funder of governmental projects or as an independent producer.

International Financial Institutions (IFIs) pushed for the dismantling of the electricity system that was established in the 1960s as the only solution to the sector's crises during this period. They argued that by liberalizing fuel and electricity prices, unnecessary energy demands, such as the increasing dependence on energy-intensive air conditioning by wealthy families, could be reduced. The same institutions proposed cutting subsidies, arguing that they were a major factor in the budget deficit.

These IFIs argued that liberalization would transform the Egyptian Electricity Holding Company (EEHC) from a net loss-making body into a

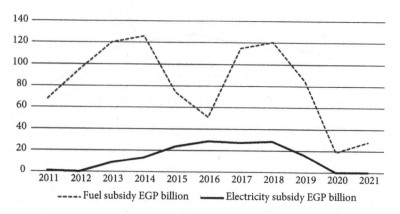

Figure 7.4   Fuel subsidy and electricity subsidy (Egyptian pound billions).

profitable one, which would encourage more favourable contractors, particularly international funders. It was argued that liberalization would also foster rapid infrastructure development, while improving efficiency and reducing energy waste. Ultimately, liberalizing the tariff of electricity produced from natural gas will increase the competitiveness of renewable energy. Indeed, this would be more attractive for the private sector, lead to more investment in solar and wind power plants, diversify electricity resources and reduce the risk of crises stemming from the scarcity of traditional fuels.

The following section will explain how international finance has been used to tackle the growing problem of power cuts, allowing for an accelerated liberalization process that has removed the electricity sector from state control.

THE ORIGINS AND DEVELOPMENT OF LIBERALIZATION

Egypt experienced the highest records of power cuts in the summer of 2014, with an average disconnected load reaching 6,050 MW, almost double the cuts in 2008. The Ministry of Electricity was unable to implement approximately one-third of power station maintenance plans due to insufficient resources. This led to an even greater inability to generate enough energy to meet the rising demand, especially during warmer weather with the greater need for air conditioning.

The state initiated an urgent plan in 2014 to address the energy crisis and strengthen electricity production. The main objective was to complete overdue works and create up to 3,636 MW of new energy capacity. Funding was secured through EPC-F[25] contracts, enabling contractors to search for funding sources.[26]

The medium term governmental plans were highly ambitious and with an expansionist vision. The state opted for the most fuel efficient strategy: it aimed to complete works on three new large-scale combined-cycle power plants. With a combined capacity exceeding 14,000 MW, these plants produced a massive upsurge of electricity production after 2014. Of course, costs were exorbitant, at six billion euros. These massive funds were arranged through the same logic of the urgent plan: project contractors acted as intermediaries who arranged loans for the state. While local and international banks were responsible for the financing, the loans were guaranteed by European support, particularly by the German export credit agency Euler Hermes and the Italian Export Credit Agency (SACE).[27] The funds for these new sources of energy were collected for the EEHC.

This period, therefore, witnessed a marked increase in EPC-F contracts. Contractors in the construction industry saw this as a golden opportunity for the government to expand the provision of financing. For instance, Osama Bishai, the chief executive officer of Orascom Construction, one of the most prominent contractors implementing the new power plants, commented on this form of financing in a media interview: "Why should the government bear the burden of debt and its service when someone else can bear it for them?"[28] Nonetheless, electricity marketization actually meant that the debt would be repaid by the consumer, not the state, as had already begun to happen after the liberalization of electricity tariffs.

In addition to opening up to external financing during this period, the state also introduced new legislation to encourage the private sector to take part in electricity production and sales to the national network, which had become more attractive with tariff liberalization. By now, the risks for the private sector were drastically reduced: it was no longer expected that a large gap would arise between the EEHC's purchase price from independent producers and its selling price to the public, which might in turn result in the EEHC's failure to pay those producers. In this context, a law[29] was passed in 2014 regulating the private production of electricity from renewable sources and its sale to the state. The law introduced a "feed-in" tariff – a tariff approved by the state as meeting the price for purchasing electricity from private companies based on the actual cost of the energy produced. A year later, another law was introduced to regulate the electricity sector. Observers have argued that this 2015 law brought about a major advance in the liberalization of the electricity sector, since it included clauses that explicitly broke the state's monopoly over electricity production. The law also gave the EEHC a period of eight years to restructure its subsidiaries so that they will be able to operate in an increasingly competitive market.[30] Furthermore, the law placed the Egyptian Electric Utility and Consumer Protection Regulatory Agency (EgyptERA), which existed since 2000 with a marginal role, at the top of the electricity sector, granting it robust powers, notably to set electricity tariffs "according to economic rules and foundations". This represented a transfer of the mandate that had been placed in the hands of the government for decades. Thus, the mandate to set tariffs was transferred to a body that is independent of the government, with a strong representation of the private sector on its board of directors, including chief executive officers of the Federation of Egyptian Industries and the Federation of Egyptian Chambers of Commerce.[31]

Although the 2015 law enables the government to continue to subsidize electricity, it was treated as an exception to the rule. It allows the Council of Ministers to bridge the difference between the cost and the sale price, through subsidies if the service is provided at a tariff that is lower than the one set by EgyptERA. This clause aims to ensure social stability in the case of price rises.[32]

Noteworthy is that the law, with its radical shift away from the stability of the 1960s, was not a product of 2015, as it has been conceived after the outage crises. According to a World Bank study, it began to be conceived by state officials in 2005, following international consultations provided to the Egyptian government on the subject. Indeed, a study by Merrill Lynch submitted to the Egyptian government in 1996 offered recommendations that are very similar to the clauses contained in the 2015 law.[33]

OUTCOMES OF OPENING UP TO EXTERNAL FINANCE

Overall, the policies of opening up to external finance, both in terms of the construction of infrastructure and legislative amendments, have led to the diversification of energy sources and an increase in power generation capacities.

Figure 7.5 illustrates a considerable shift in production levels compared to the demand, after a period of close convergence between nominal capacity (i.e. energy production capacity) and peak load (i.e. demand). Since 2013, nominal capacity has increased at a quicker rate than energy demand. As a result, some observers have argued that the recent expansions were unnecessarily large, and that the returns from these energies are unclear. Mohammed

Younes, a scholar specializing in the energy sector, has termed this as "the burden of surplus electricity".[34]

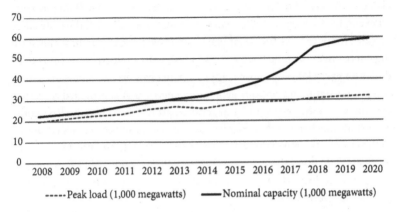

Figure 7.5 Peak load and nominal capacity (1,000 megawatts)

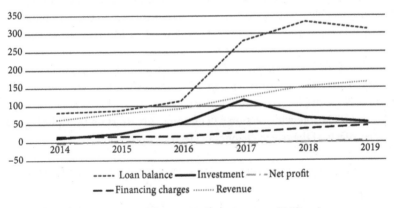

Figure 7.6 Financial statements of EEHC (Egyptian pound billions).

EEHC's financial statements show how both higher fees and openness to borrowing have increased investment in infrastructure. To be more specific, both revenue and loan balances increased rapidly after 2014.

Regarding the diversification of energy sources, the capacity of solar and wind power plants to generate electricity has grown by nearly 100 percent during 2017 to 2018 and from 2018 to 2019.[35] Such a leap was enabled by private sector investments in the energy sector, aided by the state and within the regulatory framework of the recent legislation.

At first, a traditional approach involved contractors establishing electrical plants for the state.[36] However, since the time it was initially adopted to establish gas powered stations before 2014, the state has developed a stronger interest in the generation of solar and wind energy. The second approach, also closely related to the state, is referred to as "feed-in tariff investments", an example of which is the Benban platform, through which the state allows private groups to construct solar plants while pledging to purchase the result-ing energy over the long term (i.e. 25 years), with a return of 8.4 cents to the dollar for every kilowatt. A final approach involves private renewable energy companies setting up stations generating energy that is provided directly to consumers via the system of "independent producers". There is also another important mechanism that enables various consumption sectors connected to the electricity network to establish their own plants and produce renewa-ble energy. It allows those sectors to have offsets against what they produce and consume from the network, which, in turn, helps them to reduce the cost of the electricity bill.[37]

In sum, international financing has not only succeeded in extricat-ing Egypt from the electricity production crisis, but ended up pushing for unnecessary electricity projects. However, this form of financing came with the condition of dismantling state control over the electricity sector and has increased the dominance of private companies over the public service of providing electricity. International financial institutions, particularly the World Bank, have a distinct perspective, suggesting that the commodifi-cation of electricity has introduced "social reforms" that benefit the most deprived. The following section will examine this assertion by exploring the social effects of the recent changes in electricity provision.

THE SOCIAL IMPACTS OF ELECTRICITY LIBERALIZATION

"I am against wasting subsidies" was the slogan of a government propa-ganda campaign during the period of electricity liberalization, an initiative hailed by the World Bank as reflecting the energy reform policies it had recommended to numerous countries via its Energy Sector Management Assistance Programme (ESMAP).

The impact of international finance on electricity policies in Egypt went beyond commercial finance, which sought to corporatize the sector's insti-tutions. Development organizations have also influenced the sector by their funding, arguing that the subsidized tariff benefits the rich and that elimi-nating subsidies would prevent resources from being wasted on non-eligible people. The World Bank, alongside the African Development Bank (AfDB) and the Agence Française de Développement (AFD), provided a $3.1 billion loan to Egypt to support electricity liberalization and other reforms. Additionally, scrapping electricity subsidies was part of an economic reform programme financed by the IMF in 2016 with a $12 billion loan.[38]

The World Bank's criticism of the electricity subsidy system focuses on the link between higher incomes and higher electricity consumption. As the wealthier classes use more air conditioners and electrical appliances in their homes, it is argued that they receive the biggest proportion of electricity subsidies. A World Bank report summarizes the arguments about how various social classes benefit from support, both at the level of public goods and services as well as the obligations that indirect taxation imposes on them. It shows that the upper classes' share of benefits of electricity subsidies reaches 40 percent, while the lower classes receive just 10 percent, according to the World Bank's own estimates. Therefore, the World Bank recommends avoiding this waste and redirecting financing to sectors that benefit the poor more, such as free public education.[39]

At first glance, the World Bank's arguments seem logical and may elicit a change in our perception of the energy reforms. According to it, liberalization is not a process of privatization but rather a review of the waste of financial resources, which further accentuates the state's deficit and disrupts the financing of electricity infrastructure. The reality on the ground differs from the bank's narrative, and it can be countered with the following arguments.

First, the electricity sector is distinctive from other forms of universal subsidies since high electricity consumption is generally linked to higher income groups, while low consumption is linked to lower income groups. Therefore, it is possible to differentiate the income levels of subsidy recipients. The groups with the lowest consumption levels could have been exempted from the alarming increases in electricity bills, which could have acted as an outlet for providing support to those in need. However, this was not done.

The World Bank argues that groups with lower consumption levels were paying much lower bills than the actual electricity cost. In other words, they were partially subsidized. This raises a crucial question: why were they not totally exempted from the tariff increase, that is, why was the electricity subsidy not fully maintained? In fact, during the reform period new fees were introduced under the guise of "customer service" charges, and they were subject to rapid yearly increases. These charges were imposed on all the populace, with no exemptions for low-consumption groups.

Second, the World Bank's argument does not take into account the coincidence of the electricity liberalization process with a series of impoverishing measures that have been part of the reform programme supported by IFIs since 2015. These measures included an increase in the value-added tax (VAT) and the liberalization of fuel prices. These reforms have led to a sharp rise in inflation rates, particularly since the central bank left the local currency to the whims of supply and demand in November 2016. As a result, the local currency lost more than half of its value against the United States dollar, giving inflation a strong boost in 2017.

Third, inflation has pushed vulnerable classes into further poverty as compensation measures have been severely lacking. The high poverty rates in 2018[40] show the irreversible impacts of such economic patterns. Despite the slight decline in poverty rates in 2020, they did not return to pre-reform levels.

Fourth, there are several criticisms of the means-tested assistance policy that is offered by financial institutions as an alternative to generalized forms of support, such as electricity subsidies. One example is the "*Takaful* and *Karama*" cash pension programme. These criticisms revolve around the unavailability of adequate state-level data infrastructure for targeting beneficiaries. Instead of a rapid transition that relies on targeted cash support, a gradual move away from universal forms of support would have had better outcomes. Indeed, cash support in Egypt does not come with clear requirements connecting the value of the support to inflation. Thus, its value was quickly eroded during the inflationary period that accompanied the reform process.[41]

Fifth, the World Bank's argument focuses on the distinction between the poorest and the richest beneficiaries from electricity subsidies, without addressing the position of the middle class in between. In fact, segments of the middle class are vulnerable to falling into poverty due to economic pressures, such as energy price increases. Additionally, even if one assumes that reducing electricity subsidies and implementing targeted cash support programmes for the poorest will reduce poverty in the long term, it still remains true that such policies lead to a deteriorating quality of life for the middle classes. Indeed, these policies reduce the spending capacity of the middle class in vital areas, such as health and education, and divert it to basic expenditures, such as household energy.

Lastly, there has been insufficient spending on those subsidies that the World Bank deems to be more worthy for groups with the lowest levels of income. Indeed, many human rights organizations have criticized the government's failure to provide the minimum spending on healthcare and education stipulated in the 2014 constitution.[42]

CONCLUSION

The production and pricing of electricity in Egypt has for decades been under the complete control of the state, which helped maintain a certain level of family-based subsidies. However, this financing model was not sustainable, due to the growing difference between the prices of imported petroleum products and the price paid for energy by the end users. Furthermore, the state's commitment to ensuring a low budget deficit has limited its ability to invest in electricity infrastructure. As the power outage crisis grew due to limited oil resources and inadequate infrastructure, the state moved reluctantly towards implementing measures about which it had hesitated for decades. Among these measures was the opening up to the private sector, both as electricity producers and as infrastructure financiers. Such penetration by the private sector, alongside the increased price of petroleum products, required the liberalization of electricity prices.

Within the span of a few years, Egypt has undergone significant strides in transforming its electricity production sector into one that is free from state control, although the state remains a major energy producer.

This privatization of the electricity sector stands in stark opposition to the support policies from which Egyptians have benefited for decades. The liberalization of the electricity sector is one of the main features of the energy transition process in Egypt. This transition has helped make the financing of the electricity sector more sustainable, while protecting the sector from crises relating to financing infrastructure, and reducing dependence on traditional fuels in favour of renewable energy. But is the energy transition taking place in Egypt just?

The concept of "just transition" was developed in the United States in the 1970s. It gained prominence in the following decades, with a growing demand for just policies to shift energy sectors towards less reliance on polluting sources. A just transition implies a more environmentally sustainable and more equitable energy sector, especially with regards to the various social classes who are dependent on energy.

In Egypt, international finance has principally impacted the formulation of energy policy, regardless of its social dimension. This is evident in the unfolding of the electricity liberalization process, which occurred in conjunction with austerity measures that worsened the social crisis facing the country. Examples of the latter measures include liberalizing fuel prices, increasing public transport costs and increasing the VAT. The World Bank has argued that traditional subsidy policies through electricity tariffs were a waste of resources, as the majority of high consumers were deemed to be from higher income groups. The bank believes that the best course is to end subsidies and redirect funds towards social benefits for lower income groups. However, lower income groups have not been immune from electricity tariff increases, and indeed the liberalization process has taken place at an alarming pace, placing a huge pressure on the middle classes.

Any discussion of a just energy transition in Egypt must factor in an analysis of who controls the resources of the energy system, and who benefits from their usage. While international finance ensures the financial sustainability of the energy system's infrastructure, it is also radically transforming a basic service into a commodity – and is often the first and the last voice shaping Egypt's social policy.

NOTES

1. "Development of Electricity in Egypt," The Ministry, Ministry of Electricity and Renewable Energy. "History," Development of Electricity in Egypt, http://www.moee.gov.eg/english_new/history2.aspx.
2. Law No. 129 of 1947 concerning concessions of public utilities.
3. Law No. 145 of 1948 concerning the establishment of the Cairo Electricity and Gas Administration.
4. Presidential Decree No. 86 of 1962, which incorporated the Egyptian Electricity Company into the Electricity and Gas Administration.
5. Presidential Decree No. 1471 of 1964 on establishing the General Body for the Electrification of Egypt.
6. Presidential Decree No. 470 of 1971 establishing the Public Body for the Electrification of the rural areas.
7. In 1952, the Free Officers staged a military coup against the monarchy and established a republic. Under the mandate of the second president of the republic, Gamal Abdel Nasser, the welfare state was strengthened as the state took on greater control of economic activity. This control peaked in the early 1960s with the broad nationalization of production and service activities.
8. Abdel-Khalek Gouda, MDG-Based Debt Sustainability Analysis: The Case of Egypt, *UNDP/UNDESA*, 2007.
9. Korayem Karima, "Structural Adjustment and Reform Policies In Egypt : Economic and Social Implications," UN. ESCWA. Sustainable Economic Development Division, 1997.
10. Rana Anshul and Khanna Ashish, *Learning from the Power Sector Reform: The Case of the Arab Republic of Egypt*, World Bank, 2020.
11. Law No. 100 of 1996 amended some provisions of Law 12 of 1976, thus establishing the Egyptian Electricity Body.
12. Hoda Selim, *Has Egypt's Monetary Policy Changed after the float?*, The Economic Research Forum, 2010.

13. Anshul and Ashish, *Learning from The Power Sector Reform.*

14. Shahid Hasan, Turki al-Aqeel and Hafez El Salmawy, *Electricity Sector Liberalization in Egypt: Features, Challenges and Opportunities for Market Integration* (The King Abdullah Petroleum Studies and Research Center, 2020).

15. Anshul and Ashish, *Learning from the Power Sector Reform.*

16. Mohamed Gad, Sarah Seif, Bisan Kassab, and Noha Magdy, *Aldarayib maslahat min* [Taxes for Whom?] (Al-Maraya House for Cultural Production, 2019).

17. Minister of Electricity and Renewable Energy, *Addressing Egypt's Electricity Vision* (Egypt Economic Development Conference, 2015).

18. Omar Salem, *mahataat aldawrat almurakabat alhalu alsuhraa litawfir alwaqud* [Combined Power Stations: The Magic Solution to Saving Fuel and Money] (Al Mal journal, June 2013). www.almalnews.com.

19. Mazut is a low-quality heavy fuel oil.

20. Contracts for oil extraction are usually established on the condition that the foreign partner carrying out the extraction activities obtains a share of the extracted products, while the government, at a later stage, buys this share at the price agreed upon.

21. Reuters Staff, "Egypt to Pay USD1.5B in Arrears to Foreign Oil Firms," *Reuters*, October 2, 2014.

22. Anshul and Ashish, *Learning from the Power Sector Reform.*

23. Mohamed Gad, "mustathmiru qitae albitrul fi misr mutafayilun lakinahum yantazirun mazid min alaimtiazat alhukumia" ["Investors in the Petroleum Sector in Egypt are Optimistic, But they are Waiting for More Government Concessions"], *Aswat Masriya*, February 2, 2015.

24. Akram Abdel Rahim, "mustathmirun wa'ashab mazariei: khasayir kabiratan limurabiy aldawajin bisabab ainqitae alkahraba" ["Big losses for poultry farmers due to power outages"], *Al-Masry al-Youm*, May 24, 2013; "'simu': ainqitae altayaar alkahrabayiyi yazid min alkhasayiri, wamuhawalat laeadatih ['Simo': The Power Outage Increases Losses, and Attempts to Restore It]," *Al Mubasher*, March 4, 2013; Mahmud Amin, "alkahraba' fi hukm al'iikhwan .. misr ''mudlimatun'' eali 'ahluha (tasalsul zamaniun)" [Electricity Under the Brotherhood, Egypt Blacks Out its People (Chronology)"], *Masrawy*, June 25, 2013.

25. Medhat Saad Eldin, "euqud alhandasat walmushtariat walbina' waltamwil - EPC +" ("Finance Engineering, Procurement, Construction and Financing Contracts - EPC + Finance"), *LinkedIn*, October 16, 2018, https://tinyurl.com/y9we8fk4.

26. The Ministry of Electricity and Renewable Energy, "Inauguration of the Projects of the Ministry of Electricity and Renewable Energy (Statement)," 2018.

27. Siemens AG, "The Egypt Megaproject: Boosting Egypt's Energy System in Record Time," 2018, siemens.com/egyptmegaproject.

28. Enterprise, "Five Questions for Osama Bishai," March 4, 2017, https://enterprise.press/2017/03/04/20632/.

29. Egypt Renewable Energy Law, Decree No. 203 of 2014, on encouraging the private sector to produce electricity from renewable energy sources.

30. Article 63 of Law No. 87 of 2015, which puts the Electricity Law into action.

31. Articles 4 and 5 of Law No. 87.

32. Article 41 of Law No. 87.

33. Anshul and Ashish, *Learning from the Power Sector Reform*.
34. Mohamed Younes, "eib' fayid alkahraba ["The burden of Surplus Electricity"], *Mada Masr*, July 30, 2019, https://tinyurl.com/yu5tkbk4.
35. Egyptian Electricity Holding Company, *Annual Report*, 2019.
36. Currently, there are three types of state contracting: EPC reverse auctions, which are government projects put forward by the Renewable Energy Authority with the aim of designing, supplying and installing projects owned by the authority; auctions, which are state projects that are purchased at the lowest prices; and build-own-operate (BOO) projects, which are projects offered by the Egyptian Electricity Transmission Company to investors in the sector with specific capabilities, awarded to those offering the lowest prices. Source: Renewable Energy Authority (2018) *Annual report of the Renewable Energy Authority*.
37. Mohamed Gad, "hal naeish aleasr aldhahabia liastithmarat altaaqat almutajadidati? ["Are We Witnessing the Golden Age of Renewable Energy Investments?"], *MadaMasr*, January, 5, 2021, https://tinyurl.com/248znf83.
38. World Bank, "Maximising Finance for Development in Egypt's Energy Sector", ESMAP Impact, No. 16, June, 1, 2019, https://tinyurl.com/248znf83.
39. Gabriel Lara Ibarra, Nistha Sinha, Rana Fayez and Jon Jellema, "Impact of Fiscal Policy on Inequality and Poverty in the Arab Republic of Egypt", *Policy Research Working Paper*, No. 8824 (World Bank, 2019), http://hdl.handle.net/10986/31579.
40. According to the Central Agency for Public Mobilization and Statistics's data on income, expenditure and consumption, poverty rates reached 32.5 percent between 2017 and 2018 against 27.8 percent in 2015. They then declined to 29.7 percent in 2019.
41. Mohamed Gad and Hasnaa Mohamed, "alnuqud wahdaha la takfi: tatabae athar baramij aldaem alnaqdii ealaa alfuqara" [Money Alone is Not Enough: Tracking the Effects of Monetary Support Programmes on the Poor]," Egyptian Initiative for Personal Rights, February 2019.
42. In 2021, the Egyptian Initiative for Personal Rights filed a lawsuit against the Egyptian government for going against Article 19 of the constitution by which the government is obliged to allocate four percent of the GDP to expenditure on education. The government was allocating only 2.6 percent of the GDP to education.

8

The Energy Sector in Jordan: Crises Caused by Dysfunctional and Unjust Policies

Asmaa Mohammad Amin

In February 2011, parts of the Egyptian-Jordanian gas pipeline in the Arish region in Egypt were blown up. These attacks continued over the next two years,[1] culminating in gas supplies from Egypt to Jordan stopping completely in 2013.[2] With the country's electricity supply and thus all economic activities threatened, Jordan was plunged into a crisis. The events highlighted the fact that Jordan was excessively dependent on Egyptian gas, which provided over 80 percent of the country's requirements for power generation at the time.[3] Having no ready alternatives to Egyptian gas, Jordan was forced to import large quantities of oil products at a time when global oil prices were extremely high, leading to steeply growing costs for local electricity production. Cumulative losses by the electricity authority reached $7.7 billion by 2022, which amounted to 14.3 percent of the country's public debt.[4]

The gas pipeline crisis demonstrated the fragility and dysfunctionality of Jordanian energy policies. In the years leading up to the crisis, especially around 2007, the government, supported by the International Monetary Fund (IMF), started privatizing Jordan's electricity generation. Distribution and investment companies were introduced as partners under agreements which shielded them from financial risks and guaranteed their profits. These dynamics increased the government's financial burdens and reduced performance in the public energy sector. At the same time, state planning failed to address the issue of energy security.

Although the official narrative about the gas crisis blamed the Arab Spring and the scarcity of domestic fossil resources for the increasing costs of energy production, poor management and planning contributed greatly to the crisis. The country's dependence on a single insecure source was a result of political considerations and policy choices that had eschewed the possibility of diversifying sources and establishing plans for emergencies.

As a result of the crisis, the government adopted a new strategy after 2011, based on diversification and the exploitation of local energy sources. This strategy involved multi-billion dollar contracts with multinational corporations to develop new power generation projects. One such investment is a contract signed in 2016 with the American enterprise Noble Energy and its partners in the Leviathan field to purchase Israeli gas. Despite widespread popular protests, legal issues, and doubts surrounding the project's safety and its economic usefulness to Jordan, the government forged ahead with the deal.[5] However, this deal repeated the dynamics of the pre-2011 period, in that it neglected energy security and exposed the Jordanian people and economy to the risks of dependency on foreign sources of energy.

Another post-2011 project was the construction of the oil shale-fuelled Attarat Power Plant by the Estonian company Eesti Energia (Enefit), with financing from multinationals. This was initially expected to meet 15 percent of Jordan's electricity needs, at a cost of $3 billion.[6] However, the government realized that the high cost of the project would significantly increase its annual shortfalls and resorted to international arbitration in the hopes of reducing the exorbitant cost.[7]

Finally, since 2011 the Jordanian government has also promoted renewable energy and has established a domestic market for renewables. Renewables now constitute 30 percent[8] of Jordan's total power generation, with the country promoted as a model for clean energy. However, this figure conceals a crucial fact: Jordan's renewable energy projects are dominated to a large extent by rich individuals and private companies, and policies in this area prioritize personal production instead of reducing the costs of state-level energy production or supporting economically depressed sectors, such as agriculture, and low-income citizens. Such profit-oriented policies have also resulted in significant public losses, which have been met by borrowing from the IMF and the World Bank. These lenders' remedial plans for the energy sector follow a basic recipe: privatization and liberalization of the energy market, and increasing electricity prices, irrespective of the severe socioeconomic impacts this may have.

The remainder of this chapter will explore each of these dynamics in more depth, before providing some recommendations for an alternative vision for Jordan's energy future.

BEFORE THE GAS STOPPAGE CRISIS

The World Bank and the IMF: privatization and diminished consumer subsidies

The Jordanian energy sector was partly privatized in 2007 as a condition for the loans the government received from the IMF and the World Bank to cover its public deficit. In the 1980s, decreased financial flows from the Gulf and increased public spending had resulted in a severe economic crisis in the country. Economic growth declined steeply, with the exchange rate collapsing in 1989. "Corrective" lending programmes became the only means of avoiding a worsening of the crisis. "Washington Consensus" policies adopted by the IMF and the World Bank as a quick fix for "failed states" – including Jordan, in their view – were premised on controlling public spending, liberalizing markets, removing barriers to international trade and privatizing state institutions. It was argued that this would resolve weak government performance and reduce the economic burden on the state.[9]

Privatization was institutionalized in Jordan from 1996, when the Executive Privatization Unit[10] was established in the Prime Minister's Office, in collaboration with the World Bank.[11] In the same year, a recommendation was made to convert the Jordan Electricity Authority, a public body founded in 1967 which owned and managed all activities in the sector, into a public shareholding company, the National Electric Power Company, owned by the government.[12] This new structure was more amenable to possible future privatization. A restructuring in 1999[13] further divided the company into three distinct companies: the National Electric Power Company (NEPCO), in charge of purchasing primary energy and transmission, control and interconnection; the Central Electricity Generating Company (CEGCO), in charge of electric power generation stations; and the Electricity Distribution Company (EDCO), in charge of electrical distribution. These three companies are administratively and financially independent. The Energy and Minerals Regulatory Commission (EMRC) was also subsequently set up as an independent entity regulating the relationship between different activities in the sector.[14]

The restructuring of the Jordanian energy sector was a prelude to the privatization of distribution and generation companies, a process that has displaced the public sector's role and turned companies and investors into key stakeholders in the energy sector. Despite evidence showing the efficient performance of the National Electricity Authority, the state adopted the IMF's neoliberal vision and privatized the authority in 2007. At the same time, 51 percent of the shares of the CEGCO were sold to the Emirati company Dubai International Capital (DIC).[15] One year later, 100 percent of the shares of the public shareholding EDCO and 55.4 percent of the shares of Irbid District Electricity Company – a distribution company in Northern Jordan – were sold to the Kingdom for Energy Investments Company (KEC) also owned by DIC, the Kuwaiti Privatization Holding Company and the United Arab Investors Company.[16] Subsequent generation projects were transferred to the private sector through direct proposals or competitive bidding. Distribution activity was, therefore, fully privatized, while generation activity saw a mixture of public and private ownership, with a tendency towards the latter. Transmission and fuel purchasing activity remained under the ownership of NEPCO, which represents the government in the sector, and the independent EMRC continues to regulate all of these activities.

The sector's structural problems only worsened after privatization, in that the dismantling of a single government authority led to unnecessary administrative costs in the sector and weakened its overall performance. Under privatization, contracts and agreements tend to "privatize profits and socialize risks", including by applying the "cost-plus" approach, which guarantees a fixed profit rate for companies without effective guarantees of performance and efficiency. Furthermore, generation contracts oblige the state to pay the costs of extra generation capacities even if they are not needed or used.

Ultimately, privatization has worsened Jordan's energy crisis. The Privatization Evaluation Committee, which was established in 2013 by a royal decree, released an evaluation report in 2014 on the privatization experience in Jordan. The report found that performance indicators showed a decline in the quality of services (e.g. increased power losses in privatized distribution companies), alongside an increased financial burden. The report stated: "Privatization had not achieved the desired economic goals … [of] increasing strategic investment, protecting the treasury from the consequences of increasing fuel cost, maximising the efficiency of the sector, or diversifying energy sources."[17]

Furthermore, the report argued that the majority of companies' profits (which averaged 20 percent annually at the time) were linked to the high energy prices set by the Electricity Regulatory Commission, rather than to increased efficiency or productivity, with companies achieving an unusually high rate of profit due to sales and profits being guaranteed. In its defence, the government argued that privatization occurred in collaboration with financial investors who were not specialized in energy and whose goals, therefore, centred around profitability rather than enhancing the sector's productivity. The government argued that it had to resort to these investors in some cases because strategic investors did not show an interest in the process.[18]

The second aspect of the World Bank's and the IMF's plans after restructuring Jordan's electricity sector relates to reducing consumer subsidies for electricity. As a condition for their lending to the Jordanian government after the 2011-13 Egyptian gas disruption, these institutions required that the energy mix be diversified and the electricity tariff adjusted, i.e., abolishing fuel and electricity subsidies, which they argued was necessary to resolve the debt crisis of the national energy company, NEPCO.

The IMF's conditional loan of approximately $2.06 billion in 2012,[19] followed by additional loans and grants in subsequent years, led to the implementation of a major subsidy reform programme which removed subsidies on petroleum products, increasing prices by 14 to 50 percent,[20] and involved a five-year plan to raise the electricity tariffs in five stages, starting in 2013.[21] This plan was only partially implemented as electricity prices increased three times between 2013 and 2015, but implementation was halted after NEPCO's losses diminished steeply as a result of the stabilization of international oil prices and the renewed access to gas. Nevertheless, NEPCO's shortfalls are now climbing again, with further increases forecast, which means more IMF plans and other proposals to reduce subsidies are likely to be back on the table. This could be a gateway to renewed privatization.

World Bank reports insist that the IMF's planned policies will lead to financial savings in the energy sector, which will enable investment in programmes targeting the poor and ultimately improve the living standard in Jordan as a whole. However, recent facts and figures do not support the claims that reducing subsidies raises profits: rates of economic growth[22] continued to decline even as the IMF's plans were being implemented,[23] the middle class was eroded, poverty levels rose and purchasing power decreased.[24] Even though subsidies were not abolished outright, the negative impacts of rising electricity prices on poor and middle-class people was evident.

The least expensive option first

The Egyptian gas crisis was not the first to hit Jordan's energy sector. A similar – albeit less severe – crisis occurred due to the stoppage of Iraqi oil in 2003 after the US invasion of Iraq. As with Egyptian gas, Iraqi oil was a cheap but insecure source that was heavily relied upon for power generation.

As a result, the stoppage led to energy price increases in Jordan. Nevertheless, this experience did not change the government's approach to the sector, and in the same year Egypt signed a 15-year agreement to supply Jordan with natural gas, covering 80 percent of the country's power generation requirements at a low price that offered protection against the rapidly climbing global oil prices.[25] At first, the agreement had a positive effect on energy prices and the Jordanian economy, but gas supplies began to fluctuate and decrease in 2008, triggering a renewed crisis. Furthermore, in 2010, only 60-70 percent of the agreed quantity of gas was delivered, casting doubt on Egyptian gas as a secure long term solution to Jordan's energy needs.[26] Despite these problems, the government did not search for new sources of gas, as had been explicitly recommended in its Energy Strategy 2007-20. Indeed, the sector continued to rely on Egyptian gas as a primary source for electrical generation right up until the explosions in Arish in 2011-13.

When the Egyptian gas crisis struck, to replace the now-unavailable Egyptian gas, Jordan began to import oil (and its derivatives) despite record prices. This greatly impacted the energy costs paid by the state-owned company, which increased by a staggering 129 percent, from about 9.6 cents per kilowatt hour (kWh) in 2010 to 22.5 cents per kWh in 2014.[27]

The Jordanian energy sector struggled with both high prices and inadequate supplies until the return of gas supplies in 2015, when imports began under new gas agreements mainly with Qatar.[28] These agreements made use of a Floating Storage Unit for liquefied natural gas in the Sheikh Sabah al-Ahmad Oil Terminal in Aqaba, on the Red Sea in the south of the country, which was chartered as part of an agreement with the company Golar LNG.[29]

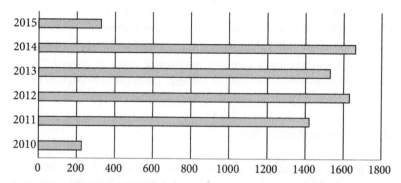

Figure 8.1 Annual government losses during the gas stoppage period (in US $ billions). Source: NEPCO annual reports. The disruption of Egyptian gas flows to Jordan since 2011 led to the accumulation of losses for the state-owned NEPCO.

At about the same time as the gas supplies began to arrive, global oil prices stabilized, causing a fall in the costs of energy production by approximately 10.3 cents per kWh in 2016.

Fear of repeating the 2011-13 crisis finally resulted in Jordan shifting away from heavy reliance on a single energy source and towards a policy of diversifying energy sources and forms of generation, as well as intensive domestic exploitation. In 2012, legislators rushedly passed the Permanent Renewable Energy and Energy Efficiency Law, after years of delay, and began issuing bids and competitive offers for renewables projects, quickly establishing a large local renewable energy sector. At the same time, billions of dollars' worth of nuclear energy and oil shale projects were also planned. Nevertheless, the high costs of these latter projects made the government uneasy, and nuclear projects in particular were abandoned before they could be implemented. All of this occurred alongside talks that led to a restoration of natural gas supplies from Egypt. Also at this time, a gas agreement was concluded with Israel that threatened Jordan's government with either rising electricity prices or periodic blackouts.[30]

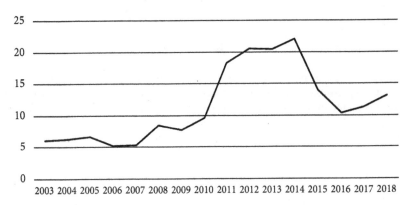

Figure 8.2 The cost of producing electric energy from 2003 to 2018 (in US cents per kWh).

The new policies and practices implemented since 2011 can be assessed by asking the following questions: Are the new energy sources sufficiently secure to protect the Jordanian people from the risk of interruptions and supply shortages? What has been the impact of these sources on energy bills and on the economy? Have these resources been properly managed at the financial and technical levels? Are all of these energy sources necessary for the country's energy security and the needs of its people?

AFTER THE CRISIS

Gas: the key word

Natural gas is the most important word in Jordan's energy security strategy. It is the most used component in domestic power generation and the most significant contributor to energy costs. Gas agreements, prices and sources are therefore critical for Jordan's energy sector.

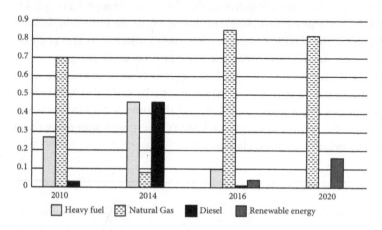

Figure 8.3   Energy mix from 2010 to 2020 (in percentage). Source: NEPCO Annual reports. Gas as a percentage of the total energy mix in 2010 before Egyptian gas was cut off; in 2014 when gas supplies were cut off; in 2016 when the exploitation of the floating storage unit began in the Port of Aqaba; and in 2020 as Jordan's reliance on gas for electricity production increased again.

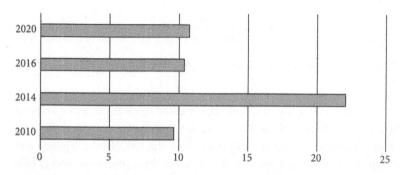

Figure 8.4   The cost of producing the kilowatt-hour (kWh) in US cents. Source: NEPCO Annual reports. The percentage of gas in the electricity mix correlates closely with the cost of production.

Twelve years after Jordan's 2011 energy crisis, it is important to reconsider current gas agreements, the most important of which is the agreement with Israel. In 2015, Jordan signed a $10 billion agreement to import Israeli gas from the Leviathan field in the Mediterranean over a period of 15 years, for electricity generation. This agreement followed another deal with the Jordanian Arab Potash Company, worth $770 million, to import gas from the Tamar field, also off the coast of Israel.[31] Not only did the government ignore Jordanian public opinion, which rejected the Leviathan agreement, but it also ignored all considerations related to the security of Jordan's energy supply, given that the owner of this energy source is a strategic enemy[32] with whom relations are marked by deep tensions and conflicts.

The Leviathan agreement, which was supposed to be kept secret, includes high penalty terms ($1.5 billion) for the buyer (NEPCO) in the event of default, with lower penalties applying to the seller (only $1.2 billion). Furthermore, on paper the seller is a company that is registered in the Cayman Islands, meaning partners in the Israeli field may be able to avoid all responsibility in the event of a default. This suggests the possibility of gas being cut off as a result of an arbitrary decision, with limited consequences for Israel but devastating impacts on Jordan. The agreement also constrains Jordan's ability to replace imports with its own production: if Jordan increases domestic extraction, it cannot reduce the contracted quantity by more than 20 percent, and that only after at least half the contracted quantity has been purchased.[33]

In its 2020 report, the Economic and Social Council (ESC) of Jordan stated that the Leviathan gas agreement is problematic because engineering, financial and supply details are cloaked in secrecy and because the contract mentions some parties without clearly identifying them. Further, the pipeline that carries Leviathan gas to Jordan is managed by the Egyptian Fajr company, without Jordanian input. As such, the details of the agreement and its implementation are beyond Jordan's control. This reliance on external partners without adequate protections or guarantees threatens the country's long-term energy security. In this respect, Article 19.4.15 of the agreement states: "Under no circumstances shall the Partners in the Leviathan Field have any obligation or liability whatsoever to the Purchaser [i.e. NEPCO] in connection with the subject matter of this Agreement."

In the absence of explicit contractual and legal guarantees ensuring Jordan's right to have its gas supply needs fulfilled, the pumping of gas and control over the quantity of gas supplied can be used as leverage by Israel. This represents a real threat to Jordan's energy security and economy, and provides an opening for foreign interference in Jordanian affairs. The ESC recommends cancelling the Leviathan deal or recontracting. If this is impossible, it argues that Israeli gas should not exceed 15 percent of Jordan's imports or consumption, so that any interruption to supply can be absorbed. Nevertheless, the $1.5 billion penalty clause for breaking the terms of the agreement, as well as additional fines for reduced purchases, call into question whether any of the council's recommendations are possible.[34]

In line with the spirit of the ESC's recommendations, some state officials, such as the ministers of energy and finance, have attempted to reduce the strategic position of Israeli gas by limiting imports from Israel to no more than 20 percent of the energy mix.[35] However, imports are likely to exceed this cap as various gas contracts, including the one with Egypt and the agreement relating to the Floating Storage Unit in Aqaba, come to an end. At the same time, the Energy Strategy 2020-30, which sets the Ministry of Energy's goal of reducing the percentage of gas-powered electricity generation from 82 percent in 2020 to 53 percent by 2030, states that nearly half of the gas used in electricity generation shall be Israeli during this period. Thus, the pricing and security of Israeli gas will remain a critical factor in years to come.

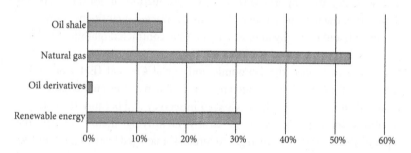

Figure 8.5 The energy mix according to the Energy Strategy 2020-30. Source: Jordan Energy Strategy (2020-30).[36]

Oil shale : A wasted black gold?

In 2012, as part of the turn to local energy sources, the state started reconsidering the question of oil shale. Under 14 previous energy ministers this discussion had remained stuck at the level of studies and consultations, due to the relatively high costs of oil shale exploitation. Studies have estimated Jordan's reserves of oil shale at between 40 and 70 billion tonnes, the sixth largest national reserves globally, with available energy estimated at 600-900

MW per site in several locations,[37] equal to roughly half of Jordan's annual electricity consumption. After extensive studies, an agreement was concluded with the Estonian National Energy Company (Enefit) in 2017 to establish an oil shale-burning power plant in the Attarat region southeast of Amman, with a total capacity of 470 MW (15 percent of Jordan's electrical needs). The Attarat Power Plant came into operation in 2021 and is currently the largest energy project in Jordan.[38]

However, in 2019, as the project was nearing completion and operation, the government announced that it was resorting to international arbitration against the company managing the plant. It cited the "obscene injustice" of the price charged for energy produced by the plant, and argued that NEPCO should be able to terminate the agreement if that gross injustice is not rectified.[39] Importantly, the World Bank had earlier asked the Jordanian government to reconsider the cost, usefulness and productivity of the Attarat project, while an IMF report[40] connected the project to increasing projected losses for NEPCO. According to the estimates of Jafar Hassan, former Minister of State for Economic Affairs, the project will increase NEPCO's annual losses, which are estimated to be more than $560 million by 2024, with total losses between 2020 and 2024 projected to reach $2 billion.[41] This is assuming that the tariff remains unchanged, and that Brent crude oil prices stay at $55 per barrel. Furthermore, government reports (NEPCO 2020) confirm that NEPCO saved $82.6 million when COVID-19 delayed the Attarat project.

According to the Jordanian government, the real cost of the Attarat project was misrepresented by the managing company[42] and the contracted tariff was too high, hence the state's decision to resort to arbitration to reduce the selling price. Government data show that the average tariff for buying

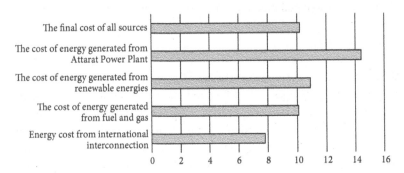

Figure 8.6   Average cost of various energy sources in US cents per kWh (2017-20). Source: NEPCO Annual reports.

electricity from Attarat is about 14 cents per kWh and the electricity is sold to the consumer at 18 cents,[43] which includes transmission and distribution costs. This makes Attarat's oil shale production the most expensive source of electricity in Jordan per kWh – more than 41 percent above the average.

While it is true that the Jordanian government signed the Attarat agreement during a moment of crisis, when global oil prices were high and the country was facing dramatic shortfalls in primary energy sources, which led it to accept the project's significant costs, official statements fail to mention that in 2016 the government had a chance to back out of the contract. At this time, global oil and primary energy prices had dropped significantly and stability had returned to local energy as production costs decreased. Nevertheless, when the Attarat company nearly failed to secure the necessary funding in 2016,[44] the government pressed forward with the deal, giving the company an additional two months to resolve its financial issues, rather than cancelling the contract altogether. As such, the government missed a real opportunity to avoid these financial burdens.

One cause of the high costs of shale oil, which makes it uncompetitive compared to other energy sources, is the immaturity of the technology involved. Shale oil also has significant environmental costs. It emits a lot of carbon dioxide in comparison to burning natural gas, let alone renewable energy production, and burning shale oil also demands large quantities of water, which is especially problematic in a country facing a major water crisis. Thus, it can be concluded that it was a strategic mistake for the Jordanian government to enter into this high-cost investment at this point in time (given that the costs of shale oil are likely to decrease over time as the technology matures), and to invest in a solution that will only exacerbate the country's water scarcity. Furthermore, the Attarat project is also a local polluter, so the cost of pollution to the healthcare sector needs to be factored in.

Renewable energy

Renewable energy currently constitutes 30 percent of Jordan's energy mix,[45] the largest share after gas and the largest domestic source. Renewable energy thus plays a pivotal role in determining energy costs and influencing the way the energy sector is organized and planned. The renewable energy sector witnessed a significant boom after the 2012 Renewable Energy and Energy Efficiency Law was passed. This boom continued until 2018, during which time Jordan's clean energy model was widely publicized for its promise of energy self-reliance and competitive prices. However, this promise quickly began to dissipate, with green energy dreams turning into a nightmare for the government institutions managing renewables. Poor planning in the new sector led to widespread problems that created a burden for producers and workers alike, leading to discussions in 2022 about the immediate closure of local renewable energy companies.

Jordan has excellent prospects for renewable generation, especially solar. Large areas of land are available in the southern and eastern regions, and the country is located on the "global solar belt", with ideal solar radiation and long hours of sunlight throughout the year.[46] Solar projects could thus provide a very reliable and secure source of energy due to the abundance of solar energy, its accessibility, its sustainability and the low cost of operation and use.

Before 2011, only modest attempts were made to include renewable energy in the energy mix, due to the high costs at that time, with renewable energy about twice as expensive as energy produced from Egyptian gas. However, following the gas stoppage crisis, diversification of energy sources was prioritized and the Renewable Energy and Energy Efficiency Law was approved in 2012. The law set out instructions, regulations and guidelines governing the renewable energy sector, in parallel with the development of renewable projects. Two distinct categories of projects were created: first, direct bidding projects and competitive bids that sell energy to the government according to specific agreements; and second, self-generation projects owned by end-users (citizens or corporations) in the form of wheeling and net metering projects (see below).

Direct bid renewables projects

With the aim of attracting foreign and domestic investment, the government invited investors and developers after 2012 to submit technical and financial proposals for renewable energy plants. In the first phase of direct bids, the government avoided the competitive bidding process and offered attractive prices of 16 cents per KWh for solar energy and 10 cents per KWh for wind, with a 15 percent premium for domestic projects.[47] This first phase of direct bids saw the establishment of plants with a total capacity of 204 MW of solar energy and 423 MW of wind energy, mainly concentrated in southern Jordan. Competitiveness increased in the second and third phases of offers, for which the government implemented a competitive bidding model. As a result, projects were awarded to proposals offering lower prices: 6.5 cents per kWh in the second phase, and 2.5 cents per kWh in the third phase (which has since been discontinued).

At all stages, power purchase agreements were signed requiring the government to purchase all of the energy produced at the agreed price for 20 to 25 years. It should be noted that such agreements stand in contrast to the situation in other countries, where a shift is taking place to variable price contracts that deliberately reduce the purchase price after investors have recovered the costs of the project. However, it is the case that funding agencies remain unwilling to accept the investment risks associated with such variable price contracts.

The high price of the first round of bids generated debate in Jordan about the viability of renewable energy, with the sector being held responsible for increasing energy prices at this time. However, official figures refute this interpretation: for solar projects, for example, first-phase projects account for just 2.5 percent of total generated energy,[48] so their impact on overall energy costs is minimal. The average cost of renewable energy per kWh is close to the average cost of gas-generated energy (the least expensive option locally), and the difference in the final cost is just 0.7cents per kWh. This indicates that renewable energy is not to blame for Jordan's rising electricity prices.

Nevertheless, objections were raised regarding Jordan's renewable energy projects, with some considering the high prices of the projects agreed in the first phase as providing a form of subsidy to investors. These objections, and the issue of increased electricity prices, led the Jordanian Ministry of Energy and Mineral Resources in 2022 to announce "corrective measures" to reduce the costs of electrical production, including renegotiating prices with renewable energy companies covering 29 projects.[49] Ultimately, however, the government made the decision in 2019 to end the bidding, with the third phase of bids discontinued. This is in spite of the fact that the estimated

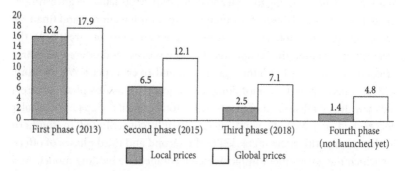

Figure 8.7 Comparison between costs of direct bids for renewable energy in Jordan with the average cost of renewable energy globally (in US cents per kWh). Source: NEPCO Annual reports.

renewable energy prices in the third and fourth bidding phases stood to be more than 9 cents per kWh cheaper than the 2021 energy prices. This would have made renewables the cheapest available energy source in Jordan, significantly cheaper than either natural gas or oil shale.

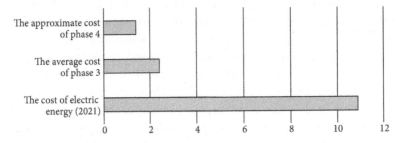

Figure 8.8   Comparison of the average total electric energy cost for 2021 with solar energy costs of the suspended third and fourth phases (in US cents per kWh). Source: NEPCO Annual reports; IRENA.[50]

By withdrawing support for renewables at this juncture, Jordan has therefore missed a great opportunity to reduce energy prices and to accelerate the energy transition. Yet, if renewable projects can reduce energy costs, why did the government stop the third phase and stop accepting offers? Several potential factors behind this decision can be identified. In the first place, Israeli gas, the oil shale project and fossil plants were meeting energy needs. Thus, further renewable energy generation projects would have led to oversupply. Second, for technical reasons (mainly due to the intermittency of solar and wind energy), the electricity network's capacity to absorb renewable energy remains limited. Third, there is also the fear of emergencies when components of the system cease to work, with wind farms and solar plants becoming unresponsive – which can be translated into a serious risk of complete shutdown. This calls for investments in infrastructure, expanded networks and increased generation capacity. This investment has been lacking due to financial and organizational challenges in the government energy sector. Lastly, there were significant problems with the way the government had invested in self-generation renewable projects (this is discussed below).

Self-generation renewable projects: net metering and wheeling

In 2012, after the introduction of the Renewable Energy Law, the government paved the way for consumers to establish renewable energy projects for their own use, under two systems: net metering and wheeling. Both systems enable consumers to produce electric energy from renewable sources, to use it for personal purposes and to exchange any surplus with government-owned energy to cover times when their own renewable sources cannot cover their own needs. The difference between the two systems is that a net metering system is installed near to where the electricity is consumed, while wheeling systems are located at a distance and are linked to the consumption location through transmission and distribution networks. In both systems, the producer-consumers are billed for the shortfall between the surplus electricity they produce and the electricity they draw from the government. In the event of the surplus electricity produced exceeding the amounts drawn, the government commits to buying it at a determined reference price. This is known as a "compensation" scheme and is a government strategy to support and stimulate the renewable energy sector, enabling consumers to enjoy "zero" bills.

Unfortunately, those in charge of regulating Jordan's electricity market, and the compensation scheme specifically, did not take into account all of the economic ramifications of the scheme. The estimated cost of 1 kilowatt produced from the net metering and wheeling systems did not exceed 5.5 cents, while the average cost for every 1 kilowatt drawn from the state was 11 cents, notwithstanding the fact that the cost is even higher during the night when most energy is drawn. The cost difference, therefore, constitutes a loss which the state is compelled to shoulder. This is one of the core organizational errors behind these projects.

Self-generation systems have achieved tremendous growth in recent years, reaching 972 MW – 39 percent of Jordan's total installed renewable energy capacity in 2021.[51] However, most of these systems belong to large users, such as telecommunication companies, banks, hotels, private hospi-

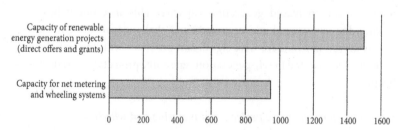

Figure 8.9 The total installed capacity of electric power generation projects from renewable energy sources until 2021 (in MW). Source: Ministry of Energy and Mineral Resources (2021 Annual Report).

tals, and large household consumers, who see renewable energy as a way to avoid the high tariffs charged by the state. Thus, alongside the growth in private renewable energy projects, there has been a decrease in energy sales to large subscribers[52] (the large industrial sector[53] especially). In 2020, industrial energy sales were just 68 percent of what they were in 2014, before the introduction of renewable energy. Although some of this is due to the closure of some factories for economic reasons, the biggest part of this decline is due to the growth in self-generation, especially renewable generation.[54] This lost revenue has had significant impacts on the public energy sector.[55] These large consumers were previously subsidizing the operation of the national grid, and thereby subsidizing poorer energy users. They are now no longer doing so. It can be argued that the state has thus incurred direct losses due to its ill-conceived, neoliberal approach to the renewable energy sector. Indeed, these richer energy users are themselves being supported by the compensation scheme. At the same time, their use of the national grid infrastructure causes additional wear and tear that is not covered by the fees they pay.

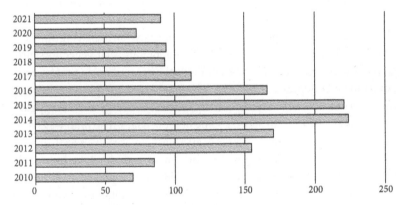

Figure 8.10 Government revenues from selling energy to major subscribers (in US $ millions). Source: NEPCO Annual reports.

Major private sector institutions and companies now control a considerable share of the renewable energy sector in Jordan. This raises crucial questions. Who are the beneficiaries of renewable energy, and who can access it? In whose interest is the transition towards renewable and clean sources, and who controls it?

The planners of Jordan's energy sector have thus far not taken the question of a just distribution of renewable energy into account. In contrast

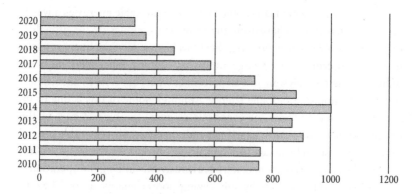

Figure 8.11 The amount of energy sold to major subscribers (gigawatt-hours GWh). Source: NEPCO Annual reports.

to the current approach, which, as we have seen, benefits large private sector players and high-income individuals, just policies would instead direct energy to vital and economically challenged sectors, such as agriculture and medium-size and small industries, and/or would use renewable generation to reduce the government's overall energy bill. For example, alternative renewable policies could have focused on replacing financial subsidies for lower income groups with autonomous solar generators, protecting them from tariff increases, or on deploying renewable generation in government buildings and facilities, which are major energy users. In 2012, the government established the Jordan Renewable Energy and Energy Efficiency Fund (JREEEF) to address some of these issues. The fund was set up to provide financing and technical assistance to various social groups, including small- and medium-size consumers and those in productive sectors, to enhance their access to renewable energy and to reduce their energy consumption. However, the fund's output was modest compared to private sector investments or existing electricity subsidies. According to the fund's website, the financial savings from the fund's projects in all sectors amounted to $9.44 million, while annual electricity subsidies are estimated at $564 million.[56]

Guaranteeing just access to renewable energy does not necessarily mean undermining the ability of the private sector to reduce its costs through self-generation, or abandoning market competition. However, it does require certain key changes to be made to the renewables regime in Jordan to ensure fair and targeted distribution of renewable energy to industrial, commercial, agricultural, water and domestic users. Key here is the setting of fair tariffs that reflect the costs generated by private renewable projects while using the national electricity and grid systems. These tariffs must enable investment in power networks to improve their reliability and capacity to absorb self-generated electricity in a useful way. If this were to happen, renewable self-generation would allow the government system to provide better and cheaper energy access for the Jordanian people. Furthermore, a new billing system needs to be put in place that disposes of net metering, so that the system reflects the true value of the energy drawn from the network, instead of the current unfair exchange whereby the state loses tens of millions of dollars by effectively buying self-generated electricity at deeply unfavourable rates. This net billing system must be established in a way that benefits all parties. Lastly, large consumers should continue to support the provision of energy to poorer users in various forms. These can be direct cash transfers or through a system that requires self-generation projects to produce electricity in excess of their need, so the surplus quantities of renewable energy generated are supplied to the grid.

In sum, any system supporting renewable energy production should ensure justice and access to adequate energy for all.

The trajectory of the renewable energy sector: rise and fall... and rise again?

In 2019, the government announced that it would stop issuing licenses for renewable energy projects above 1 MW. This was presented as a temporary suspension, while the government studied the capacity of the grid.[57] However, even before this decision was made there had been clear signs that the government was reducing its commitment to new renewable projects, and also struggling to meet consumer demand for private solar licences. New projects had faced delays in receiving approval, approvals were made contingent on scaling down generation capacity, and onerous technical requirements were imposed that increased the costs.

Tracking the volume of renewable energy investments in Jordan over time reveals the level of growth in the sector, and the extent to which this has been affected by government decisions. Between 2012 and 2020, some $3.09 billion was invested in renewable energy. Investment peaked in 2016, at $955 million per annum,[58] with foreign investment accounting for about 75 percent of this.[59] The period between 2015 and 2018, corresponding with the second phase of direct bids for renewables projects, represented a high point of growth. The most significant development in this period was the large scale licensing of net metering and wheeling projects. This produced a renewable energy boom, and the sector was seen as a promising area for investment. However, this growth slowed abruptly with the 2019 decision to suspend approval of third-phase bidding and to stop licensing projects exceeding 1 MW. The situation was further exacerbated by another decision to suspend wheeling plants on Fridays and Saturdays during COVID-19. Investments thus fell sharply in 2020, to just $16.8 million. This decline posed a threat not only to the renewable energy sector as a whole, but also specifically to its local workforce.

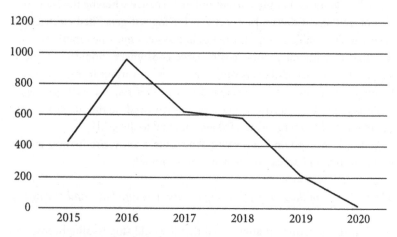

Figure 8.12   Renewable energy investments in Jordan 2015-20 (US $ millions).

However, a change occurred after Bisher al-Khasawneh's new government assumed power in October 2020, with the renewable energy sector again experiencing growth. In 2021, the value of investments reached $307 million. Furthermore, in 2022 the government decided to resume licensing renewable energy projects larger than 1 MW, albeit with strict new regulations.[60]

While this return to growth in the sector is positive, it should be noted that there have been no radical amendments to the billing and tariff systems, and no policy reforms. A modest fee (equivalent to $2.8 per kWh of renewable energy) was imposed in the domestic sector only (specifically, those owning renewable energy systems), overlooking other sectors, such as banks, private hospitals, factories and telecommunications companies, for which a revised tariff structure is equally needed. As a result, the problems that have hitherto plagued the sector may expand in the future.

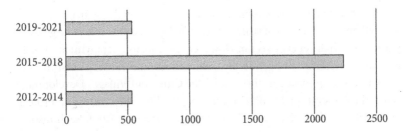

Figure 8.13 The volume of investments in the renewable energy sector (US $ millions).

THE FUTURE OF THE ENERGY SECTOR IN JORDAN

Mainstream media in Jordan have presented a rosy image of Jordan's rapid transition to renewable energy, but this narrative glosses over the considerable obstacles still facing the country's electricity system, such as the inability of the power grid to absorb more energy. This problem arises because the very rapid growth in new energy sources has not been accompanied by matching growth in infrastructure and the development of the power grid.

Any proposed solutions to the energy crisis in Jordan need to be realistic and implementable. At the same time, popular, civil and trade union forces should push for national policies that make energy sovereignty a top priority. Despite the unavoidable ramifications of the unjust contracts entered into by previous Jordanian administrations, it is possible (and necessary) to start gradually building the capabilities of the sector towards achieving its independence. To do this, the sector should continue to expand at the local level, thereby reducing both energy production costs and the need for large projects that are funded by the private sector and foreign investments. This will benefit NEPCO financially and will also reduce its need for international funds. However, this will be possible only by moving towards more renewable energy sources, and by putting in place a framework to treat these sources as public goods, rather than commodities.

Developing a national strategy for building Jordan's renewable energy capacity is the first building block in this trajectory. The next priority should be to develop decentralized small-scale self-generation systems for individuals and small institutions that can meet their needs equitably and directly while also reducing the load on the national grid. This will improve people's lives by lowering their energy costs while simultaneously reducing the burden of subsidies on the state. Such decentralized systems are a crucial means of achieving development and can be effectively deployed and strengthened on a national scale. This can help reduce the role of large companies in Jordan's energy sector. However, the effective deployment of these systems requires a strong role for public funds and financial institutions, such as has been played in the past by the Cities and Villages Development Bank, the Agricultural Credit Corporation, the Rural Electrification Project, and Jordan's Renewable Energy and Energy Efficiency Fund. Such institutions could act as the largest funders of renewable energy systems and could seek partnerships with engineering and commercial sectors, and with local expertise, to establish new systems and projects.

Jordan's energy sector has suffered in the past due to reliance on sources (local and imported) in excess of its needs, often at unaffordable prices, which has raised the cost of energy domestically and caused financial losses for the public energy sector. This could be remedied through a comprehensive strategy that stimulates the demand for electricity in key sectors where this would be socially beneficial, including by electrifying the activities of government ministries. For example, transport (public and private), which consumes approximately 49 percent of Jordan's primary energy imports,[61] could be electrified. This and other similar changes would reduce the state's energy bill, optimize the use of local resources and stimulate much needed economic growth.

Another area of use for Jordan's existing generation capacity is sea water desalination. By powering desalination plants, electricity could be used to address the country's water poverty. Here it is important to take note of the recent agreement between Jordan and Israel, sponsored by the UAE,[62] under which Jordan will purchase 200 million cubic metres of desalinated water from Israel each year, in exchange for electricity produced by a 200 MW solar power plant in Jordan. This makes Jordan dangerously dependent on Israel, especially alongside its above-noted reliance on Israeli gas (see Manal Shqair's chapter in this book). It is essential to put pressure on the state to cancel this agreement, and to strive for domestic synergy between the energy and water sectors, including by using modern renewable energy technology for water desalination and restructuring water networks to waste less energy and water. This would enhance the role of clean energy in Jordan and provide for its water needs into the future.

Finally, creating a common Arab and regional energy market remains an important, and as yet unfulfilled, project for the region. This would contribute to resolving the energy crises, not only in Jordan, but across the region. Strengthening current electricity interconnection projects with Egypt, Palestine and Iraq, and establishing new and expanded projects to connect the region, would enhance the stability of power systems in Jordan and abroad. Further, such a move would reduce the technical problems currently facing the sector and increase the capacity of renewable energy networks. This would be of great economic benefit as it would exploit all the contracted generation capacities and would allow for the mutually beneficial sale of energy to neighbouring countries, such as Lebanon, that are suffering from acute energy shortages.

NOTES

1. After the outbreak of the Arab Spring in 2011, armed groups in Sinai repeatedly attacked the pipelines transporting gas to Israel and Jordan.
2. *Al Jazeera*, "tawaqquf 'imdadat alghaz almisri lil'urdun" ["The Suspension of Egyptian Gas Supplies to Jordan"], July 7, 2013, https://tinyurl.com/3d4k8m7m.
3. *Ibid.*
4. *Shaheennews*, "Electricity Company Scored JD5.5 Billion in Losses, Noble Was "Last Option": NEPCO," January 6, 2020, https://shahennews.com/eng/archives/5302.
5. ESC, *State of Country Report - 2020*, https://tinyurl.com/mdeuzywb.
6. *The Jordan Times*, "Oil Shale Power Station to Generate 15% Of Jordan's Electricity Needs by Next May," May 8, 2019, https://tinyurl.com/bce4wzhf.
7. *The Jordan Times*, "Attarat Project takes Centre Stage at Lower House Oversight Session," January 27, 2021, https://tinyurl.com/2swva2ru.
8. *Ammon News*, "Rawashdeh: Renewable Energy Constitutes 30% of Local Energy Mix," November 15, 2022, https://en.ammonnews.net/article/61521.
9. Taher Kanaan and Hazem Rahahleh, *The State and Market Economy: Readings in Privatization Policies and its Arab and International Experiments*, 2016. Washington Consensus is a set of economic policy recommendations for developing countries, and Latin America in particular, that became popular during the 1980s.
10. Executive Privatization Commission, "General Information", https://tinyurl.com/c2m2s7ce.
11. World Bank, *Second Economic Reform and Development Loan*, Report No. P-7004-JO, November 14, 1996, https://tinyurl.com/bdku4s5k.
12. "National Electricity Company," Government Agencies, The Official Site of the Jordanian e-Government, https://tinyurl.com/2p9cff67.
13. *Ibid.*
14. "Development of Electricity," NEPCO Home page (Arabic), https://www.nepco.com.jo/sectordevelopment.aspx.
15. Suleiman al-Khalidi, "UPDATE 1-Jordan Sells 51 pct of State Generation Firm," *Reuters*, September 20, 2007, https://tinyurl.com/4dapn63e.
16. The Kingdom for Energy Investments Company "About Us", https://tinyurl.com/29fbr439.

17. Tariq Al-Da'jah and Radad Snowy Al-Qarrala, ""alghad" tanshur nasa taqrir taqyim altakhasia ["Al-Ghad" publishes the privatization evaluation report]," *Alghad*, March 30, 2014, https://tinyurl.com/5n8uye3y; "Report of the Privatization Evaluation Committee," Jordanian Prime Minister's Office, https://tinyurl.com/3h4rxbeb.

18. *Ibid.*

19. IMF, "IMF Executive Board Approves US$ 2 billion Stand-By Arrangement for Jordan," Press Release No. 12/288, August 3, 2012, www.imf.org/en/News/Articles/2015/09/14/01/49/pr12288.

20. Jodi Rudoren, "Riots Erupt Across Jordan Over Gas Prices," the *New York Times*, November 13, 2012, https://tinyurl.com/3xmcfhht.

21. World Bank, *Hashemite Kingdom of Jordan – Promoting Poverty Reduction and Shared Prosperity: Systematic Country Diagnostic*, Report No. 103433-JO, February 2016, https://tinyurl.com/2txa75ea.

22. World Bank, *Jordan Economic Monitor – Fall 2022: Public Investment - Maximizing the Development Impact*, January 19, 2023, https://tinyurl.com/rav5wbe3. Hala "The World Bank Expects Jordan's Economic Growth to Decline in the Medium Term" (in Arabic), January 25, 2023, https://tinyurl.com/vtm4pts8.

23. Central Bank of Jordan, Financial Stability Report, 2015, https://tinyurl.com/25zthej9.

24. *Ammon News*, "We stand between two pressures – Minister of Finance," January 9, 2023, https://en.ammonnews.net/article/62873.

25. European Investment Fund, "EUR 100 mio Support for the Regional Gas Pipeline between Egypt and Jordan," Press Release, June 7, 2004, https://tinyurl.com/d7v84h9r.

26. Allan Martinez Venegas, "Jordan 's Energy Security: Impact of Dependency on Unstable Foreign Sources on Social Stability and Policy Alternatives," *SIT Digital Collections* Spring 2013, https://tinyurl.com/56zs9a72.

27. National Electric Power Company (NEPCO), *Annual Report*, 2010, https://tinyurl.com/yc6fyujn.

28. Economist Intelligence Unit, "Jordan Targets Energy Diversification," October 27, 2016, http://country.eiu.com/article.aspx?articleid=1054753689.

29. Golar Eskimo, Time Charter with Jordan, November 2015, https://tinyurl.com/3yucpfh.

30. Abdullah al-Shami, "alghaz min 'iisrayiyl: kadhibatan bimilyar dinar 'urdny" ["Israeli Gas: A Million Dinars Jordanian Lie']," *7lber*, December 14, 2014, https://tinyurl.com/bdzkb8v5.

31. Mohammad Ghazal, "Potas Company To Import Israeli Gas at Preferential Prices," *Jordan Times*, February 19, 2014, https://bit.ly/3ZwIZzx.

32. Jordan has fought three wars with Israel: the 1948 war, the 1967 war and the 1968 Battle of Karameh. A peace agreement was signed between the two countries in 1994, but since Israel still seeks to solve the Palestinian question at Jordan's expense by settling Palestinians in Jordan, Jordanians continue to consider Israel an enemy. National and humanitarian reasons are also a consideration here, particularly Israel's continued occupation of Arab lands.

33. Hisham al-Bustani, "safqat altbeyt wal'iidheani: 'abraz bunud atfaqyt alghaz almustawrad min "iisrayiyl" ["A Deal of Dependence and Submission: The Most Important Terms of the Agreement on Gas Imported from Israel"], *7iber*, August 4, 2019, https://bit.ly/3y63MOU.

34. ESC, *State Of Country Report*.

35. Al-Shami, "Israeli Gas".

36. Ministry of Energy and Mineral Resources, *Summary of Jordan Energy Strategy 2020-2030*, https://tinyurl.com/4ryfrb3e.

37. Japan International Cooperation Agency, *Project for the Study on the Electricity Sector Master Plan in the Hashemite Kingdom of Jordan – Final Report*, February 2017, https://openjicareport.jica.go.jp/pdf/12283693_01.pdf.

38. Attarat news, "Attarat Mining Company Officially Opens its Oil Shale Laboratory," *Attarat*, December 8, 2019, https://attaratpower.com.jo/5445-2.

39. Ahmed Al-Naimat, "wazirat altaaqati: "kulfat mashrue aleitarat 'aqala mimaa qil lina" ["Minister of Energy: The Cost of the Attarat Project is Less Than What We were Told"], *Al-Mamlaka*, June 16, 2021, https://tinyurl.com/54w2tpsk.

40. IMF, *Jordan: Second Review Under the Extended Arrangement Under the Extended Fund Facility, Requests for a Waiver of Nonobservance of Performance Criterion, an Extension of the Arrangement, and Rephasing of Access-Press Release; Staff Report; and Statement by the Executive Director for Jordan*, Country Report No. 19/127, May 15, 2019, https://tinyurl.com/2pscff3e.

41. Jaafar Hassan, *The Jordanian Political Economy: A Structure in the Womb of Crises* (Dar Al-An Publishers, Amman, 2020).

42. Al-Naimat, "Minister of Energy."

43. *Hala*, "zawati: 200 milyun dinar khasayir 'alkahraba alwatania' fi 2021 baed bad aleamal bi al-Attarat" ["Zawati: 200 Million Dinars, the National Electricity Losses In 2021 after the Start of Attarat"], December 23, 2019, https://tinyurl.com/ydxphw9d.

44. Mohammad Ghazal, "Company Developing First Oil Shale Plant Seeks Extension on October Deadline," the *Jordan Times*, October 1, 2015, https://tinyurl.com/yck53kdw.

45. *Almmon News*, "Rawashdeh,"

46. "About Renewable Energy Sector," Energy and Minerals Regulatory Commission, https://emrc.gov.jo/Pages/viewpage?pageID=111.

47. Projects that use locally manufactured system components are usually solar panels.

48. *Addustour*, "zawati: taerifat alshira' min aleitarat tablugh 100filis/kW[h]" ["Zawati: The purchase tariff from Attart is 100 fils/kWh"], December 2, 2021, https://tinyurl.com/34e3mms4.

49. *Jordan News Agency*, "alkharabishah: nadrus muqtarahat limurajaeat atifaqiaat shira' altaaqa" ["Al-Kharabsheh: We are Studying Proposals to Review Energy Purchase Agreements"], July 2022, https://tinyurl.com/2spyxanh.

50. International Renewable Energy Agency. *Renewable Power Generation Costs in 2021*, 2022, https://tinyurl.com/ypm8x8ty.

51. Ministry of Energy, *Annual report*.

52. The major subscribers are the large industrial sector, Jordan's Radio and Television Corporation, and Queen Alia International Airport.
53. The large industrial sector is made up of potash and phosphate companies, and cement and chemicals factories.
54. The National Electricity Company (2022) "A Decrease of 3.6% in Purchases of the Large Industrial Sector" (in Arabic), https://bit.ly/41Cv39a.
55. The National Electricity Company: annual reports from 2015 to 2018.
56. *Ammon News*, "Zawati: 400 Million Dinars to Subsidize Electricity Bills Annually" (in Arabic), 8 January, 2020, https://tinyurl.com/2n8dxr3z.
57. Ministry of Energy, "waqaf masharii tawlid alkahraba' lifatra muaqatat li'asbab tandimia" ["The Temporary Suspension of Electricity Generation Projects for Regulatory Reasons"], no date, https://tinyurl.com/yvwhfun8.
58. *Bloomberg*, "Top Five Emerging Markets for Clean Energy Investment," January 21, 2020, https://tinyurl.com/ykywebad.
59. *ClimateScope*, "Clean Energy Investments", https://tinyurl.com/bdesp7ts.
60. *Jordan Times*, "Kharabsheh Announces Criteria for Renewable Energy Projects Surpassing 1 MW Capacity," July 5, 2022, https://tinyurl.com/46p22x2d.
61. Ministry of Environment, Facilitative Sharing of Views, November 11, 2021, https://tinyurl.com/2p96h5f8.
62. Bruce Riedel and Natan Sachs, "Israel, Jordan, and the UAE's Energy Deal is Good News," November 23, 2021, *Brookings*, https://tinyurl.com/yc779h5s.

9

Renewable Energy in Tunisia: An Unjust Transition

Chafik Ben Rouine and Flavie Roche

Tunisia has achieved a high electrification rate, increasing from 21 percent at the creation of the Tunisian Company of Electricity and Gas (STEG) in 1962 (six years after independence) to 99.8 percent nowadays.[1] However, the country's energy sector is currently facing several issues. Highly dependent on fossil fuels, accounting for 97 percent of electricity production, the energy sector is facing increasing demand while the already limited national resources are diminishing. Between 2010 and 2018, the national production of primary energy fell by 36 percent. Over the same period, the demand increased more than twofold. This context has led to the expansion of the primary energy balance deficit, which increased from 15 percent in 2010 to almost 50 percent in 2018, simultaneously enhancing Tunisia's energy dependence: more than half of the natural gas consumed is imported from Algeria. All of this has resulted in a steady rise in electricity prices for consumers.[2]

Tunisia signed the Paris Agreement in 2015 and is committed to its Nationally Determined Contribution (NDC), which aims to reduce the country's greenhouse gas emissions across all sectors by 41 percent by 2030 compared to 2010 levels, despite the fact that Tunisia accounts for only 0.07 percent of global emissions.[3] The planned reductions include a 46 percent decrease in emissions in the energy sector. Achieving these objectives would help Tunisia accomplish its ambition of reducing its energy deficit.

While promoting the diversification of its energy mix by developing renewable energies, in the last few decades, Tunisia's climate policies failed to bring about the necessary changes. This is because its policies remain embedded in a capitalist framework that imposes a quest for endless growth and prioritizes private profit above all else, resulting in an "energy expansion, rather than an energy transition".[4]

An alternative to the models followed so far is needed, one that would incorporate social and environmental goals within climate policies. This could be achieved through a public goods and ownership approach, provided the latter are carried out by *accountable* institutions, namely within the framework of energy democracy. This implies a scheme that involves "genuine popular participation and accountability".[5] To tackle this challenge and to bring about actual solutions, the concept of a "just transition" has been introduced into global discussions around the energy transition. This concept advocates a fair shift to an economy that is ecologically sustainable, equitable and just for all its members. It asserts that transforming the way we use and think about energy requires deep transformations in every sector, and that the energy transition should be implemented carefully to avoid reproducing or deepening existing inequalities. The idea of a just transition, therefore, emphasizes the issues of democracy and sovereignty over public goods and the environment, among others.[6] Indeed, focusing on communities' interests in the design of an energy transition must involve taking steps away from the current financial, profit-based system and include consideration of other dimensions. According to the just transition framework, a real solution cannot simply tackle only one aspect of the problem of climate change – for instance, the sources of energy – while overlooking the social and environmental sectors that may depend on those sources in various ways. The idea is to get away from narrow visions and goals and to consider the way in which renewable energies should be developed.

In 2017, in reference to the necessity of a just transition, the Movement Generation argued that "Transition is inevitable. Justice is not". However, considering recent developments, including the fact that the COVID-19 pandemic has failed to bring about any significant change in our system, it is becoming increasingly obvious that the *status quo* could well last much longer than we might have expected, while its adverse repercussions grow exponentially. In this context, there is no room left for a wait-and-see attitude: immediate action must be taken towards both an effective and just transition.

While Tunisia is now entering a new energy transition in line with its international commitments, almost no debate has occurred at the national level on the redistributive aspects of this transition, which raises serious concerns and crucial questions. For instance, who will benefit and who will lose from this transition? Who controls the knowledge and technology that will be used to implement this transition and to what extent will the transition deepen Tunisia's historical dependency on imperial powers? Will this energy transition open the door to the liberalization and the privatization of the energy sector? To what extent will this energy transition help to address the issues of structural unemployment and inequality in the country? Will this transition facilitate an increase in democratic control over natural resources, or will it exacerbate capitalistic land-grabbing at the expense of local communities?

This chapter seeks to provide some reflections to help answer these questions. It aims to explore the concept of a just transition in the context of Tunisia. In the first part, it introduces Law No. 2015-12 on renewable energy and its implications. It then examines the extent to which the transition to renewable energy represents a convincing development opportunity for Tunisians themselves. Last, it questions the impacts of renewable energy development on people's rights and on the environment.

THE RENEWABLE ENERGY LAW:
A TURNING POINT IN TUNISIA'S ENERGY TRANSITION

The Tunisian Solar Plan: a renewal of the trend towards dependency as strategic orientation

As part of the country's energy transition strategy, Tunisia launched an updated version of the Plan Solaire Tunisien (PST) in 2015.[7] The plan was originally published in 2009 with the aim of increasing the ratio of renewable energy from three percent in 2016 to 30 percent in 2030.[8] This required the production of an additional 3,815 MW from renewable energy. According to the PST, 46 percent of new renewable energy will be produced by wind turbines, 39.6 percent by solar photovoltaic panels, 11.8 percent by concentrated solar power[9] and 2.6 percent by biomass.[10]

The PST's intermediary goals were updated after a conference in December 2017 on accelerating the implementation of renewable energy projects.[11] This policy followed a regional, if not global, trend towards the expansion of renewable energy, partly through Public-Private Partnerships (PPPs), justified by the lack of sufficient governmental resources to build power plants.[12] For example, Morocco has been following a similar path since 2009, when its solar plan was presented by King Mohammed VI.[13] However, the promotion of PPPs as a substitute for public procurement and public debt is misguided as this kind of partnership aims to derisk public projects, allowing profits to be privatized and losses socialized.[14] The PST requires around eight billion euros of investment from 2015 to 2030, includ-

204 • DISMANTLING GREEN COLONIALISM

ing 6.3 billion for equipment and 1.7 billion for the development of the power grid.[15] According to the plan, two thirds of this funding will come from private sources, predominantly foreign investment, and one third from public sources. Most of these financing needs focus on importing knowledge and expertise (i.e. technologies, equipment and patents) and will accelerate Tunisia's current path of dependency. Such dependency will take the form of deepening Tunisia's foreign debt in order to finance the imported technology, which is subject to monopoly conditions and intellectual property rights. In this context, the PST was designed in a way that would reinforce power dynamics whereby a country in the global South would need to borrow more so as to import technology and knowledge from the global North in order to transition to renewable energy. Through this plan, Tunisia is continuing to promote an economic model that is led by foreign investment as the primary way of financing its development. While some parts of the funding needed for Tunisia's renewable energy plan may come through foreign investment – or even climate finance and debt – no efforts have been made to explore ways of producing and controlling the knowledge necessary to achieve some parts of the PST in order to reduce knowledge and capital dependency on industrialized countries.

The 2015-12 law: liberalization, privatization and lack of state control

Since 2009, steps have been taken towards progressively liberalizing the Tunisian energy sector: Law No. 2009-7 of February 9, 2009 introduced private sector electricity production from renewable energies for self-production by firms.[16] This was followed by Decree No. 2009-2773, concerning the conditions of surplus selling to the national company STEG. A big step was taken in 2015 with Law No. 2015-12,[17] relating to the production of electricity from renewable resources. This law opened up the power grid to private companies, enabling them to produce energy, primarily for domestic use and export, through an authorization regime for projects of between one and ten megawatts, and a concession regime for projects of more than ten megawatts. These liberalization measures, which put an end to STEG's monopoly, aimed to make the regulatory framework more attractive for foreign investors.[18] Subsequent decrees and regulations specified conditions and procedures for the achievement of these projects,[19] including connection to the national grid,[20] and providing standard contracts for firms to start producing under the above-mentioned regimes. In this context, the idea that the energy sector functions most efficiently when it is managed by private companies, as opposed to the ineffectiveness of public management, still prevails today in Tunisia, despite a serious lack of independent studies on the impact of liberalization policies on the electricity production sector.[21]

As a matter of fact, the claim that private companies provide better services for a lower price has not been confirmed by the facts. On the contrary, while PPPs are sought after by states for development reasons, private companies tend to prioritize corporate profit above all under these contracts, and this aspect of divergent interests has often been overlooked. These partnerships often induce increased prices, along with labour violations, declining service quality and a failure to implement an ambitious climate strategy. The Tunisian law concerning PPPs, enacted in late 2015, does not provide sufficient tools for the state to address the negative impacts of these types of projects, and to ensure the protection of public and citizens' interests. No right to compensation for affected communities is envisaged, nor are there mechanisms for government control and supervision – to prevent green grabbing, for instance.[22] Moreover, civil society and local communities have limited access to information about PPP proposals and are not encouraged to participate in discussions.[23] Therefore, PPPs raise financial issues for the government, as much as they represent a threat to the efficient delivery of services and genuine democratic control over projects.

The influence of international interests in the context of the policy-making process

The energy transition in Tunisia is being promoted by international actors, some of whom are connected to previous projects that aimed at developing renewable energy in northern Africa for export to Europe. One of them, Desertec, was centred around an "unrestricted flow of cheap natural resources from the global South to the rich industrialized North, maintaining a profoundly unjust international division of labour", as Hamza Hamouchene has described it.[24] Two companies, Nur Energy (based in the United Kingdom) and Zammit Group (based in Malta) are the main stakeholders of the TuNur project, which aimed in its early days to establish a giant solar power plant in the region of Kebili, with the purpose of exporting the produced electricity to Europe through cables under the sea. This project mobilized a powerful lobby that sought to include provisions relating to exports in the renewable energy legislation, against resistance from the state electricity monopoly.[25] The role of international actors in influencing domestic policies has been well-documented in the field of renewable energy, especially as regards the relationship between Tunisia and Germany in this sector. Germany, which is a pioneer in the area, sees in Tunisia a high potential to be tapped. Accordingly, since the 2012 German-Tunisian partnership on energy, Germany has been providing technical and financial support through industrial investments and the establishment of institutes and foundations in Tunisia. The latter seek, among other things, to influence political parties by promoting "green" development ideas.[26] These actions, carried out in the context of bilateral cooperation, have had some repercussions in relation to the Tunisian regulatory framework.

In fact, some recommendations made by the German Agency for International Cooperation (GIZ) and Desertec Industry Initiative (Dii) seem to have anticipated some of the measures contained in the 2015 law. The official motives of German cooperation are said to be beneficial for the development of Tunisia, particularly in relation to employment.[27] Germany's actions in Tunisia fit into the context of European Union (EU) activities in this area. A 2015 communication by the European Commission about the strategy for an Energy Union clearly expresses the EU's wish to encourage and develop renewable energies, notably through international cooperation with countries outside the EU area.[28] This would be done within the framework of the Energy Charter Treaty (ECT), which was established in the early 1990s. In fact, the European effort to involve Tunisia in this process dates back to 2013, when the country was approached by the ECT secretariat, through the mediation of the German embassy, to join the treaty within the context of its "MENA Project" of expansion in the region. The country's membership of the ECT is still under discussion.[29] The ECT includes provisions on foreign investments in the energy sector, including in relation to investor-state dispute settlement (ISDS). This tool gives corporations the power to sue governments when they consider that states' policies are detrimental to their profits, even if those policies aim to foster an energy transition or social rights that are in the public interest. ISDS claims have already delivered billions of dollars of taxpayers' money to big corporations and the mere threat of ISDS, therefore, constrains states in their policy design, thus interfering with democratic processes.[30]

In line with the principles set out in the ECT, the EU is seeking to deepen liberalization in order to standardize the Tunisian legislative framework through negotiations around the Deep and Comprehensive Free Trade Agreement (DCFTA). This liberalization offensive would undermine the state's capacity to regulate – sometimes against investors' interests – and, therefore, would facilitate the introduction of European investors, who benefit from the EU's extensive subsidy programmes, into the Tunisian market. This would eventually open the way for exports, thus ensuring energy security for Europe, rather than for Tunisia.[31] For European companies, accessing the Tunisian market means increased cost-effectiveness and competitiveness because of lower salary and fiscal charges and the transfer of environmental costs. The pressure exercised on Tunisia, and the lack of consultation with civil society in the DCFTA negotiation process, have been already pointed out.[32]

Progress of, and resistance to, the privatization process

Prior to Law No. 2015-12, the production of electricity – excluding the self-production regimes – was the monopoly of the STEG. This public company had already embarked on several investments to develop the production of electricity from renewable energy. For instance, two wind power plants belonging to STEG were established in the north of Tunisia before 2015: a 54 MW plant in Sidi Daoud and a 190 MW plant in Bizerte.[33] However, the company's executive considers these projects very expensive. In an interview with *Nawaat* magazine, Taher Aribi, the former chief executive officer of STEG, stated: "To invest in such projects, we are obliged to sign debt agreements. Clean electricity production projects cost three times as much as a conventional plant. Our financial capacity is fragile for investment, borrowing or guarantee remittances".[34]

Since the liberalization of renewable electricity production under the authorizations and concessions regimes, the proportion of private investment has increased. According to 2018 figures, 42.5 percent of the electric power coming from planned wind and solar energy projects will be produced as a result of PPPs. However, it should be mentioned that not all of those power plants are operating yet.[35] In parallel, STEG has been developing its two photovoltaic (PV) power plants in Tozeur (Tozeur I and Tozeur II), each with a capacity of 10 MW.

Because of the lack of available information on the progress of renewable energy projects, it is hard to define the extent and conditions under which the sector is currently developed. For instance, the information published on the Ministry of Industry, Mines and Energy website states that "in 2017, STEG began building the first 10 MW PV plant in Tozeur [Tozeur 1] which has been put into operation on 10 March 2021. A second 10 MW extension plant at the same location (Tozeur 2) has been put into service on 24 November 2021."[36] However, a press article published on November 5, 2021 reports that the operationalization of the plants is just starting. The article mentions that the delay was due to financial issues faced by Tozeur 1 and to postponements in the shipping of equipment related to the COVID-19 pandemic for Tozeur 2.[37] Both plants are now operational and have been officially inaugurated in March 2022.

Another power plant has been built in Tataouine and was ready for operationalization in June 2020. However, the trade union – *Union Générale des travailleurs Tunisiens* (UGTT) – has blocked the plant's connection to the national grid,[38] claiming that the process will eventually lead to the privatization of STEG.[39] In July 2020,[40] Mongi Marzouk, Minister of Industry, Mines and Energy, posted on Facebook a message accusing the General Federation of Electricity and Gas (FGEG), a branch of the UGTT, of "sabotage" against the operationalization of the 10 MW PV power plant in Tataouine, which has been built under the authorization regime by the Tunisian Enterprise of Petroleum Activities (ETAP), a public company and a subsidiary of the Italian company Eni. However, the FGEG's opposition to the project needs to be read in the context of its opposition to privatization in general. As a matter of fact, the UGTT's opposition to the PPPs in particular, and to privatization in the production of electricity in general, is not new.

As early as January 2014, the FGEG spoke out against the bill prepared by the Ministry of Industry and adopted by the government which would eventually become the 2015-12 law. It criticized the decision-making process behind the bill, asserting that it was drawn up without involving the UGTT, or STEG's executives and engineers. The FGEG's secretary-general noted that the project was launched in haste and without referring to studies prepared in advance or to a general national energy strategy. On March 27, 2018, a call for the non-privatization of the electricity production sector was reiterated by the FGEG. Later, on February 26, 2020, a few months before the blocking of the Tataouine power plant by the UGTT, the government issued a decree authorizing the creation of self-production companies for the production of electricity from renewable energies and defining the conditions for the transport of electricity and the sale of excess energy to STEG. Abdelkader Jelassi, Secretary-General of the FGEG, has then expressed the categorical opposition of the federation to the privatization of electricity production in Tunisia. These policies were depicted by the FGEG as paving the way for private and foreign investment, favouring investors' profits over the STEG. The FGEG had stated that it would protest against this orientation because the production of electricity by private individuals and its direct sale to customers would disrupt the electricity network and impact the distribution of electricity, making it inaccessible to certain categories of the population. It also rejects the commodification of electricity, which affects national security and STEG's public status.

IMPACTS OF THE CURRENT ENERGY TRANSITION: A FAIR SHIFT FOR TUNISIA'S DEVELOPMENT AND PEOPLE'S RIGHTS?

A real development opportunity for the Tunisian renewable energy sector?

Tunisia is implementing a training scheme in the energy sector that has been adapted to renewable energies. Academic and professional programmes have been designed and are being delivered by public and private universities, including engineering schools. The National Agency for Energy Management (ANME) has also started offering training and certification programmes. These efforts have led to the development of human resources that are capable of providing companies with the skills required to help implement national renewable energy programmes with greater "competitiveness" (i.e. more cheaply). However, local skills and expertise are insufficient to enable local enterprises to conceive, carry out and maintain large-scale wind and solar power plant projects. In addition, the stagnation of the wind farm at Bizerte since 2012 has led to the dissipation of previously acquired expertise.[41] In parallel, a number of operators have emerged that give substance and structure to the renewable energies sector that is being developed: government institutions, manufacturers and suppliers of equipment, installation and maintenance companies, design agencies, etc. Also, building on its prior industrial experience, Tunisia has the capacity to develop partnerships with foreign manufacturers to produce renewable energies equipment. Indeed, in regard to PV, national firms are engaged in assembling some modules imported from China, Germany, Japan, Italy, Spain and France.

In the case of wind turbines, there is strong potential for industrial integration: a Tunisian private company, SOCOMENIN,[42] which originally specialized in metal construction, is producing wind turbine towers, and local industry is also capable of manufacturing turbine components in the mechanic, electric and electronic industrial sectors, including by adapting the production line where appropriate. In addition, related logistics, transport, construction, exploitation and maintenance activities can all be carried out by domestic enterprises. However, despite these advantages, the Tunisian renewable energy manufacturing sector remains unable to support the development of major projects. Tunisia lacks certain raw materials and intermediate technologies that are essential to the development of such projects. These include silica, PV cells, electrical wires, alternators for wind turbines and wind turbine controllers.[43] Equipment and intermediate technologies which are not produced locally must be imported, resulting in a dependency on foreign suppliers. In fact, the reality is that this sector has thus far grown mostly thanks to residential PV installation programmes – 90 percent of Tunisian renewable energy sector companies work in the PV sub-sector. As a consequence, the market is mostly developed in the area of PV installation. According to the preliminary findings of a 2019 GIZ survey, out of 150 enterprises in the sector, more than 85 percent were installers, a third were suppliers of PV components and 20 were design agencies, while there were only two project developers, two PV panels manufacturers and one training agency. Also, when we look at calls for proposals and tenders related to renewable energy authorizations and concessions in the period 2017-19, we find that development corporations were only just emerging at this time.[44]

In addition, in spite of the existence of some national actors, Tunisia's willingness to attract foreign investors tends to exclude local companies and Tunisian developers: for instance, the government prioritizes foreign companies with a background in developing projects of the same scale with the same technology.[45] Indeed, the selection of projects is based on the prior experience of the developer or its subcontractors, and on the consistency and feasibility of the project, which *de facto* gives preference to foreign investors from countries that are leading the way in the development of renewable energy projects and which have stronger financial resources.[46]

Under the authorization regime (10 MW projects), out of the 22 projects which have benefited from an agreement in principle after the three rounds of calls for tenders launched between 2017 and 2019, only half have Tunisian project leaders and only four projects are exclusively led by Tunisian firms. By comparison, five projects exclusively involve French firms and three German ones.[47] As regards concessions for solar energy production, all five projects (accounting for a total of 500 MW) are awarded to foreign firms. The Norwegian company SCATEC Solar has won tenders for three projects, accounting for a total of 300 MW.[48]

Thus, if the Tunisian-led renewable energy sector has some assets for the development of local projects, it remains too weak to carry out the expected large-scale projects in the current context. Therefore, to reduce its dependency, Tunisia would be wise to promote small-scale projects at the household or community level that would be more suited to local expertise, and less intensive in terms of capital and knowledge requirements.

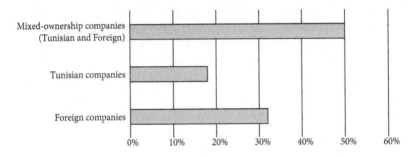

Figure 9.1 Percentage of projects obtained by companies in tenders between 2017 and 2019 under the concession regime according to nationality.

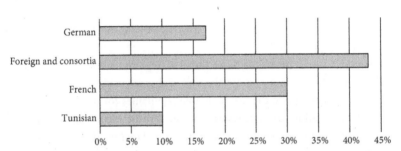

Figure 9.2 Share (in percentage) in the total power (MW) of electricity production projects obtained through the authorisation regime by companies according to their nationality.

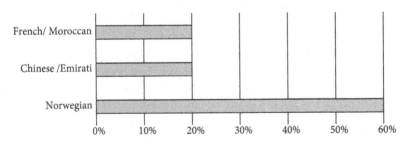

Figure 9.3 Share (in percentage) in the total power (MW) of electricity production projects obtained through the concession regime by companies according to their nationality.

Challenges to promoting local development and to reducing regional inequalities

In order to ensure that the development of renewable energy in Tunisia is beneficial for the local economy, the 2015 law was followed by several other laws and decrees. These have included laws to create an incentives framework for investments in renewable energies. Examples are Law No. 2016-71 of September 30, 2016 concerning investments in the field of renewable energies, and the subsequent government decree No. 2017-389 of March 9, 2017 concerning financial incentives to encourage targeted regional development and local employment generation through renewable energy projects. These laws and decrees also create tax benefits to encourage companies to invest in marginalized regions[49] and reinvest part of their profits.[50] However, several developers and investors have faced difficulties finding funding and have pointed out some regulatory and bureaucratic challenges to participating in calls for tenders. This is also linked to the plurality of institutions involved.[51]

Fiscal and financial incentives seek to bring development to marginalized regions, where most of the renewable energy projects are located.[52] However, the effective development of those targeted regions needs to be assessed, including by taking into account the risk of dispossession of communities. Indeed, when analysing the list of eligible companies for the installation of solar panels within the framework of the Prosol Elec's self-production project,[53] the companies based in more developed regions stand out. Indeed, out of 380 Tunisian firms, only 40 are based in the targeted regions,[54] with most of the companies being based in the Tunis and Sfax areas.[55] This means that the more developed regions are reaping most of the benefits of the development of this sector by accumulating more profit and generating more employment, at the expense of other regions that need this the most.

To make an accurate estimation of job creation, both direct and indirect employment must be taken into account. In the case of renewable energy projects, direct jobs cover activities in the areas of energy production, installation and construction, and maintenance, while indirect jobs include sales, engineering and research, training, etc. The forecast of employment creation in the field of renewable energies is around 3,000 jobs per 1,000 MW produced annually with solar PV energy. The number of additional jobs for the whole renewable energy sector in Tunisia is estimated to be between 7,000 and 20,000.[56]

As most of these jobs are needed only for the construction and commis-sioning phases of projects, they are not permanent and last only a few years, with an average of five temporary jobs at this stage for 1 MW of renewable energy. The projects' maintenance requires very few employees, at an average that drops to two permanent jobs per one megawatt.[57] Therefore, large-scale PV and wind energy projects may not be best suited to providing numerous long-term employment opportunities. In addition, job creation has to include the stimulation of all branches of the sector. In this respect, the local production of the technologies required for renewable energy projects would offer strong potential for new job creation, since low dependency on imports means more employment.[58]

Accordingly, despite an official focus on marginalized regions and local employment generation, there is a potential risk that the current framework might end up seizing land in the least developed areas in order to exploit their renewable resources, without proper compensation for local commu-nities, thus maintaining an internal colonial dynamic.[59]

Local communities' social and environmental rights: emerging concerns in the light of the Borj Salhi village mobilization

The government's 2018 strategic vision for the energy sector promoted fair energy distribution and good governance by securing fair access to energy in every region under the best conditions, developing a social responsibility policy, creating a regulatory authority and implementing a transparent process.[60]

Additionally, under the 2015-12 law, the first step required for an electric-ity production project using renewable energy, within the framework of the authorization regime, is conducting a feasibility study. This study must include environmental and social impact assessments. The environmental impact assessment must be carried out by a consulting firm and must include at least a basic description of the initial condition of the site, the characteri-zation of the site and a description of neighbouring areas, an estimation of the future impact of the project on local flora and fauna, and an estimation of the visual and acoustic impact.[61] However, despite this framework, social and environmental measures seem not to be always followed.

In 2000, the first wind turbine project in Tunisia was established around 70 kilometres from Tunis, in the northeast of the country, and was followed by other phases of installation in 2003 and 2009. This renewable energy power plant, which includes around 40 wind turbines, is providing electricity to 50,000 Tunisians. However, in Borj Salhi, the village in which the 2009 extension was implemented, villagers currently do not benefit from a connection to the high-voltage grid and cannot access STEG electricity meters, and their deteriorated grid experiences frequent outages. For more than a decade the nearby village community has denounced this power plant project, which is owned by STEG. Indeed, the 2009 extension provoked a social mobilization of the villagers living near the plant. The proximity to the wind turbines is one of the first reasons for their discontent: the closest turbine is located less than 50 metres away from one residence, causing discomfort for villagers, especially because of the consequences of the constant noise for their health and animals. From an environmental impact perspective, the landscape modifications have led to soil erosion and a dieback of olive trees. Among other issues raised by the villagers is the lack of maintenance of the wind turbines by STEG, which leads to technical accidents.

At the heart of the discontent is the lack of inclusive decision making, which could have taken into consideration the consequences for local land and inhabitants, and ensured the local people's ownership of the project. After a meeting in March 2021 between the villagers and STEG, the latter announced itself "ready to assume [its] full responsibility and to end this ten-year conflict". However, the case remains open as no further action has been taken by STEG.[62]

The example of the village of Borj Salhi demonstrates that public awareness, local communities' participation and rights, as well as environmental sustainability are not yet guaranteed in the face of other interests. The impacts on the social and environmental rights of local communities should be closely monitored, both on paper and in practice, in future projects.

CONCLUSION

This overview of developments in Tunisia's renewable energy sector highlights several incompatibilities with a just transition model. First, we identified limits in terms of democratic decision making, because of the influence of a number of foreign actors and a lack of consultation with other stakeholders, such as the UGTT or local communities. This led to the introduction of the 2015 law, which encourages privatization, including in the form of PPPs, placing Tunisia firmly within the global neoliberal scheme regarding the development of renewable energy. This has opened the door to neocolonial initiatives, such as Desertec and TuNur, which prevent government control over renewable energy projects and, by extension, impede Tunisians' sovereignty over their own resources.

Moreover, this path reinforces financial and knowledge dependency on foreign actors through foreign direct investments and technology imports, instead of Tunisia investing in energy sovereignty through local development of the renewable energy sector. This means that the current short term strategy adopted by successive governments over the last decade, by opting to invest in PPPs rather than public services, has focused more on attracting private, and especially foreign, investors and securing their profits rather than on promoting local development, regardless of the long term financial burden this may entail. As a result, communities' rights are neglected, with effects ranging from inadequate access to electricity to land dispossession, specifically for people in already marginalized regions.

This framework continues to favour actors from relatively privileged regions, whereas impoverished areas are further marginalized and dispossessed of their resources. It seems that, once more, most of the dynamics are profit-driven and short term, which also explains why the provisions regarding the protection of the natural environment are insufficient. With the priority given to the realization of large projects at all costs, insufficient consideration is given to local people's needs and the environment in which these projects are being established, and not enough attention is paid to the conditions required for their economic integration into the national economy. Although there is a lack of access to information and insufficient investigation and fieldwork that could provide more insight into ownership, activities, and winners and losers, certain instances, like the Borj Salhi case, still expose significant deficiencies in the existing framework. This means that there may be other similar examples that have not yet caught the public's attention.

From a local perspective, a just transition would provide households and communities with the means to produce their own electricity, based on small scale PV projects, which would reduce capital and knowledge requirements and promote the development of job generating activities. Worldwide, many civil society actors have embarked on a phase of "remunicipalization", i.e. reclaiming public ownership of services in order to build "community led and climate smart" public services by regaining control over local resources. Privatization should, therefore, be avoided in the first place.[63] Local governments could promote the installation of small-scale PV units by local companies, to break with the current reinforcement of regional inequalities. The management of these projects at the local level would also give local communities more ownership, rights and power to control and oversee the means of production in the energy sector.

NOTES

1. "STEG in Numbers," STEG, https://tinyurl.com/4xrd63yn.
2. Tractebel Engie, "*Projets d'énergie renouvelable en Tunisie - Guide détaillé*," GIZ and ANME, *May 2019, https://tinyurl.com/k3prf6zp.*
3. "NDC Registry," United Nations Framework Convention on Climate Change, P6 NDC Tunisie. https://unfccc.int/NDCREG.
4. Daniel Chavez, Sean Sweeney and John Treat, *Energy Transition or Energy Expansion?* Transnational Institute and Trade Unions for Energy Democracy, October 2021, https://tinyurl.com/mr43sss5.
5. Chavez, Sweeney, Treat, *Energy Transition or Energy Expansion?.*
6. "Frameworks," Just Transition, Movement Generation Justice & Ecology Project, accessed March 11, 2022, https://movementgeneration.org/justtransition/.
7. Agence Nationale de Maitrise de l'Energie, *Nouvelle Version du Plan Solaire Tunisien,* September 2015, https://tinyurl.com/yc6fdy7c.
8. Sana Kacem, "La Stratégie de la Maîtrise de l'Energie et le Plan Solaire au Niveau National," Direction des Energies Renouvelables, accessed March 11, 2022, https://tinyurl.com/3cm8h9c7.
9. Concentrated solar power uses solar thermal energy to generate power by using mirrors or lenses to concentrate a large area of sunlight onto a receiver. The concentrated light is converted to heat which drives a heat engine (e.g. steam turbine) connected to an electrical power generator (source: *Wikipedia*).
10. Ezzedine Khalfallah and Néji Amaimia, "Efficacité énergétique et énergies renouvelables (Chapter 9)," *Rétrospective du secteur tunisien de l'énergie*, 2018.
11. Tractebel Engie, "*Projets d'énergie renouvelable en Tunisie - Guide détaillé*".
12. "PPP Reference Guide – Introduction," Library, World Bank, https://ppp.worldbank.org.
13. Lazhar Rachdi, *La centrale solaire de Ouarzazate: un modèle à suivre?* Tunisian Observatory of Economy, Note de Decryptage, May 18, 2016.
14. Jihen Chandoul, *Note de synthèse à propos du projet de loi sur les PPP en Tunisie*, Tunisian Observatory of Economy, policy paper, June 4, 2015.
15. Agence Nationale de Maitrise de l'Energie, *Nouvelle Version du Plan Solaire Tunisien.*
16. The self-production regime allows the residential sector, any local authority and any public or private company, connected to the national electricity network in medium voltage or high voltage (MT-HT) and operating in the industrial, agricultural or tertiary sectors, to produce electricity for their own consumption from renewable energy and to be able to sell the surpluses to STEG.
17. Republic of Tunisia, "Law No. 2015-12 of 11 May 2015," *Official Journal of the Republic of Tunisia* 38, May 12, 2015, http://www.igppp.tn/sites/default/files/Loi percent202015-12.pdf.
18. Ophélie Julien-Laferriere, *Coopérations et diplomaties économiques concurrentes: le rôle de l'Allemagne dans la nouvelle politique énergétique de la Tunisie*, Analytical note, Tunisian Observatory of Economy, January 31,2017.

19. Decree No. 2016-1123 of August 24, 2016 laying down the terms and conditions for the implementation of projects for the production and sale of electricity from renewable energy.

20. Order of the Minister of Energy, Mines and Renewable Energies of February 9, 2017, on the approval of the standard contract for the transmission of electrical energy produced from renewable energies for own consumption, connected to high and medium voltage networks and the purchase of the surplus by STEG.

21. Imen Louati, *ALECA, Production d'électricité et Energies renouvelables : Quel avenir pour la STEG et la transition énergétique en Tunisie?*, (Briefing paper 8), Tunisian Observatory of Economy, 2019.

22. Green grabbing is a concept that was coined to designate "all the activities you can see where ecosystems are being put up for sale". Source: *Green grabbing: The Social Costs of Putting a Price on Nature*, Transnational Institute, 2012, https://www.tni.org/en/article/green-grabbing (retrieved 11 March 2022).

23. Jihen Chandoul and Cécilia Gondard, *Des projets qui ne décollent pas, Défis à relever et leçons à tirer des partenariats public-privé en Tunisie*, Tunisian Observatory of Economy and Eurodad, 2019.

24. Hamza Hamouchene, "Desertec: The Renewable Energy Grab?," *New Internationalist, Mach 1, 2015*, https://newint.org/features/2015/03/01/desertec-long.

25. Megan Darby, "Giant Tunisian Desert Solar Project aims to Power EU," *Climate Home News*, 4 August 2017, https://tinyurl.com/528kpucd.

26. For instance, the programme Policies for Future.

27. Julien-Laferrière, "Coopérations et diplomaties économiques concurrentes."

28. European Commission, "Energy Union Package: Communication from the Commission to the European Parliament, the Council, the European Economic and Social Committee, the Committee of Regions and the European Investment Bank – A Framework Strategy for a Resilient Energy Union with a Forward-Looking Climate Change Policy," 2015.

29. Energy Charter Secretariat, "Report on policy on consolidation, expansion and outreach (CONEXO) for 2013," 2013, https://tinyurl.com/3baccxfd.

30. Lora Verheecke, Pia Eberhardt, Cecilia Olivet, Sam Cossar-Gilbert, *Red Carpet Courts: 10 Stories of how the Rich and Powerful Hijacked Justice*, Transnational Institute, Friends of the Earth International, Friends of the Earth Europe and Corporate Europe Observatory, 2019, https://www.tni.org/en/redcarpetcourts; Lavinia Steinfort, Eberhardt and Olivet, *One Treaty to Rule Them All: The Ever-Expanding Energy Charter Treaty and the Power it gives Corporations to Halt the Energy Transition*, Transnational Institute and CEO, 2018, https://www.tni.org/en/energy-charter-dirty-secrets.

31. Louati, *ALECA, Production d'électricité et Energies renouvelables*.

32. Laila Riahi and Hamza Hamouchene, *Deep and Comprehensive Dependency: How a Trade Agreement with the EU could Devastate the Tunisian Economy*, Tunisian Platform of Alternatives and Transnational Institute, 2020.

33. Nour El Houda Chaabane, "Tunisie: Enjeux énergétiques, l'éolienne entre besoins et appréhensions," *Nawaat*, September 12, 2014, accessed March 11, 2022, https://tinyurl.com/4k95san6.

34. Chaabane, "Tunisie: enjeux énergétiques."

35. Republic of Tunisia, *Accélération des projets de production d'électricité à partir des énergies renouvelables*, Ministère de l'Energie, des Mines et des Energies Renouvelables, ANME and PNUD, 2018, https://tinyurl.com/4sf28fzj.

36. "Projets de la STEG," Ministry of Industry, Mines and Energy, https://tinyurl.com/ywemewww.

37. "Les essais de mise en service de la centrale de Tozeur ont commencé," *WebManagerCenter*, November 5, 2021, https://www.webmanagercenter.com/wmc-avec-tap/.

38. "Tunisie: Le projet de raccordement de la centrale électrique de Tataouine est saboté (Mongi Marzouk)," *Trustex*, July 29, 2020, https://tinyurl.com/24k5ywak.

39. Sarra Abdou, " La production d'hydrogène pour libérer les développeurs d'énergies vertes en Tunisie," *WebManagerCenter*, October 21, 2021, https://tinyurl.com/2mjh4kcs.

40. Mohamed Khalil Jelassi, "Exploitation des énergies renouvelables: pourquoi ça traîne encore?," *La Presse*, July 29, 2020, https://tinyurl.com/29snxa5c; "Centrale solaire de Tataouine: Marzouk 'étonné' de l'opposition face au projet," July 28, 2020, https://tinyurl.com/2p835kbp; "Le ministre de l'Energie fustige un 'sabotage' du projet de raccordement de la centrale électrique de Tataouine," *African Manager*, July 28, 2020, https://tinyurl.com/yckx6n5a; "Tunisie: grève générale dans le secteur de l'électricité," *Directinfo*, March 30, 2018, https://tinyurl.com/8wumbtue; "Tunisie: loi sur la production d'électricité à partir des énergies renouvelable," *Directinfo*, January 4, 2014, https://tinyurl.com/yw7fbdyj.

41. Nafâa Baccari-Anme, "État des lieux et Perspectives du Développement de l'Éolien en Tunisie," *Agence Nationale de Maîtrise de l'Energie*, March 11, 2020, https://tinyurl.com/mwxc4pz4.

42. "Wind Towers," SOCOMENIN, https://www.socomenin.com.tn/eoliennes/.

43. Tractebel, "*Projets d'énergie renouvelable en Tunisie - Guide détaillé*".

44. *Ibid.*

45. Julien-Laferrière, "Coopérations et diplomaties économiques concurrentes."

46. "Cadre réglementaire pour l'acquisition de l'énergie solaire en Tunisie : Sommaire pour les communes," United States Agency for International Development and Ministry of Local Affairs and the Environment, April 2020, https://tinyurl.com/bdf4398n.

47. "Énergies renouvelables," Ministry of Industry, Mines and Energy, https://tinyurl.com/mpaevn9e.

48. *Ibid.*

49. "Regional Development Zones," Specific incentives, the Agency for the Promotion of Industry and Innovation and the Ministry of Industry, Mines and Energy, http://www.tunisieindustrie.nat.tn/fr/dr.asp.

50. "Cadre Incitatif," STEG Energies, http://www.steg-er.com.tn/cadre-incitatif/index.html.

51. Khalfallah and Amaimia, "Efficacité énergétique et énergies renouvelables," p. 449.

52. "Énergies renouvelables," Ministry of Industry, Mines and Energy, https://tinyurl.com/mpaevn9e.

53. Production of electricity for their own consumption from PV solar energy by the residential sector and establishments and groups operating in the industrial, agricultural or tertiary sectors, while benefiting from the right of sale of surpluses of electrical energy produced to STEG.
54. Jendouba, Beja, Kasserine, Gafsa, Tozeur, Kebili, Tataouine, Gabes, Kairouan, Sidi Bouzid, Kef.
55. ANME, Projet Prosol Elec-Liste des sociétés installatrices éligibles, March 18, 2020, https://tinyurl.com/2wp8um2t.
56. Dr. Isabel Schäfer, "The Renewable Energy Sector and Youth Employment in Algeria, Libya, Morocco and Tunisia," AfDB, https://tinyurl.com/2s7yx8ff.
57. According to the results of a 2019 call for projects, the Ministry of Industry, Mines and Energy, within the framework of authorizations.
58. Schäfer, "The Renewable Energy Sector and Youth Employment in Algeria, Libya, Morocco and Tunisia".
59. Malek Lakhal, "Interview avec Sghaier Salhi : Les non-dits de la Tunisie postindépendance," Nawaat, April 5, 2018, https://tinyurl.com/3snnhadw.
60. Khalfallah and Amaimia, "Efficacité énergétique et énergies renouvelables," pp. 448-49.
61. Tractebel, Projets d'énergie renouvelable en Tunisie - Guide détaillé.
62. Aida Delpuech and Arianna Poletti, "Projets d'énergie renouvelable en Tunisie - Guide détaillé," Inkyfada, April 20, 2021, https://inkyfada.com/en/2021/04/20/wind-turbines-cap-bon-tunisia/; Forum Tunsien pour les Droits Economiques et Sociaux, Revue semestrielle de la justice environnementale. Droits, responsabilité sociétale, souveraineté alimentaire et développement durable. Troisième partie, July 2021, https://ftdes.net/rapport-semestriel/.
63. Satoko Kishimoto, Lavinia Steinfort and Olivier Petitjean, The Future is Public: Towards Democratic Ownership of Public Services, Transnational Institute and others, May 12, 2020, https://www.tni.org/en/futureispublic; Reclaiming Public Services: How Cities and Citizens are Turning Back Privatization, June 23, 2017, https://www.tni.org/en/publication/reclaiming-public-services.

10

The Moroccan Energy Sector: A Permanent Dependence

Jawad Moustakbal

Although Morocco gained its independence in 1956, the country's energy sector has remained dependent on fossil fuels and on the private sector. Between 2017 and 2020, fossil fuel imports accounted for around 90 percent of the total primary energy supply and 80 percent of the electricity supply,[1] while the private sector currently controls 84 percent of electricity production[2] and almost all energy distribution.

The ambitious renewable energy plan launched by the Moroccan government in 2009, which aims to cover 52 percent of installed electricity capacity by 2030, could have significantly reduced the country's dependence on the largely imported fossil fuels. However, the liberal policies adopted by the government for the entire energy sector, including renewables, as well as the associated public-private partnerships (PPPs), have exacerbated both the debt crisis and the dependence on the private sector.

This chapter aims to explore the historical, economic and political roots of Morocco's energy dependency, which threatens what remains of Moroccan sovereignty and aggravates social inequalities. It also shows how in this situation, it is the most deprived populations in the country that pay for the political and economic choices made by a colonized elite,[3] which serves as the partner of transnational corporations and international banks.

THE ENERGY SECTOR: FROM COLONIAL CONTROL
TO NEOLIBERAL MEASURES

At the beginning of the twentieth century, the main motivation of French colonialism with regard to electricity production in Morocco was to facilitate the extraction of raw materials, mainly phosphates, to improve mine productivity and to electrify the railway network that was used to transport raw materials to the metropole.[4] The aim was to electrify what was considered to be "the useful Morocco".[5] Concessions were granted for the production and distribution of electricity. It was replaced in 1924 by *Énergie Électrique du Maroc* (EEM), created on January 30 of that year by the *Compagnie Générale* du Maroc, itself created in February 1912 by a consortium of French banks led by *Banque de Paris et des Pays-Bas*.[6]

Despite Morocco's independence in 1956, the management of the energy sector, as with other strategic sectors, such as agriculture, industry, trade and drinking water supply, only came under the control of the Moroccan state in 1963 with the creation of the *Office National de l'Électricité* (ONE). According to its founding decree, this public body is responsible for "public service, production, transport and distribution of electrical energy".[7]

Throughout the 1960s and 1970s, Morocco chose oil as its basic primary energy resource despite the fact that it did not have any: oil accounted for more than 80 percent of the energy mix in 1980.[8] From the mid-1980s, and following the 1973 oil crisis, ONE decided to increase the share of coal in the country's energy mix.

In the mid 1990s, despite ONE's positive track record in extending the electricity network to rural areas and in providing good quality public service, evidenced in part by the absence of blackouts in major cities,[9] the government decided to adopt the neoliberal paradigm with regard to the energy sector. Prompted by international financial institutions, it began to dismantle, privatize and liberalize the distribution and production of electricity, to the benefit of large transnational corporations.

Morocco began to privatize its most profitable public enterprises and liberalize strategic sectors of the economy as part of the Structural Adjustment Programme (SAP) imposed by the international financial institutions (IFIs) following the debt crisis of the 1980s, when the country was no longer able to repay its debts and requested debt rescheduling. In 1983, the International Monetary Fund (IMF) and the World Bank required the implementation of the SAP.[10] The energy sector was among the first to be affected, with the privatization of the oil refining industry and the introduction of private production in the oil business. The *Société Anonyme Marocaine de l'Industrie du Raffinage* (SAMIR) was privatized in 1997 to the benefit of the Swedish-Saudi group Corral Petroleum Holding, mainly owned by Saudi billionaire, Mohammed al-Amoudi. In the same year, drinking water and electricity distribution services, rainwater and wastewater collection, and public lighting in the Greater Casablanca region, with 4.2 million inhabitants, were assigned to a single operator, *Lyonnaise des Eaux Casablanca* (Lydec), a subsidiary of the French company *Lyonnaise des eaux*.[11] A large array of legislation and propaganda supported this first wave of privatizations, which was portrayed as indispensable for the "modernization" of the Moroccan economy and to benefit from the technical support of transnational corporations. However, the economic, social and ecological impacts of these privatization operations proved to be disastrous.

Taking again the examples of SAMIR and Lydec, the privatization of the former led to the biggest bankruptcy in the country's history,[12] with a debt of four billion euros and over 800 workers and their families left destitute. Various reports, including official ones, such as the 2014 Court of Auditors report, reveal numerous violations of fundamental rights committed by the concession holder (Lydec) with the collusion and/or silence of local authorities and elected representatives. These abuses included depriving people of their right to electricity and water connections and increasing the cost of these services, contrary to the provisions of the concession contract. In economic and financial terms, Lydec failed to fulfil the investment programme agreed in the contract and also transferred money in foreign currency in violation of the contract, with 160 million euros being transferred to shareholders in the form of dividends and 100 million euros being transferred to its headquarters as hidden profits in the form of expenses for "technical assistance" during the first decade of the concession.[13]

From the legislative perspective, two major acts, passed during that decade, deeply impacted the energy sector, alongside Law No. 39-89, allowing the transfer of public enterprises to the private sector. These were Decree No. 94-503 of 1994, which ended ONE's monopoly and allowed private electricity producers to enter the market, and Law No. 1-95-141 of 1995, which allowed the liberalization of the petroleum products market.

RENEWABLE ENERGY IN MOROCCO:
A "GREEN" NEOLIBERALISM

While Morocco has an ambitious programme and billions of Moroccan dirhams invested in the development of renewable energy, particularly solar, and hosts one of the world's largest Concentrated Solar Power (CSP) plants,[14] the country's energy mix in 2011 remained dominated by fossil fuels, which accounted for 92.36 percent of the mix, mainly for transport, while coal constituted up to 57.82 percent of electricity production.[15] Moreover, the transport sector was the largest energy consumer in Morocco, accounting for 38 percent of the country's total consumption.[16] Almost entirely dependent on fossil fuels, in 2018 it was responsible for about 50 percent of the national energy budget, i.e. more than four billion euros, and represented 20 percent of the trade balance deficit.

In terms of electricity generation, renewables accounted for 19.81 percent of national production in 2021, with 12.37 percent coming from wind, 2.93 percent from hydro and 4.41 percent from solar. Coal was the main source of electricity (57.82 percent), followed by natural gas (eleven percent).[17]

A critical analysis of the main legislative and institutional reforms supporting the development of renewable energies in Morocco raises the question of whether these reforms have not mainly served as an excuse to further liberalize and privatize the energy sector. Indeed, Law No. 13-09 of 11 February, 2010, liberalized the renewable energy sector by allowing private companies to compete in both the production and the export of renewable electricity through the national grid.[18]

The law on PPPs came into force in August 2015, although the PPP model had long been tested through contractual forms, such as concessions outside any normative framework before the law enshrined this orientation,[19] allowing private operators to position themselves as independent power producers. These power purchase agreements (PPAs) made under PPPs oblige the state-run ONE to buy the electricity produced at an agreed price for a period of 25 to 30 years.[20]

This model, and the 2015 law that followed, derive from the French PPP law of 2004. They notably take up the concept of "availability-based payments", whereby ONE, a public institution, is obliged to buy the entire output that private concessionaires produce, regardless of actual needs. That energy, whether of fossil or renewable origin, therefore takes priority over that of public power stations.[21] In the event of a sharp drop in electricity demand, since ONE is forced to consume private concessionaires' production first, public power plants are shut down to avoid a blackout, which entails considerable extra costs for the state (i.e. taxpayers).[22]

This type of partnership thus constitutes a swindle that benefits banks and private operators. On the one hand, these are protected from any potential loss, regardless of its nature (fluctuating commodity prices, infrastructure, public service provision, climatic risks, financial risks, etc.); on the other, the profitability of their investments is fully secured, as payment is guaranteed even if energy is not needed or used. This is a typical model whereby profits are privatized and losses and risks are borne by taxpayers.

ENERGY GOVERNANCE IN MOROCCO

Who decides?

Morocco's energy sector is governed largely by autocratic instruments, with strategic decisions made beyond democratic control. The creation of the Moroccan Agency for Sustainable Energy (Masen) in 2010 and the appointment of Mustapha Bakkoury, a former president of the Authenticity and Modernity Party (PAM), as its director are cases in point. The PAM was founded by Fouad Ali El Himma, a friend and advisor to the king. In 2015, Masen was given authority over the entire renewable energy sector, with Bakkoury as chief executive officer, effectively marginalizing ONE.[23] However, in March 2021, Bakkoury was unexpectedly banned from leaving the country,[24] as part of an investigation into allegations of mismanagement and embezzlement during his time at Masen. Despite media coverage of the case, no official explanation was given at the time.[25]

Local communities and parliamentarians, as well as engineers and technicians from public companies in the fields of production, management, transport and the maintenance of energy facilities, have always been marginalized from all discussions regarding Masen's renewable energy projects. Consultation with them would have helped avoid major technical errors and made it possible to better monitor the private "partners" who, on their side, were surrounded by experts defending their interests. A specialist in the sector who requested to remain anonymous has stated:

> "Since renewable energies have become a strategic sector, the agency [Masen] has taken over all of the sustainable development prerogatives. It has become all powerful. As in any big royal project, silence prevails: everyone knew that the projects were behind schedule and costing too much, but no one dared to ask for accountability".[26]

Who benefits from it?

In 2018, local citizens ran an unprecedented boycott campaign against three companies whose owners are known to be closely linked to the royal family: Danone, Sidi Ali and above all Afriquia. The latter is owned by powerful billionaire Aziz Akhannouch, appointed head of government by the king in September 2021. Following this civil disobedience action, in 2019 the *Conseil de la Concurrence* (Competition Council) carried out an in-depth study of the oil sector which found evidence of malpractice. The report found that instead of promoting competition – its advocates' main justification – the liberalization of the sector in 2014 had created an oligopoly at all levels: from import, to storage and sale, and to distribution and consumption. With Afriquia leading the way, the top five operators captured 70 percent of the market in 2017, three of which held a 53.4 percent share.[27]

This oligopolistic situation increased with the closure of SAMIR in 2015, even though the company had provided 64 percent of the demand for refined products and had a large storage capacity (two million cubic metres). "The energy bill has thus risen sharply, the trade balance deficit has worsened and small and medium-sized structures have been weakened to the benefit of the largest players."[28]

Full private sector control of electricity

According to official government data,[29] while the target of generating 42 percent of electricity from renewable sources by 2020 was not met, the target of increasing the share of private concessions in electricity production was exceeded. At the end of 2021, the private sector controlled more than two thirds (71.8 percent) of electric power production in Morocco.

The ruling elite has made private concessionary electricity production, whether of fossil or renewable origin, a fundamental tenet and element of the energy system. This benefits first and foremost French (Engie), Spanish (Gamesa), Saudi (ACWA), Emirati (Taqa) and German (Siemens) transnational corporations, usually in cooperation with national companies owned by the royal family (Nareva) or by powerful and politically connected families, such as the Akhannouch and Benjelloun families (Green of Africa). An example of this is the international solar energy tender for the design, financing, construction, operation and maintenance of the Noor Midelt I project of 800 MW, which was awarded in May 2021 to a consortium led by *EDF Renouvelables* (France) and including Masdar (United Arab Emirates) and Green of Africa (Morocco).[30] It is worth noting that Green of Africa is owned by three of the richest families in Morocco: Benjelloun (Financecom and BMCE group), Amhal (Omafu and Somepi group) and Akhannouch (Akwa Group). Before being appointed head of government by the king in September 2021, Aziz Akhannouch had held the post of Minister of Agriculture and Fisheries for over 15 years.

In the field of wind energy, Nareva, a company belonging to the royal group Al Mada[31] is taking the lion's share via its subsidiary *Énergie Éolienne*

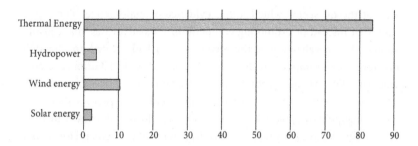

Figure 10.1 Distribution of electricity production by source of energy (in percentage). Source: 'Energy Sector – Key Figures – April 2021', Ministry of Energy, Mines, Water & Environment.

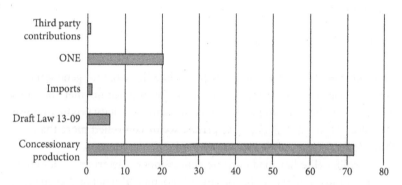

Figure 10.2 Distribution of electricity production by type of producer (in percentage). Source: 'Energy Sector – Key Figures – April 2021', Ministry of Energy, Mines, Water & Environment.

du Maroc (EEM). It currently owns five "merchant plant" wind farms under Law No. 13-09, with a total capacity of over 500 MW, the electrical energy from which is sold directly to industrial customers.[32] Nareva also owns the Tarfaya park, one of the largest in Africa, in a joint venture with the large French company Engie. The energy produced in this 300 MW park is sold exclusively to ONE under a 20-year electricity purchase and supply contract.[33] In 2016, Nareva was declared the successful bidder for the huge 850 MW Integrated Wind Project, consisting of Midelt (210 MW), Boujdour (300 MW), Jbel Lahdid (270 MW), and Tiskrad, in Tarfaya, (100 MW). Nareva won this project by partnering with wind turbine manufacturer Siemens Gamesa Renewables (Germany and Spain).

It should be stressed that while the Al Mada group presents itself as a leader in the field of sustainable development, it is responsible for the destruction and the pollution of several ecosystems. As the author has previously explained: "Not only has its sugar producing company Cosumar been involved in pollution disasters, but its mining branch Managem in its 'Imider' silver mine, located in the south of Morocco, has seen the contamination of aquifers, and there is still an ongoing conflict with the local population over water resources."[34]

In Morocco, among other countries, those who benefit from green projects generally have a long track record of polluting and destroying ecosystems. Reorienting part of their investments towards renewable energy is in reality just another, often even more profitable, way of generating profits and dispossessing local populations of their territories.

Who pays the price?

Both as taxpayers and as consumers, the population bears the financial consequences of a system that is designed to be totally inequitable and to benefit exclusively private investors. The concession contracts signed in the 1990s and early 2000s, in particular the PPAs, obliged ONE not only to buy energy from private operators according to availability and at prices that are higher than the effective selling prices for distribution and consumption, but also to bear the cost of fluctuations in the prices of raw materials, in particular coal.

This plunged ONE into an unprecedented structural financial crisis, with the government then bailing it out, through the signing of a contract that allowed ONE to increase consumer prices. As a result, consumer bills rose by 20 percent in 2014.[35] As recent renewable projects are all based on similar 30-year contracts, this situation of massive public investment without any guarantee of lower electricity prices for the population is likely to be repeated.

Masen's decision to use CSP technology, which was made without consulting any public body, including ONE, has proven disastrous, with a cost price per kilowatt-hour (KWh) of 1.62 dirham for Noor I, 1.38 dirham for Noor II and 1.42 dirham for Noor III, while each kWh is sold to ONE at 0.85 dirham. Masen thus records an annual deficit that is estimated by the Economic, Social and Environmental Council (CESE)[36] at 80 million euros for the Noor I, II and III plants.

The issue of debt and financing is fundamental. All recent power generation projects, including the so-called "green" projects, are financed by loans from international private banks, multilateral banks, the IMF, the World Bank, the AfDB, and French, German and Japanese development agencies.

In the solar energy sector, Masen's debts are guaranteed by the state. It uses these funds to develop the infrastructure needed for project development, such as roads, hydraulic infrastructure, fences, lines and transformer stations for the transport of energy. Furthermore, it uses the funds to finance its participation in the special purpose companies created for specific projects (Noor Ourzaztae, Noor Midelt, etc.).[37]

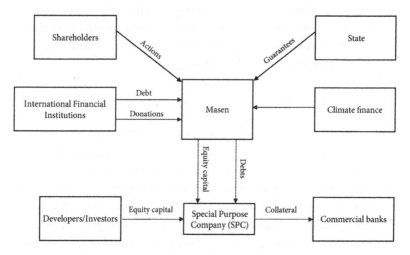

Figure 10.3 Typical financial set-up of Masen-led projects. Source: Masen promotional presentation, 2016.

Lenders remain the key players in these projects and have the final say on all strategic decisions. Consequently, it is only logical that the lenders' nationalities correspond to that of the companies involved in the project, whether as operators (French Engie, German Siemens, etc.) or as equipment suppliers (French Alstom, Japanese Mitswi, etc.).

Safi Energy Company, co-owned by Nareva (35 percent), France's Engie (35 percent) and the Japanese trading house Mitsui (30 percent) has been granted a 30-year concession for the Safi thermal power plant project, with a capacity of 1,369 MW (i.e. 25 percent of national demand) and a total investment of 2.3 billion euros.[38] The project was financed mainly by the Japan Bank for International Cooperation, Attijariwafa Bank and *Banque Marocaine pour le Commerce et l'Industrie* (BMCI), the Moroccan subsidiary of the French bank BNP Paribas. These loans will add to a public debt that approached 100 percent of GDP by the end of 2021.[39] Debt service absorbs more than a third of the state budget and amounts to almost 10 times the national health budget.[40]

SOME AVENUES FOR A JUST ENERGY TRANSITION IN MOROCCO

The liberal model has failed both economically and ecologically, particularly with regard to energy and climate justice. Governmental reports, including that of CESE, partially acknowledge this failure, while continuing to advocate for more liberalization, dismantling of state enterprises and privatization of the whole energy sector.

There will be no just transition as long as the energy sector remains under the control of foreign transnational companies and a local ruling elite that is free to plunder the state and generate as much profit as it wishes, within a culture of authoritarianism and nepotism. The debt system and PPPs are a major obstacle to any model of popular sovereignty, including energy sovereignty.

A just energy transition requires the local population's sovereignty at every stage of the production process: design, implementation, operation, storage and distribution. The energy sector must be considered a public service, co-managed by the workers involved and the local populations who agree to share part of their territories (land, water, forests, etc.) for the collective interest. In this framework, local populations should also benefit from preferential tariffs, if not wholly free electricity services. The current bureaucratic forms of government must be replaced by local and decentralized forms of governance.

Prioritizing decentralized solutions and projects also means bringing energy production as close as possible to users, in order to avoid "Joule-effect" losses[41] and to limit transport costs. This also means designing smaller-scale projects for which the necessary funds can be obtained locally or with the help of the state.

Regional integration schemes must also emerge based on the principles of solidarity and the common good. Such integration would be much more suited to the physical nature of electricity: the necessary balance between production and consumption means that the holders of surplus energy have as much interest in giving it up as those who need it have in receiving it, in order to avoid a general shutdown of the electricity supply.

In response to the neoliberal tyranny and the current imbalance of power which benefits the ruling classes, various forms of discontent and protest by local populations are rising. These seek to build alternatives to private profit obtained at the expense of the community and to neocolonialism, of which privatization is both an instrument and a symbol. If we genuinely want to build a fairer and more democratic society, it is essential that we heed these initiatives, support them and link energy transition to socioeconomic issues.

NOTES

1. International Energy Agency. *Energy Policies Beyond IEA Countries – Morocco*, 2019, https://tinyurl.com/53tk7pre.
2. Economic, Social and Environmental Council. *Accélérer la transition énergétique pour installer le Maroc dans la croissance verte.* 2020, www.cese.ma.
3. An elite that has internalized Western superiority and its contempt for indigenous cultures, and therefore for its own culture.
4. Saul Samir. "L'électrification du Maroc à l'époque du protectorat," *Outre-mers* 89, no. 334-335 (1st semestre 2002), pp. 491-512, https://doi.org/10.3406/outre.2002.3952.
5. "Les représentations politiques de la montagne au Maroc." *Revue de géographie alpine* 89, No. 2 (2001), pp. 141-44, https://doi.org/10.3406/rga.2001.4637.
6. Adam Barbe, "*Dette publique et impérialisme au Maroc (1856-1956)*" (La Croisée des chemins, Casablanca, 2020), p. 220.
7. Dahir (decree) No. 1-63-226 of August, 5 1963 concerning the creation of the *Office National de l'Électricité.*
8. Mohamed Tawfik Mouline, "*La sécurité énergétique du Maroc: état des lieux et perspectives,*" Royal Institute for Strategic Studies, Energy Research Institute, conference, Beijing, March 6, 2012, https://tinyurl.com/2r4e2f9z.
9. Office National de l'Électricité, "L'électrification rurale au Maroc : une expérience à partager," http://www.one.org.ma/FR/doc/cier/Brochure_PERG.pdf.
10. Najib Akesbi, "Programme 'd'ajustement structurel' du FMI," Africa Development/Afrique et Développement 10, No. 1/2 (1985), pp. 101-21, http://www.jstor.org/stable/24487208.
11. In 1997, the Lyonnaise des Eaux merged with Suez Environnement, which became Suez SA in July 2015.
12. El Mehdi Berrada, "Samir: retour sur la plus grosse faillite de l'histoire du Maroc," *Jeune Afrique*, May 25, 2020.
13. The Court of Audit, "Delegated Management of Local Public Services: Summary Report" (October 2014), https://tinyurl.com/bdcwjyxn.
14. CSP is a solar energy technology where electricity is generated when the concentrated light (by mirrors on a receiver) is converted to heat, which drives a turbo alternator (usually a steam turbine).
15. Our World in Data (2021). Morocco: Energy Country Profile, https://ourworldindata.org/energy/country/morocco.
16. "Transport," Moroccan Agency for Energy Efficiency, https://www.amee.ma/en/node/119.
17. Our World in Data (2021).
18. "Publications," Department of Economic Studies and Financial Forecast, Ministry of Economy and Finance, 2014, http://depf.finances.gov.ma/etudes-et-publications/.

19. Zineb Sitri, "Partenariats public-privé au Maroc : soubassement juridique d'un mode de gouvernance alternatif", *Les Études et Essais du Centre Jacques Berque* 26, 2015, http://www.cjb.ma/.

20. These are energy sales contracts between the electricity producer and the state energy distributor. In the case of Morocco, it is ONE that commits to buy its energy over a defined period. This type of contract is required by private producers and donors in order to secure their income for the duration of the contract and to protect themselves from potential price fluctuations or/and a drop in energy demand.

21. Information obtained directly from ONE officials during a visit to the Mohammedia thermal power plant in spring 2017.

22. Each time a thermal power plant is shut down, restarting it is expensive, first because it takes a long time to heat up and requires a large amount of fuel, and, secondly, in terms of maintenance, as the equipment's lifespan is altered by stopstart cycles.

23. Bmourahib, "Masen ou la montée en puissance de Mustapha Bakkoury," *Telquel*, December 28, 2015, https://tinyurl.com/mrymn28j.

24. Kenza Filali, "Comment Mustapha Bakkoury s'est brûlé les ailes à Masen," *Le Desk*, March 30, 2021, https://tinyurl.com/38cuwssv.

25. Estelle Maussion, "Maroc : que cache la disgrâce de Mustapha Bakkoury?" *Jeune Afrique*, April 2, 2021, https://tinyurl.com/2mcrkejx.

26. Ghalia Kadiri, "Au Maroc, les ratés de la stratégie solaire", *Le Monde*, May 6, 2021, https://tinyurl.com/bddwfuvr.

27. Economic, Social and Environmental Council, 2020.

28. *Ibid.*

29. Ministry of Energy, Mines, Water and Environment, "Energy Sector – Key Figures," April 2021.

30. Masen, "Masen Announces the Winning Bidder for the Noor Midelt I Solar Project", press release, May 21, 2019, https://tinyurl.com/yxfrypd2.

31. Formerly Société nationale d'investissement (SNI) and Omnium Nord-Africain (ONA).

32. Fahd Iraqi, "Maroc : comment Nareva s'est imposée dans le secteur de l'énergie," *Jeune Afrique*, June 13, 2018, https://tinyurl.com/3vre35hx.

33. Nareva, "Our Assets and Projects," www.nareva.ma/en/project/wind-farm-tarfaya.

34. Jawad Moustakbal, "On the Perspective of Ruling Classes and the Elite in Morocco on Global Environmental Issues," *CADTM*, October 12, 2016, https://www.cadtm.org/On-the-perspective-of-ruling.

35. Yassine Majdi, "Le prix de l'électricité augmente à partir du mois d'août," *Telquel*, July 24, 2014, https://tinyurl.com/yhhccy5j.

36. Economic, Social and Environmental Council, 2020.

37. "Documents", Masen presentation, African Union, 2016, https://tinyurl.com/t94w6vec.

38. "Maroc: Safi Energy mobilise 2,6 milliards de dollars pour sa centrale électrique," *Jeune Afrique*, September 18, 2014, https://tinyurl.com/ybmsnb9y.

39. S. Es-Siari, "Ratio de dette publique : le Maroc frôle-t-il les 100% ?" *EcoActu*, July 19, 2021, https://tinyurl.com/yc5b28r9.

40. ATTAC&CADTM Morocco, "Maroc : Auditer la dette pour l'annuler," September 12, 2018, www.cadtm.org/Maroc-Auditer-la-dette-pour-l-annuler.

41. The Joule effect refers to the loss of energy in the transporting of electricity from point A to point B. This loss increases as the distance between the two points increases.

PART III

Fossil Capitalism and Challenges to a Just Transition

11

A Transition to Where? The Gulf Arab States and the New "East-East" Axis of World Oil[1]

Adam Hanieh

In early 2023, the world's largest privately owned oil and gas companies began to release their financial results for 2022. ExxonMobil led the way, recording the largest profit in the company's history at $55.7 billion. Shell, the UK-Dutch conglomerate, then followed, also achieving a historic milestone in its 115-year existence, with profits of nearly $44 billion, over twice the amount earned in 2021. All told, the five leading Western oil "supermajors" – ExxonMobil, Shell, Chevron, BP and TotalEnergies – reported a combined total of $200 billion in profits, an eyewatering $23 million for every hour of 2022. That same year, the ten largest climate-related disasters were estimated to have cost the world around $170 billion, including $30 billion in the devastating floods that killed over 1,700 people and displaced more than seven millions in Pakistan.[2] With around half of ExxonMobil's 2022 profits easily able to cover the costs of Pakistan's disaster, the real winners and losers of the climate emergency could not have been more starkly illustrated.

On 12 March 2023, however, these record-breaking profits were overshadowed by the release of another set of financial results: those of Saudi Arabia's national oil company, Saudi Aramco. Coming in at just over $161 billion, Aramco's 2022 profit not only exceeded the combined results of Shell, BP, ExxonMobil and Chevron, but was the largest profit recorded by any company in the world, in any business, *ever*. Aramco's results powerfully underlined a major shift that has taken place in the control of world oil over recent decades: the seemingly unstoppable rise of national oil companies (NOCs) run by governments in the Middle East, China, Russia and other large oil-producing states in the global South. Collectively, these firms have now developed into huge, diversified corporations that have overtaken the Western supermajors in key metrics, including oil production, reserves, market capitalization and export quantities. The big Western firms continue to be strongly represented in the US, Canada and Western Europe, but their overall global influence has been much diminished by the rise of the NOCs.

Given these new realities, my goal in this chapter is to examine the particular role and weight of the six Gulf Arab states (Saudi Arabia, Kuwait, United Arab Emirates, Qatar, Bahrain and Oman) in the world oil industry. Of course, as the location of some of the largest reserves in the world, the Gulf states have long been leading exporters of oil and natural gas. But for much of the twentieth century, the oil industry in the Gulf was controlled mostly by American and European oil firms, who paid royalties and other fees to the region's ruling monarchs in return for access to oil. Following the nationalization of oil throughout the 1970s and 1980s, Gulf governments assumed direct control of upstream production, with NOCs such as Saudi Aramco, the Abu Dhabi National Oil Company and the Kuwait Petroleum Corporation taking over the exploration, extraction and export of the Gulf's oil supplies. As with the earlier evolution of Western oil, these Gulf NOCs are now present in territories far beyond their national borders, with involvement in activities that stretch across the entire oil value chain. Moreover, as the COP27 and COP28 climate talks vividly demonstrate, accompanying this expansion of the Gulf's oil industry is the region's increasingly conspicuous place in international discussions about climate change.

In what follows, I argue that the rise of the Gulf needs to be understood through the significant changes that have taken place in global capitalism over the last two decades. Since the early 2000s, the emergence of China and Asia in general as the geographical centre of global commodity production has shifted how fossil fuels and their various products circulate through the world economy. Key to this is a new hydrocarbon axis linking the oil and gas reserves of the Middle East with the production networks of China and Asia.[3] This "East-East" hydrocarbon axis has been associated with a sizeable rise in the levels of wealth amassing in the Gulf, with flows of so-called "petrodollars" having a major impact on political and economic structures in the Gulf and the wider Middle East. A range of interdependencies between business and state elites in the Gulf and Asia are also developing alongside this eastward shift in the oil industry – these are not restricted to the export of crude oil but extend to "downstream" sectors, such as refining and petrochemicals. All told, this new axis of world oil serves to embed the Gulf at the core of contemporary "fossil capitalism".[4]

Climate activists need to give much greater attention to these shifts in the world oil industry and the role of the Gulf states. Aramco's extraordinary profits and the relative eclipse of Western supermajors mean that a major obstacle to ending the world's dependence on fossil fuels is now located outside the core Western markets. The dangers of ignoring these trends are indicated in the Gulf states' explicit plans to massively expand oil and gas production over the coming decade, through a series of what have been called "carbon bombs",[5] while simultaneously seizing the market opportunities presented by the new "low-carbon" technologies currently under development. Consequently, both in the Middle East and globally, the character of any "green transition" will be significantly determined by the actions and policies of these states. Without understanding the changes to the control and structure of the oil industry – and strategizing effectively around them – it will be impossible to build successful campaigns to halt and reverse the effects of anthropogenic climate change.

FROM THE SEVEN SISTERS TO OPEC

Oil did not definitively displace coal as the world's primary fossil fuel until the 1950s, but the early decades of the twentieth century were crucial in shaping the industry's later structure.[6] Stretching across the roughly 70-year period between 1870 and the eve of the Second World War, a handful of large oil companies emerged in the US and Western Europe. More than in any other comparable industry, these firms were marked by an extreme degree of vertical integration, through which crude oil was transferred internally within the same company to be refined and sold. Vertical integration enabled the largest firms to exert pressure on competitors and shift profit-making activities down the value chain, depending on price fluctuations and market demand.[7] Rapidly expanding beyond their domestic markets, these vertically integrated firms came to control a densely interlocked network of oil fields and circulatory infrastructure stretching across the globe. By the mid-twentieth century, just seven of these firms dominated virtually all the world's production and trade in oil.[8] They were dubbed the "Seven Sisters" by their industry rivals in the 1950s, and the leading oil firms that today remain at the centre of global debates around energy use and the climate transition – ExxonMobil, Chevron, BP, Royal Dutch Shell and so forth – are their direct descendants.[9]

These seven Western firms remained the controlling force in world oil well into the 1970s, but they were not themselves equally balanced. Despite the considerable international presence of major European players, such as Royal Dutch Shell (UK-Dutch) and BP (UK), the industry steadily gravitated towards a more American-centric landscape in the first half of the twentieth century. One reason for this was the presence of large oil reserves inside the US itself, which established the country as the core hub of global crude production and consumption for much of the century.[10] American oil firms also held a dominant position in the big Latin American oil-producing states, such as Venezuela. The global strength of these American oil giants reflected the broader consolidation of US power during this period, as an oil-fuelled global capitalism became increasingly synonymous with an American-centred one.

Following the Second World War, US oil companies finally broke into the main oil-rich areas of the Middle East, ending the previous stranglehold of European firms. Nonetheless, burgeoning anti-colonial and radical nationalist movements in the main oil-producing states in the Middle East and Latin America began to upset the control Western oil firms held over oil production, refining, pipelines and pricing.[11] These movements eventually culminated in the formation of the Organization of the Petroleum Exporting Countries (OPEC) in 1960, initially made up of Saudi Arabia, Venezuela, Iraq, Iran and Kuwait. At that time, the five countries constituting OPEC produced around 37 percent of world crude and most oil outside of the US. Over the following decade, the organization's membership continued to expand. Today, most leading oil producers (with the significant exceptions of the US, Russia and Canada) are OPEC members.

With the establishment of OPEC, governments across the Middle East and Latin America gradually nationalized oil resources, and state-owned oil companies took over much of the exploration and production of crude outside of the US. The largest Western firms retained their dominance in downstream refining and marketing of oil but increasingly had to contend with powerful non-Western NOCs in the upstream sectors across the main oil-producing states. Crucially, Western firms gradually lost their ability to set the *price* of oil, which increased dramatically in 1973-74 and again in 1978-1980. Rising oil prices, coupled with the changes in the structure of ownership in the oil industry, massively increased the financial surpluses (subsequently dubbed "petrodollars") flowing to oil-producing states, especially those located in the Gulf.[12] By the end of the 1970s, Western firms owned less than a third of the crude oil located outside of the US.

Reflecting the pressures of OPEC competition and a decline in oil prices from the mid-1980s, a major wave of corporate consolidation began to take place among Western oil firms. The most important example of this was the merger of the two US oil giants Exxon and Mobil in 1999, creating Exxon-

Mobil, the biggest private company in the world.[13] At the time, this was the largest industrial merger in history, surpassing an earlier oil sector deal – BP's acquisition of the American firm Amoco in 1998 – which had previously held that record. Other significant corporate consolidation at this time included Chevron's takeover of Texaco in 2001, and the merger of Conoco Inc. and Phillips Petroleum Company to create ConocoPhillips in 2002. Outside the US, the large French oil firm Total merged with Petrofina in 1999, and then later took over Elf Aquitaine to create Total SA (now known as TotalEnergies). The net result of these mergers was a reconfiguration of the Western oil industry around a handful of firms that are dominant today: ExxonMobil (US), BP (British), Royal Dutch Shell (British-Dutch), Chevron (US), Eni (Italy), TotalEnergies (France) and ConocoPhillips (US).

This wave of industrial consolidation was accompanied by other important changes in how Western oil companies functioned. As the largest privately owned firms in the world, the oil supermajors were deeply implicated in the wider turn to financialized capitalism that took place through the 1980s and 1990s, particularly in US financial markets. Of particular note was their increasing emphasis on share buybacks and the prioritization of dividend payments to shareholders – a feature of Western oil firms that has continued through to the current day.[14] With reduced access to conventional onshore oil fields (now controlled by the largest non-Western NOCs), Western oil majors moved towards environmentally risky, technologically intensive oil production in areas where oil was difficult to extract (e.g. deepwater drilling and fracking for shale resources) and continued to emphasize downstream activities, especially the production of petrochemicals. Several of the Western supermajors also sought to project themselves as "energy companies", and even began distancing themselves (misleadingly) from oil in their corporate branding.[15]

CHINA, WORLD OIL AND THE GULF'S POLITICAL ECONOMY

Beginning in the late 1990s, these structural features of the world oil industry were deeply shaken by China's opening to the world economy and subsequent positioning at the centre of global manufacturing. Fed by foreign capital flows seeking to take advantage of the country's enormous pools of cheap labour, the emergence of China as the "workshop of the world" created a boom in the global demand for energy, with the world's annual oil consumption increasing by around 30 percent between 2000 and 2019.[16] In 2000, China accounted for just six percent of world oil demand; by 2019, the country was consuming around 14 percent of the world's oil, more than anywhere else except the US. With China's manufacturing zones sitting at the core of a wider regional production network, the demand for oil and other raw materials increased significantly across Asia as a whole. By 2019, Asian oil consumption stood at close to one-third of the world's total, more than that of Europe, Russia, Africa, and Central and South America combined.[17]

Despite being one of the world's largest oil producers – ranking fifth in the world in 2010 – China's sizeable reserves were insufficient to meet the country's soaring demand. As a result, China's rise not only drove an increase in global oil consumption, but also had a considerable impact on the volume and direction of the world oil trade. Fully dependent on oil supplied from elsewhere to supplement domestic reserves, China's new position in the global economy pulled the export of oil away from the West and towards the East. By 2019, about 45 percent of all the world's oil exports were flowing to Asia – with more than half of these destined for China alone.[18] Most of these oil supplies originated in the Middle East, with the Gulf states and Iraq collectively providing almost half of China's oil imports by 2020 (up from around one-third in 2001).[19] Once again, this demand for Middle East oil was a pan-Asian trend – around 70 percent of all crude oil exports from the Middle East (primarily from the Gulf) are currently destined for Asia.

The large increase in China's oil consumption, and the oil consumption of Asia more widely, played a substantial role in driving a surge in global oil prices during the first two decades of the new millennium (although this was not the only reason for this price increase).[20] From an average monthly price of around $25/barrel in January 2000, global oil prices rose steadily over subsequent years, eventually peaking at just under $150/barrel by mid-2008. A short sharp downturn followed the global economic crash of 2008, but oil prices resumed their upward trend from January 2009, fluctuating around $100/barrel for most of the period between 2011 and mid-2014.[21] Importantly, over this period, oil sat at the centre of a broader boom in global commodity prices, including the price of metals, food and fertilizers. Much like the experience of the oil shocks of the 1970s, for poorer countries that are dependent on food and energy imports, these rising prices had profoundly negative implications.

For oil-producing states, however, this almost 14-year stretch of ever-increasing exports and rising prices was an immense boon.[22] For the Gulf states in particular, the surge in prices brought trillions of dollars of surplus capital into the region: a petrodollar bonanza that transformed the Gulf into one of the world's "new powerbrokers", according to the global consultancy firm McKinsey.[23] But these pools of surplus capital did not remain solely in the hands of the governments of the Gulf states. As has been the case historically, much of this newfound wealth was redirected towards the Gulf's private sector, helping support the accumulation of the large capitalist conglomerates that dominate the region's political economy.[24] This occurred through various mechanisms, including awarding lucrative state contracts for construction and real estate development, fostering joint ventures and partnerships between private and state firms, and state-owned banks providing generous loans to big private firms. Additionally, Gulf stock markets became an important route for domestic capital accumulation, with shares of large state-owned companies partially listed on these markets, thereby allowing wealthy citizens access to a portion of the revenues earned by these firms. The most notable example of the latter was the landmark listing of 1.5 percent of Saudi Aramco on the Riyadh stock exchange in 2019. First mooted by Saudi Crown Prince Mohammed bin Salman in 2016, this was the largest share offering in history. With the value of the company priced at just under $2 trillion, Aramco overtook Apple to become the most valuable company in the world.

Gulf petrodollars also found their way into international markets. In years past, the Gulf's surplus capital had been invested primarily in North America and Western Europe, playing a pivotal role in the development of the global financial architecture. During the oil boom of the new millennium, Western states remained an important destination of Gulf investments, but a growing portion of the Gulf's private and public wealth also targeted neighbouring Arab countries, attracted by the investment opportunities that sprung up following the adoption from the early 2000s of structural adjustment packages by many governments in the region.[25] This internationalization of Gulf capital afforded state and private conglomerates based in the Gulf a dominant position in key economic sectors throughout the Middle East, including real estate and construction, logistics, banking and finance, agribusiness, retail and infrastructure.[26]

In all these ways, Asia's voracious appetite for energy was intimately linked to the emergence of a regional economy in the Middle East centred around the tempo of capital accumulation in the Gulf.

REFINING AND PETROCHEMICALS

When thinking about these geographical shifts in world oil trade, it is essential to recognize that crude oil is a commodity that had little practical use prior to its transformation into various kinds of liquid fuels or raw materials. For this reason, when mapping the emerging patterns of control in oil, it is critical to consider the "downstream" segment of the oil industry, particularly the all-important stage of refining. For most of the twentieth century, the downstream segments of the world oil industry were almost completely run by the largest Western oil firms – indeed, it was through their control over refining and the marketing of oil that these firms managed to maintain their global dominance following the OPEC countries' nationalization of crude oil reserves in the 1970s. Ownership of the world's refineries was thus concentrated in the hands of a very small number of firms, led by the Western supermajors. In 1999, for instance, just 15 companies held around 40 percent of all the world's refining capacity, with Royal Dutch Shell, Exxon and BP-Amoco occupying three out of the top four positions.[27] Today, this longstanding Western domination of refining has been eroded substantially. Around half of the top 15 companies in the world are now NOCs, with the first, second and fourth spots taken by Chinese and Saudi companies (Sinopec, Chinese National Petroleum Corporation and Saudi Aramco). Only ExxonMobil remains within the top four global refiners. The geographical concentration of refining has also shifted, reflecting the eastward orientation of crude exports. In the early 1990s, nearly half of the world's refining capacity was located in North America and Europe – this has now fallen to around one-third. In contrast, Asian refining capacity tripled between 1992 and 2020, with the absolute number of oil refineries in the region growing more than 2.5 times. By 2020, Asia's share of global refining capacity stood at 37 percent – more than North America and Europe combined.

The only other region of the world that has seen growth in its share of world refining capacity is the Middle East, where absolute capacity more than doubled between 1992 and 2020, and which now holds a 10 percent share of the world's total refining capacity. Quite remarkably, two-thirds of all oil refineries that have been built over the past five years, and over 80 percent of those currently under construction, are in the Middle East and Asia.[28] As with exports of crude oil, the growth of Middle Eastern and Asian refining is closely tied to production networks in China and the wider Asia region. Crude oil is either extracted in the Middle East and exported to China or another Asian country for refining, or it is extracted and refined in the Middle East and then exported to Asia. In this manner, the refined fuels and chemicals produced from Middle East oil enter into Asian production chains, where they are transformed into commodities that are consumed globally. Within this axis, the refining process is dominated by large NOCs headquartered in the Middle East, China and the wider Asia region, with Western firms holding a relatively marginal position.

A crucial part of these Asian production networks are petrochemicals, which form the basic input into plastics and other synthetic materials.[29] With the growth of China's manufacturing dominance, the country's consumption of petrochemicals has skyrocketed, and much of this demand is met by petrochemical plants located in the Gulf. Most significant here is ethylene, often described as "the world's most important chemical", which is essential to the manufacture of packaging, construction materials and automobile parts.[30] Between 2008 and 2017, the Gulf's share of ethylene production capacity grew from 11.5 percent to 19 percent. In this period, the Gulf rose from being the world's fourth-ranked producer of ethylene to being in second place, just behind North America (whose global share of ethylene production capacity fell from 27 to 21 percent).[31] This vital chemical is manufactured in massive integrated refineries and petrochemical complexes across Saudi Arabia, the UAE and other Gulf states, and then exported eastwards – indeed, just under half of all China's ethylene imports now come from the Middle East. China's emergence as the "workshop of the world" would not have been possible without these flows of refined chemical products from the Middle East to Asia.

These trends have elevated Gulf-based firms to the centre of the world petrochemical industry. Most important here is the Saudi Basic Industries Corporation (SABIC), which now ranks as the fourth largest chemical company in the world by sales (up from 29th in 2000).[32] SABIC was established by Saudi Royal Decree in 1976, with the goal of utilizing the country's crude oil and gas to manufacture basic chemicals for a range of industries, including automobiles, agriculture, construction and packaging. In the early 2000s, the company began to grow internationally through investments in Europe and the US. A major milestone was the acquisition of the plastics division of the US firm General Electric in 2007, which enabled the company to take substantial steps into advanced petrochemical production. Since that time, SABIC has expanded still further, and it now has activities in more than 50 countries across the world.

For most of its history, SABIC was 70 percent owned by the Saudi state, with the remaining 30 percent listed on the Saudi stock exchange. In 2020, however, the state's share of SABIC was taken over by Aramco, in a notable restructuring of the Saudi oil industry that illustrated the strong push towards vertical integration in the Gulf. Similarly, the leading petrochemical firms in the UAE, Kuwait, Qatar and Oman are all subsidiaries of state-controlled NOCs. These state-run petrochemical firms are closely connected to privately owned Gulf conglomerates through joint ventures and strategic partnerships, as well as the partial listing of companies such as SABIC on Gulf stock markets.[33] In this manner, the petrochemicals sector serves as another important conduit for private wealth accumulation in the Gulf.

NEW "EAST-EAST" INTERDEPENDENCIES

These patterns confirm the strong interdependencies that are emerging between the Middle East (especially the Gulf region) and Asia (especially China) in the oil sector. But this encompasses much more than the simple export of Middle East crude to Asia – rather, it is a process involving a considerable increase in *cross-regional* investments between the two regions. These investments come from both the large Gulf and Asian NOCs, as well as major privately owned conglomerates located in both areas. Through these flows of capital, there is an extensive intermeshing of all steps in the oil value chain: refining, petrochemical production and the onward circulation of oil products to the consumer. As such, Gulf hydrocarbon interests are embedded *inside* Asian production networks, and *vice versa*. At the political level, these linkages have also been accompanied by the development of much closer ties between the two regions, represented in a raft of recent bilateral agreements, high-level governmental visits and various other diplomatic initiatives.

To get a clearer picture of these capital flows and their implications, it is essential to look at all aspects of the hydrocarbon circuit – upstream, downstream, and activities such as transportation, drilling, storage and the laying of pipelines. Across these oil-related activities, China made more than $76 billion in outward investments globally between 2012 and 2021.[34] The first phase of these Chinese investments (2012-16) followed the announcement of the Belt and Road Initiative (BRI), and focused mainly on North America, Western Europe and Russia/Central Asia. Following 2016, however, there was a substantial reorientation in Chinese overseas oil investment. Between 2017 and 2021, more than 30 percent of Chinese investments in oil-related activities went to the Middle East region, greater than any other region and a five-fold increase in the Middle East's relative share compared to the 2012-16 period.

This investment has given Chinese firms a prominent role in oil-related industries across the Middle East. In the UAE, for example, Chinese firms are leading partners of the state-owned Abu Dhabi National Oil Company (ADNOC), and hold major stakes in onshore and offshore oil fields. In Iraq, a privately-owned Chinese firm now operates one of the largest oil fields in the world, the "supergiant" Majnoon oil field. In Kuwait, a subsidiary of the Chinese oil firm Sinopec has become the largest oil drilling contractor, controlling 45 percent of drilling contracts in the country. The largest deal involving China in the Middle East oil sector was finalized in 2021: this concerns Chinese participation in a multinational joint venture (JV) that owns a 49 percent equity stake in Aramco Oil Pipelines Co., a company that will have rights to 25 years of tariff payments for oil transported through Aramco's crude pipeline network in Saudi Arabia.

At the same time as this influx of Chinese investment into the Middle East is taking place, the Gulf states have become the primary foreign presence in the Chinese oil sector, through numerous JVs with Chinese entities. These projects aim to secure a market share for the Gulf's crude exports and include refineries, petrochemical plants, transport infrastructure and fuel marketing networks. An important example of this is the Sino-Kuwait Integrated Refinery and Petrochemical Complex, a 50:50 JV between Sinopec and the Kuwait Petroleum Corporation that is the biggest refinery JV in China, incorporating within it the country's largest petrochemical port (completed in May 2020). Both the refinery and port are viewed as an integral component of China's BRI, enabling China to import crude oil from the Gulf to manufacture fuels and other basic chemicals that are then exported to neighbouring Asian countries. For its part, Saudi Arabia's significant presence in China is evident through several large JVs between Saudi Aramco and Chinese firms in the refining and petrochemical sector, as well as a network of over 1,000 service stations in Fujian province, which was the first province-level fuel retail JV in the country. These partnerships involve both Chinese NOCs, such as Sinopec, and leading privately owned refining companies in China (which control around 30 percent of China's crude refining volumes). Qatar is also a prominent Gulf investor in China's energy sector, focusing particularly on securing markets for its liquefied natural gas (LNG) exports.

This expansion of the Gulf's hydrocarbon industry into China is part of a broader involvement by the Gulf states in the oil-related sectors of other Asian countries. Indeed, between 2012 and 2021, nearly half (by value) of all foreign investments from outside Asia into Asian oil-related assets came from the Gulf, including the four largest deals during this period.[35] Through these investments, Gulf firms have sought to expand their production of refined oil products and basic chemicals within Asia itself (utilizing crude feedstocks imported from the Gulf), which are then circulated within Asia by the trading arms of Gulf firms. Key regional targets for this downstream diversification of Gulf oil firms are South Korea, Singapore, Malaysia and Japan. Across these four countries – which each possess established industrial capacity that is often closely linked to the accumulation of domestic capitalist groups – Gulf firms have fully or partially acquired leading companies, and have also undertaken other kinds of partnerships, such as JVs.

Unsurprisingly, the chief Gulf firm in this respect has been Saudi Aramco, which now has a notable presence in key Asian states. In 2015, for example, Saudi Aramco acquired control over the South Korean firm S-Oil, which is the third largest refining company in the country (with about 25 percent market share) and which operates the sixth largest refinery in the world (located in Ulsan, South Korea). This acquisition enabled S-Oil to expand its petrochemical capacity in Korea, and the firm is now a top producer of various refined fuels and basic chemicals that Saudi Aramco's regional trading arm (Aramco Trading Singapore) then exports to other Asian countries. Also in South Korea, Saudi Aramco became the second largest shareholder of Hyundai Oilbank in 2019, following the purchase of 17 percent of the company's shares. Hyundai Oilbank is the fourth largest refining company in Korea, and is majority owned by the Hyundai industrial conglomerate. In Malaysia, Saudi Aramco is currently building a refinery and petrochemical plant that is projected to be the largest downstream petrochemical plant in Asia; the project is a 50:50 JV with the Malaysian NOC Petronas. In Japan, Saudi Aramco became the second largest shareholder in Idemitsu Kosan in 2019 – the firm is the number two refiner in Japan, controlling roughly one-third of the domestic oil products market through six refineries and a network of 6,400 retail service stations.

CONFRONTING THE CLIMATE EMERGENCY: TAKING THE MIDDLE EAST SERIOUSLY

With Gulf NOCs and other capitalist firms increasingly located *directly* inside Asian production networks – and not simply acting as suppliers of crude – we need to rethink how we approach the geographies of the global fossil fuels industry. It is not enough to focus solely on reducing the direct consumption of fossil fuels or carbon emissions in traditional Western centres. Global commodity production, including much of what ends up being actually consumed in North America and Western Europe, remains grounded in an axis of fossil capitalism that connects the oil fields, refineries and factories of the Middle East and Asia. The profound interdependencies established across this axis are a significant component of capital accumulation in both regions and help support the power of state and business elites. From an ecological perspective, these "East-East" interdependencies serve to re-embed fossil fuels at the core of global production chains, constituting a sizeable barrier to any successful green transition.

Such global shifts explain why the Gulf states have no intention of reducing their production of fossil fuels any time soon. Rather, as capitalist states, their strategic interests lie in the continuation of an oil-fuelled world for as long as possible. The Saudi energy minister, Prince Abdulaziz bin Salman, put this perspective bluntly in 2021, pledging that "every molecule of hydrocarbon will come out", amid plans to increase the kingdom's oil production capacity by more than eight percent by 2027, reaching 13 million barrels per day.[36] With this objective in mind, Saudi Aramco has invested more in oil field expansion in 2022 than any other company in the world. Such moves have led the *Financial Times* to note that Aramco "is doubling down" on oil, aiming to be "the last oil major standing" and "betting that it can continue to do what it does best: pump oil for decades to come and gain even more market power as other producers cut back".[37] All the hydrocarbon-rich states in the Gulf have signalled their intention to follow the same course.

This does not mean, however, that the Gulf monarchies deny the reality of climate change or stand apart from the global rush towards the new "green" technologies. Indeed, precisely the opposite is the case. All the leading Gulf NOCs have expressed support for the Paris Agreement goals and have endorsed their countries' net zero pledges.[38] They are also investing massively in hydrogen, carbon capture and solar, with the explicit goal of becoming world leaders in these technologies (see Christian Henderson's chapter in this book). Most visibly, the Gulf states have taken a prominent position in regional and international forums, such as COP27 and COP28. At the COP27 meeting in Egypt in 2022, for instance, the largest national pavilion was that of Saudi Arabia, followed by those of the UAE, Qatar and Bahrain. Measuring 1,008 square metres, the Saudi pavilion was exactly double the size of the pavilion housing the entire continent of Africa – a part of the world that is most directly under threat from the effects of climate change. COP28 will take place in the UAE.

All of this shows that the Gulf states see no contradiction between an embrace of "low-carbon solutions" and pursuing the path of accelerating fossil fuel production. Importantly, however, this is not just a rhetorical exercise in greenwashing: to a significant degree, the expansion of the renewable sector is a necessary step towards enabling the Gulf states to sell *more* oil and gas. With very high levels of energy consumption at home, the domestic substitution of oil and gas with alternative energy sources means that more fossil fuels can be made available for export. Indeed, such reasoning is explicitly behind Saudi Arabia's plan to generate half of the country's electricity from renewables by 2030 (which would be faster than most other parts of the world, including the European Union). As Prince Abdulaziz bin Salman put it, such a shift to renewables is envisioned as a "triple-win situation": greater oil exports, cheaper energy bills at home and the prestige of meeting emissions targets.[39]

The technologies and energy infrastructure associated with the green transition are also providing lucrative opportunities for Gulf-based firms, including NOCs such as Saudi Aramco. In December 2022, Saudi Arabia became the first country in the world to commercially ship a cargo of "blue hydrogen" – the shipment was destined for South Korea, raising the prospect that the East-East axis of world oil will soon take a renewables turn.[40] The UAE, Bahrain, Oman and Kuwait are all planning huge hydrogen sites on their territories, which will make the region one of the largest producers of hydrogen in the world.[41] Similarly, carbon capture and solar energy are receiving major investments from Gulf governments (again, often channelled through NOCs). All the Middle East's leading renewable energy companies, such as Masdar (UAE), ACWA Power (Saudi Arabia) and Nebras Power (Qatar), are Gulf-based. Through these companies and their dominance of the emerging renewables markets, the Gulf will take a commanding role in steering the form of any "green" transition in the region.

By appearing to transform themselves into key actors in the fight against a warming climate, the Gulf states obscure their ongoing centrality to a globalized fossil capitalism. This is the real goal behind their leadership in the deliberations at both COP27 and COP28 – it is a means of shaping the world's response to climate change and of resisting any move away from an oil-centred global order. But these realities also firmly tie political struggles in the Middle East to our planetary future. With the Gulf monarchies sitting atop the region's extreme inequalities in wealth and power, popular movements aimed at challenging these regimes and winning social and economic justice across the region need to be understood as crucial allies of global *ecological* struggles. A perspective on the climate crisis that ignores the Gulf and the politics of the wider region – concentrating its fire solely on Western governments and the Western oil industry – is not only out of step with the realities of world oil, but inadequate to the enormous challenges at hand.

NOTES

1. Parts of this chapter draw upon Adam Hanieh, "World oil: Contemporary Transformations In Ownership and Control," in Greg Albo, Nicole Aschoff and Alfredo Saad-Filho (eds.) *Socialist Register 2023 Volume 59*, (Merlin Press, 2022).

2. Christian Aid, *Counting the Cost 2022: A Year of Climate Breakdown*, https://tinyurl.com/5bhn5344.

3. Unless otherwise noted, "Asia" is used throughout this chapter in reference to the major oil consuming countries of East and Southeast Asia, i.e. China (inc. Hong Kong), Japan, South Korea, Taiwan, Indonesia, Malaysia, Philippines, Singapore, Thailand and Vietnam.

4. Andreas Malm, *Fossil Capital: The Rise of Steam Power and the Roots of Global Warming* (Verso, Brooklyn, 2016).

5. Damian Carrington and Matthew Taylor, "Revealed: The "Carbon Bombs" Set to Trigger Catastrophic Climate Breakdown," *The Guardian*, May 11, 2022.

6. Many illuminating accounts of this history are available. See, especially, Timothy Mitchell, *Carbon Democracy* (Verso, New York, 2011); Matthew T. Huber, *Lifeblood* (University of Minnesota Press, Minneapolis, 2013); Giuliano Garavini, *The Rise and Fall of OPEC in the Twentieth Century* (Oxford University Press, London, 2019).

7. An early and brilliant analysis of the Western oil industry is presented in John Malcolm Blair, *The Control of Oil* (Pantheon Books, New York, 1976).

8. In 1949, around two thirds of the world's known crude reserves, and more than half of the world's crude production and refining were controlled by these companies. Outside the US, these companies held more than 82 percent of all known crude reserves, 86 percent of crude production, 77 percent of refining capacity and 85 percent of cracking plants (used in the manufacture of petrochemicals). Between them, it was estimated that they owned at least half of the world's tanker fleet in 1949, and around two thirds of privately controlled tankers (see Federal Trade Commission, United States Congress, Senate Select Committee on Small Business, Subcommittee on Monopoly (1975) *The International Petroleum Cartel* (US Government, Washington).

9. The original "Seven Sisters" were the Anglo-Iranian Oil Company (first called Anglo-Persian, now BP), Royal Dutch Shell, the Standard Oil Company of California (now merged into Chevron), Gulf Oil (now merged into Chevron), Texaco (now merged into Chevron), the Standard Oil Company of New Jersey (Esso, later Exxon, now part of ExxonMobil) and the Standard Oil Company of New York (Socony, later Mobil, now part of ExxonMobil).

10. See: Huber, *Lifeblood*; Blair, *Control of Oil*.

11. Christopher R. W. Dietrich, *Oil Revolution* (Cambridge University Press, Cambridge, 2017); Garavini, *The Rise and Fall of OPEC*.

12. Adam Hanieh, *Money, Markets, and Monarchies: The Gulf Cooperation Council and the Political Economy of the Contemporary Middle East* (Cambridge University Press, Cambridge 2018).

13. United States Congress, House Committee on Commerce, Subcommittee on Energy and Power (1999) *The Exxon-Mobil Merger: Hearings before the Subcommittee on Energy and Power of the Committee on Commerce, House of Representatives, One Hundred Sixth Congress, first session, 10 and 11 March* (US GPO, Washington), p. 4.

14. In 1982, a rule change implemented by the Securities and Exchange Commission permitted companies to repurchase their own shares on the open market within certain limits, dependent on trading volumes (often financed through debt). The reduction in the number of shares led to an increase in companies' stock prices, allowing senior executives to make millions through the exercise of their stock options. US oil companies were at the forefront of this practice; indeed, between 2003 and 2012, ExxonMobil was the largest stock repurchaser on US financial markets. See: William Lazonick, "Profits without Prosperity", *Harvard Business Review*, September 2014.

15. Perhaps the most notorious example of this was British Petroleum's rebranding of itself as BP in 2000, under a new tagline "Beyond Petroleum" and with a new green sun-burst logo. At the time, BP remained the second largest oil company in the world and spent more on the corporate rebrand than it spent on renewable energy.

16. Figures in this paragraph are drawn from ENI, *World Energy Review*, 2021, https://tinyurl.com/54ynz8h8.

17. See footnote 3 for a definition of Asia. ENI, *World Energy Review*.

18. BP, *Statistical Review of World Energy 2020*, 2020, p. 30, https://tinyurl.com/3aw4xdaz.

19. *Ibid*.

20. There is significant debate about this boom in oil prices, which was closely connected to the financialization of oil and the role of financial markets in the determination of oil prices. See: Mazen Labban, "Oil in Parallax: Scarcity, Markets, and the Financialization of Accumulation," *Geoforum* 41, No. 4 (2010), pp. 541-552; and Adam Hanieh, "The Commodities Fetish? Financialisation and Finance Capital in the US Oil Industry," *Historical Materialism* 29, No. 4 (2021), pp. 70-113.

21. Prices then fluctuated between $30 and $70 a barrel until the COVID-19 pandemic induced a massive demand shock through 2020. Since that time, prices have risen significantly with the lifting of pandemic restrictions and the war in Ukraine.

22. High oil prices also had major implications for Western oil firms. Most importantly, they incentivized the production of so-called "non-conventional" oil and gas supplies – reserves that are more difficult and significantly more expensive to extract than conventional fossil fuels. Of particular importance here was US shale, crude oil that is held in shale or sandstone of low permeability and which is typically extracted through fracturing the rock by applying pressurised liquid (hence the term "fracking"). In this context, the oil price rise between 2000 and 2014 helped to attract large investments into North American shale field development, with the production of US oil shale growing more than ten-fold between 2007 and 2014.

23. McKinsey Global Institute, *The New Power Brokers: How Oil, Asia, Hedge Funds, and Private Equity are Shaping Global Capital Markets*, McKinsey & Company, 2007, https://tinyurl.com/2p8n7pu5.
24. Hanieh, *Money, Markets, and Monarchies*, Chapter 3.
25. Commonly cited figures for the 2000s indicate that around 50-55 percent of all Gulf investments went to US markets, 20 percent to Europe, 10-15 percent to Asia and 10-15 percent to the Middle East and North Africa.
26. Hanieh, *Money, Markets, and Monarchies*.
27. See Hanieh, "World oil", for a detailed analysis of these trends in refining and sources for the figures in this section.
28. International Energy Agency, "World Energy Investment 2020," p. 48, 2020, www.iea.org/reports/world-energy-investment-2020.
29. For a discussion of petrochemicals and their place in the study of oil, see Adam Hanieh, "Petrochemical Empire: the Geo-Politics of Fossil-Fuelled Production", *New Left Review*, 130, July-August 2021, pp. 25-51.
30. Approximately 75 percent of the global demand for ethylene comes from these three manufacturing activities. Gulf Petrochemicals and Chemicals Association (2019) "Ethylene a Litmus Test for the Chemical Industry", p. 2.
31. Deloitte, "The future of Petrochemicals: Growth Surrounded by Uncertainty," Deloitte Development LLC, 2019, p. 4, https://tinyurl.com/3ha2ee5m.
32. C&EN, "C&EN's Global Top 50 Chemical Firms for 2022," https://tinyurl.com/mrk4b2my.
33. Hanieh, *Money, Markets, and Monarchies*.
34. The figures in this paragraph have been calculated by the author, drawing on Orbis Database, which is available at https://tinyurl.com/998yv68b. They include investment in both upstream production as well as refining, petrochemical production, and oil and gas services (drilling, pipelines, etc.).
35. Calculated by the author, drawing on Orbis Database. Asia is defined here as China (including Hong Kong), Taiwan, Korea, Malaysia, Indonesia, Japan, Thailand, Singapore and the Philippines.
36. Javier Blas, J. "The Saudi Prince Of Oil Prices Vows to Drill 'Every Last Molecule'", *Bloomberg*, July 22, 2021, https://tinyurl.com/53s2w6fm.
37. Tom Wilson, "Saudi Aramco Bets on Being the Last Oil Major Standing," *Financial Times*, January 23, 2023, https://on.ft.com/3W5eAWU.
38. With the exception of Qatar, all Gulf states now have net zero emissions target dates.
39. Andrew England and Samer Al-Atrush, "Saudi Arabia's Green Agenda: Renewables at Home, Oil Abroad", *Financial Times*, November 22, 2022, https://tinyurl.com/yeyud3xb.
40. Blue hydrogen is a form of hydrogen production derived from natural gas. It is viewed by many in the environmental movement as a trojan horse of the fossil fuel industry because of the likelihood that it will significantly increase global demand for gas.
41. Qatar's strategy appears to be one of focusing on its plentiful LNG supplies, which will be exported abroad for the production of blue hydrogen.

12

The Challenges of the Energy Transition in Fossil Fuel Exporting Countries: The Case of Algeria

Imane Boukhatem

THE NEED FOR AN ENERGY TRANSITION IN ALGERIA

Fossil fuel extraction in oil rich countries is a principal contributor to greenhouse gas (GHG) emissions. After South Africa and Egypt, Algeria has the third highest GHG emissions in Africa.[1] However, in 2020 Africa accounted for only 3.8 percent of global carbon dioxide (CO_2) emissions, the smallest share of any region in the world.[2] Most of the fuel produced in Algeria is exported and burnt elsewhere, producing additional CO_2. According to the Intergovernmental Panel on Climate Change (IPCC) Special Report on global warming, achieving emissions reductions that can limit warming to 1.5°C requires rapid and deep transitions in energy systems.[3]

Algeria has a population of over 44 million people and is the largest country in Africa, extending from the Mediterranean coast, where most of its inhabitants live, to the Sahara Desert, which covers more than four fifths of its land area and has the highest surface temperatures in the world.[4] When the country gained independence in 1962, the economy was primarily rural and based on agriculture, with Algeria's production sent to France, the former colonial power, to supplement production there. However, major oil and gas reserves were discovered in Algeria's Sahara in the late 1950s. Thereafter, the Evian Accords (1962) and the Franco-Algerian Agreement (1965) provided a framework for Algerian-French energy cooperation and management up until the nationalization of hydrocarbon resources in 1971, when Algeria gained control of its hydrocarbon industry.[5]

According to *"La Charte Nationale Algérienne"* of 1986, nationalization was viewed as a significant act of political independence, and Algeria's oil and gas resources are now seen as part of its national wealth, which should be shared among the people, in the form of providing the funding for social services, such as free healthcare and education.[6]

Hydrocarbon exports have long played an important role in both Algerian politics and economy. Increased hydrocarbon exports financed President Houari Boumediene's industrialization agenda from 1965 to 1978. Then, following liberalization and the shift to a market economy from the early 1980s, Algeria's manufacturing expertise was undermined and its industrial potential was ultimately destroyed, and it became a relatively basic oil and gas exporter. Oil and gas now represent 93.6 percent of the total national export revenue and about 50 percent of the national budget.[7] Meanwhile, since the early 1980s Algeria's food production has fallen far short of self-sufficiency[8] thanks to liberalization and privatization.[9] Nevertheless, living standards have risen to a level commensurate with middle income countries and Algeria's GDP per capita reached $3,815.25 in 2020,[10] making it one of the five richest countries in Africa.

In regard to energy use, Algeria expended enormous efforts to provide cheap and reliable electricity to its population, achieving a 99.8 percent electricity access rate in 2020.[11] However, the country currently faces a triple challenge in its energy sector: economic dependence on hydrocarbon revenues, growing domestic electricity demand, and long-term fossil fuel export agreements that the country must honour in order to avoid sanctions, court cases and fines. At the same time, with Algeria's population growing rapidly at an average rate of two percent per year and forecast to reach 53 million by 2030,[12] natural gas exports have been reduced significantly to meet the fast-growing domestic demand for electricity.[13]

In light of this urgent situation, Algeria is faced with the need to rapidly transform its energy sector, and to do so while focusing on social justice. However, oil export revenues are a major obstacle to a just energy transition. Oil and gas revenues have played, and continue to play, a critical role in meeting the basic needs of people in Algeria, from food to healthcare and education, and in providing them with a standard of living that exceeds that of many countries in the region. A transition which undermines people's rights to food, health, education, livelihoods and development cannot be considered "just". At the same time, powerful political actors capture a large share of the oil and gas revenue and benefit disproportionately from the present extractivist economy.[14] Thus, there are major socioeconomic, institutional, political and policy obstacles to an energy transition in Algeria. Moving away from an economy that is centred on fossil fuel exports will require a dramatic social and economic transformation. This chapter highlights the opportunities, challenges and potential injustices involved in a green energy transition in Algeria.

For countries dependent on exporting fossil fuels in the global South, including Algeria, reducing GHG emissions has not yet taken priority over socioeconomic development. From a justice perspective, these countries have historically been disadvantaged by colonialism and have not benefited significantly from industrial development, nor are they historically responsible for causing the climate crisis. Yet they are the most affected by climate change and the ongoing effects of different forms of the predatory extractivism into which they have been locked. In countries like Algeria, whose economies have been built on the export of primary commodities, such as fossil fuels, they face a double burden: the direct impacts of climate change and the need to reduce and ultimately eliminate fossil fuel extraction.

Key actors within the Algerian government and energy sector are advocating a diversification of the energy system, but their motivation is economic rather than environmental concerns. In particular, the transition promoted by these Algerian elites is primarily motivated by a desire for economic diversification to free the country from dependence on fossil fuel revenues and to protect the national economy from the volatile fossil fuel market. At the same time, green energy development has so far been driven by a desire to sustain the current rentier system by replacing fossil fuels with returns from renewable energy exports.

Despite facing several crises in the last decades, Algeria has earned a reputation as a reliable and significant exporter of gas to Europe, ranking third after Russia and Norway. In response to the Ukraine crisis, Algeria offered to increase its gas exports to Europe, as a form of support to the continent. For instance, the state-owned Sonatrach and the Italian Eni oil and gas companies signed a contract to pump an additional nine billion cubic metres of gas between 2023 and 2024.[15] As Algeria's current gas reserves cannot meet European demand during the conflict in Ukraine, there may be pressure on the government to develop and export unconventional gas resources or, at the very least, to drill more gas wells. However, since the EU aims to become carbon neutral by 2050, this rise in gas demand will not last long and, therefore, if the country invests in new oil and gas resources to meet the expanding gas demand from Europe, there is a significant risk of lock-in and of stranded assets.[16]

At the same time, Algeria's internal gas consumption situation faces rapidly growing demand and, possibly, declining production.[17] Consequently, to protect the country's future capacity for gas exports, Algeria's ruling classes are seeking to develop renewable energy as a substitute for domestic gas usage.

Increasing the share of renewable energies in the energy mix, while saving gas for exports, would also be an appealing scenario for those who currently profit from the rentier system. In the short or medium term, this would ensure a continuous source of revenue and, consequently, socioeconomic and political stability. However, this is not a realistic option in the long run: a growing number of scholars and experts recognize that addressing the climate crisis will require leaving a significant proportion of known coal, oil and gas reserves in the ground, and with Europe aiming for carbon neutrality in the coming decades, and enforcing carbon taxation, Algerian hydrocarbons will not bring lucrative returns. Furthermore, establishing an energy transition with the goal of saving gas for exports will merely perpetuate the country's rentier and extractivist economic model, which has failed to deliver the progress to which the country aspires. At the same time, it would contribute to further deepening the climate crisis, which is already likely to threaten the region's very existence.

Instead of this elite vision for an energy transition, the transition to a sustainable energy system must be accompanied by long term economic, social and environmental changes, and must be founded on the principles of social and economic justice. For instance, workers' and their families' quality of life should be improved through fair compensation, respectful balance between work and life, and the creation of a healthy working environment. A just redistribution of Algeria's national resources will be an important tenet of such a transition as today many Algerians simply do not benefit from the country's wealth. In addition, democratization and empowerment of citizens to decide their own energy future is a necessary form of energy democracy and should be part of the transition.

To summarize, a just transition in Algeria should be developed with the goal of lowering emissions, protecting the environment, respecting people's rights to resources and a liveable environment, and preserving natural resources, including water and land, for future generations. It must also improve the quality of life of Algerians by promoting social and economic justice, fair distribution of wealth, and energy democracy, rather than simply generating revenue from renewable energy exports. To this end, proposals for an energy transition should address how energy is used and by whom, not only where it comes from.

ALGERIA'S CLIMATE AND ENERGY POLICY

In 1993 Algeria ratified the United Nations Framework Convention on Climate Change (UNFCCC), and in 2005 the country signed the Kyoto Protocol. Since then, all of Algeria's socioeconomic plans have included climate change mitigation and adaptation measures.[18] By 2030, the Algerian government plans to reduce its emissions by seven percent unconditionally or by 22 percent with support from the international community.[19]

Developing a strategic action to tackle global climate change and to promote sustainable development in the country can attract international financial support for Algeria. This has so far been lacking compared with the support given to neighbouring countries.[20] However, the issue of funding climate action in Algeria goes beyond the technical questions involved in developing a plan to access funds. It is uncertain whether Algeria will abandon its oil and gas industry in the absence of firm national and international commitments to finance an energy transition. But if global climate action requires that Algerian oil not be extracted and exported, and if we agree that the Algerian people should not be required to pay for global climate action, how will current export earnings from oil and gas be replaced, and who will pay for this?

Furthermore, aside from existing international commitments regarding the export of fossil fuels, climate policies have so far received little institutional support or attention in Algeria. Although the National Agency for Climate Change (ANCC) was established in 2009, it remains understaffed and institutionally weak. At the same time, organizational changes and frequent reshuffles of departments and ministries have caused confusion and disrupted longer term work programmes. In May 2015, for example, a partial change in government led to the termination of the tasks related to sustainable planning and protection of the environment then being undertaken by the Ministry of Land Use Planning, Environment and Tourism. The environmental question was subsequently assigned to the Ministry of Water Resources. Two years later, the Ministry of Environment and Renewable Energies was created and given responsibility for the environment and renewable energy.[21] Then, in 2019, responsibility for renewable energy and the environment was divided, with a new Ministry of Energy Transition and Renewable Energies given sole responsibility for energy transition. In the same year, the ministry in charge of the environment launched its first project relating to a climate law, with support from the German Agency for International Cooperation (GIZ).

After years of unsuccessful implementation of renewable energy policies in the context of sustainable development, the Algerian government has now realized the significance of incorporating state owned oil, gas and electricity companies in this process. Without the financial resources of the national oil company, the National Company for Research, Production, Transport, Processing and Marketing of Hydrocarbons (Sonatrach), the state owned utility, the National Company for Electricity and Gas (Sonelgaz), and technical and managerial energy expertise, the Renewable Energy and Energy Efficiency Programme (2015-30) faces the risk of failure. Cognisant of this fact, in 2020 the Ministry of Energy and Mines launched a new Renewable Energy Programme (2020-30).

Algeria primarily uses petroleum, petroleum products and natural gas to meet around 98 percent of its domestic energy demand. Algeria does not have any nuclear capacity, nor does it have any substantial hydroelectric, coal-fired or installed renewable capacity. However, in a bid to meet the increasing domestic demand and to diversify its energy mix, the country is slowly moving towards integrating more solar and wind-based generation.[22]

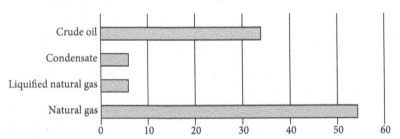

Figure 12.1   Algeria's primary energy production in terms of quantity (in percentage). Source: Ministry of Energy, 2020.

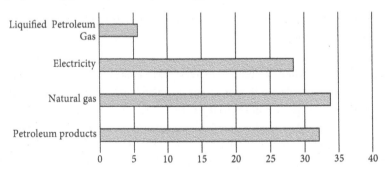

Figure 12.2   Algeria's final energy consumption in terms of quantity (in percentage). Source: Ministry of Energy, 2020.

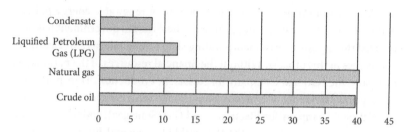

Figure 12.3 Algeria's primary energy exports in terms of quantity (in percentage). Source: Ministry of Energy, 2020.

THE RENEWABLE ENERGY SECTOR IN ALGERIA

Algeria is well positioned in terms of its geography and climate to take advantage of opportunities for renewable generation. It enjoys 2,000 to 3,000 hours of sunshine each year in its desert, which covers 80 percent of the country's surface area. This whole area, therefore, has the potential to generate more than 169,400 terawatts per hour, which is 5,000 times the yearly national electricity consumption.[23] Due to Algeria's proximity to European energy import centres, geographical vastness and reputation as a reliable energy exporter, export scenarios for renewable energy are also quite viable. Solar and wind power dominate the National Renewable Energy and Energy Efficiency Programme, accounting for 85 percent of the total projected capacity by 2028.[24] The *Algerian Renewable Energy Resource Atlas*, produced by the Centre for the Development of Renewable Energy (CDER), was released for the first time in 2019. It features a set of geographical representations that showcase Algeria's energy potential, which includes solar, wind, geothermal and bioenergy.[25]

National Programme for the Development of Renewable Energy and Energy Efficiency (PNEREE)

Algeria has launched a number of renewable energy programmes since 2011, but renewables still make only a small contribution to meeting domestic energy needs. On February 3, 2011, the government approved the first national programme for the development and promotion of renewable energy and energy efficiency (PNEREE), which aims at achieving a capacity of 22,000 MW from renewable electricity production by 2030, with 10,000 MW committed for export, in order to achieve 40 percent renewable energy of the total electricity mix.[26]

The 2015 update of the PNEREE was driven primarily by significant reductions in investment costs for electricity production from diverse renewable sources, especially photovoltaic (PV). As a result, the concentrated solar power (CSP) portion was reduced to less than a third of what had originally been planned in the earlier version of the programme (from 7,200 MW to 2,000 MW). Regarding solar PV, its share was increased by nearly five times (from 2,800 MW to 13,575 MW).[27]

However, neither the 2011 programme nor its 2015 update have actually been implemented to any significant degree. By 2020, Algeria had installed only around 425 MW renewable energy, which is very far from the target capacity of 4,375 MW.[28] Achieving the intended capacity has been hampered by a lack of coordination in the execution of the PNEREE, including the 2015 update, a lack of monitoring and evaluation, and, most importantly, a lack of serious political commitment to the energy transition.

National Energy Transition Programme of 2020

After failing to achieve the targets set out for the 2015-30 programme, in 2019 the government underlined its determination to catch up in the renewable energy sector. The National Energy Transition Programme of 2020 was then formulated; it aims to achieve a capacity of 16,000 MW by 2035, based on solar PV. According to the Ministry of Energy, this target is a crucial plank of Algeria's energy transition. It is planned that PV plants connected to the national grid will generate 15,000 MW by 2035, with the first tranche of 4,000 MW expected to be finished by 2024 and the remaining 11,000 MW expected to be deployed by 2030. To put this into action, the project Tafouk I was announced in May 2020, with the goal of generating 4,000 MW of solar PV capacity.[29]

Recent developments in the renewable energy sector

In 2021, Algeria announced a plan to achieve 15,000 MW of renewable energy capacity by 2035. The new *Société Algérienne des Énergies Renouvelables* (SHAEMS), a joint venture shared between the two public companies, Sonatrach and Sonelgaz, has been tasked by the minister of energy transition and renewable energies with processing the call for tenders. SHAEMS will also make an investment in each project company, either on its own or in partnership with other public and private entities. If the bid is successful, it will result in a 25-year power purchase deal.[30] An industrial local content component is not required in this bid, but the use of locally produced equipment is incentivized. Moreover, the 51/49 rule, which limits the part of foreign investment in any project to 49 percent, has also been dropped for the renewable energy sector, confirming the neoliberal orientation of the current Algerian government (see more on this below).

CHALLENGES OF AND BARRIERS TO THE ENERGY TRANSITION IN ALGERIA

The first obstacle facing the exploitation of Algeria's massive solar capacity is the distance between demand centres and supply hubs. Demand centres are located in the north, where urban density precludes the creation of huge projects. However, the supply side lies in the Sahara, in the country's southern half, where sun irradiation and geographical space are abundant. Added to the issue of remoteness, climatic conditions, among other considerations, mean that the cost of building solar PV installations is 30 percent higher in Algeria than the global average.[31]

Beyond governance barriers and towards energy democracy

The lack of a long term energy strategy is a main governance hurdle facing the energy transition: renewable energy initiatives are inefficient, fragmented and lack coordination. Algeria's energy sector has been slow to adapt to the world's critical need for renewable energy, driven by climate change. As summarized earlier, Algeria has announced ambitious plans, but little has been accomplished due to poor management, the lack of a unified energy strategy and insufficient political will. While energy policies and regulations in the sector are either inspired by those in foreign jurisdictions or generated domestically, in each case their implementation is impeded by bureaucracy and corruption.

The energy sector is centralized in Algiers and is led by the energy and mining ministries, and by the oil/gas and electricity monopolies Sonatrach and Sonelgaz. In this top down and highly authoritarian centralized governance system, ideas from the grassroots are unlikely to be heard and accepted. Added to this, Algeria adopts a republican governance model, which means that the elected body is the sole decision-making authority. The president appoints provincial governors, exhibiting a top-down organization in which policy innovations at the local level remain uncommon. Because civil society is weak and fragmented, with little interest in climate and energy issues, the fossil fuel industry's agenda is largely unaffected by the influence of popular action. There is thus a need to adopt a more flexible, participatory and transparent policy approach, so that Algerians can participate in discussions of, and offer solutions to, the country's energy problems. Furthermore, including individuals in energy decision making would boost their sense of ownership of public energy assets, which could lead to a shift in the population's behaviour towards more responsible and solution-oriented attitudes. Rebuilding confidence between the government and the population by increasing openness, accountability and, most importantly, respecting citizens' decisions would be a first step towards energy democracy in Algeria. More decentralized alternatives might offer people the power to choose how they generate, consume and exchange energy, while maintaining the state's critical role as manager, controller and legislator.

Algeria suffers from acute water scarcity, which threatens the country's food security, among other consequences, such as risking the collapse of agriculture and the displacement of local communities. For example, in August 2021 huge wildfires burnt tens of thousands of hectares of woodland in the north of the country, killing at least 90 people. However, despite the obvious catastrophic impacts of climate change on the country in recent years, climate change is rarely addressed in the country's energy and environmental plans, in part due to the institutional weakness of the Ministry of the Environment. There is, therefore, an urgent need to include the climate variable in future energy policies and scenarios.[32]

Financing the energy transition

Despite the falling cost of solar and wind technologies, renewable energy projects remain capital intensive. Financing the transition is thus a huge challenge for Algeria. Three national funding options seem to be available: public funds, domestic private funds and foreign direct investment.

In the current context of economic crisis, public funds are increasingly solicited to address what are considered to be more pressing socioeconomic grievances, and thus can offer only limited support for the transition in the short term. However, a thorough examination of the country's economic position during the last two decades reveals there has been massive waste and poor management of public finances, in addition to entrenched corruption. For any meaningful and just energy transition in Algeria, the state needs to be democratized and the endemic corruption must be rooted out. The state should also play a bigger role in the transition, should commit more public funds and should argue for significant climate finance on the international stage. Additionally, it is important to remember that when oil prices rise, as they are now, it offers the possibility of establishing sovereign funds that can enable national funding for the modernization of energy systems.

Further important financing avenues, such as climate reparation funds and the payment of ecological debt by the global North, could be effective in helping to achieve a fair energy transition for Algeria among the global South economies dependent on oil. However, doubts remain about whether such actions are likely to occur. For example, although the $8.5 billion pledge to support South Africa's phasing out of coal was made at the 2021 United Nations Climate Change Conference, COP26,[33] this has unfortunately not yet happened. Yet we cannot talk about a just energy transition without raising the question of the payment of climate debts and reparations by the rich North to Southern countries, not as new loans but rather as transfers of wealth. This is not just an ethical or moral imperative; it is also a historical responsibility. The industrialized West must pay its fair share in aiding poorer countries that are less responsible for climate change and that are often more vulnerable, with their adaptation plans and green transitions. Unfortunately, current processes do not allow for such assistance for middle-income and oil-producing countries like Algeria, as countries with more complicated energy issues, such as problems with accessing clean energy, are prioritized. This reflects how international climate policies are still trapped in an "aid and development" framework, which fails to recognize the real need to leave existing gas and oil reserves in the ground and what it entails in terms of significant revenue losses. Halting extraction will require global redistributive policies that directly address the needs of oil exporting countries, including middle-income ones like Algeria.

Turning to the domestic situation, Algeria currently lacks the political will to finance the energy transition. For instance, a special fund was established in 2015 to finance renewable energy programmes, but this has not been effectively utilized due to the lack of an executive decree providing a legal framework for using this fund. The country's sovereign policy bans foreign debt and limits international financing in vital and strategic sectors, such as energy, to 49 percent, compared to 51 percent for Algerian partners. However, under pressure from the local and global capitalist energy lobby, which argues that Algeria's "insecure" and "rigid" regulatory framework discourages foreign investment, the minimum sovereign criterion of 49 to 51 percent has been dropped for renewable energy projects. The private investor class in Algeria, which is mainly comprised of family funds, prefers to participate in projects with quick returns, where investors can recoup their capital fast, unlike renewable energy projects, which take a long time to repay investments. Furthermore, after a decade of empty promises and unmet commitments, investors in Algeria have lost confidence in the country's renewable energy schemes. Since PNEREE was announced in 2011, some private investors have nevertheless been working to establish a solar PV value chain to assist the programme.[34]

Despite the big announcements made by Algeria's political elites about the deployment of renewable energy, very little has been accomplished in terms of raising finances for the energy transition from the three sources discussed above. At the same time, the country missed the crucial opportunity offered by the oil bonanza of 2004-14, when oil and gas prices were very high, to use the significant revenues generated to industrialize and diversify its economy, embark on a solid energy transition and create green jobs. Instead, the super-profits made during this period were siphoned off by a predatory and corrupt elite.[35]

Finally, one worthwhile avenue that Algeria can pursue is strategies that encourage citizens to invest in small community-based and integrated community energy projects. This form of direct energy democracy can offer an opportunity for job creation and community empowerment.

Expertise and technology barriers

Algeria lacks expertise in green technologies; its historical experience is with oil and gas technologies. This is due mostly to the lack of political interest in green technologies that has been shown in recent years, as well as the Algerian economy's de-industrialization since neoliberal reforms began in the 1980s. The process of liberalization and the transition to the market economy was accompanied by the elimination of both theoretical and practical knowledge in industry, culminating in the liquidation of institutes specializing in crucial areas, such as the energy, steel and textile industries.

Following a campaign against technical secondary education, technical industrial branches that had contributed to the training of engineers and senior technicians for decades were liquidated.[36] The number of experts available to support the energy transition programme, particularly in terms of renewable energy, energy efficiency, and all economic and industrial activities related to it, is still far below the critical mass required.[37] There is thus a need for more applied research and practical training. Algeria needs to receive technology transfers and managerial expertise from nations that have managed the transition effectively. Such cooperation should aim to achieve human and material capacity building in Algeria. This will require overcoming the current monopolized technological boundaries and the intellectual property regime enforced by free trade agreements and the international financial institutions.

One of the most difficult aspects of any energy transition in the global South is establishing control over technology (technology transfer) and industrialization: this is required in order to achieve the level of economic integration needed for the development of a thriving green economy with green jobs. This necessitates a paradigm shift away from neoliberalism and towards greater state involvement and investment, as well as some climate funding from the most developed countries.

Enforcing a local content policy is crucial to the development of a solid and autonomous renewable energy sector in Algeria. This means mandating that firms operating in the country use goods and services produced or supplied domestically. The bidders who are chosen for developing renewable energy projects must play their part in an industrial strategy whereby solar energy components are developed locally. Such a strategy should aim to create a local solar energy economy while also lowering project costs by avoiding the costs of imported materials.[38] Although such a strategy would benefit the local economy, and particularly the job market, it will be extremely challenging to implement in Algeria because the local industrial value chain is still not fully formed: it is currently in its infancy, and much local production does not meet international standards. In this context, foreign investors are currently lobbying for the abolition of the local content requirement because they see it as an investment barrier that hinders the generation of lucrative profits for themselves. However, while abandoning such measures would perhaps attract more funds and boost the energy sector, it would not benefit the Algerian economy, Algerian industry and the Algerian job market. As such, it is crucial to look at effective solutions for creating and enhancing technology and expertise, by going beyond the Western enforced intellectual property rights regime and technology monopolies, through entering into win-win partnerships with countries in the global South, such as China.

Access to energy and the matter of subsidies

Algeria's energy market is still dominated by the government. However, the state-owned utility's unsustainable economic model and its mismanagement have led to calls for privatization and the end of subsidies. In order to adapt to the current drastic changes in the energy landscape, the Algerian utility Sonelgaz needs to adopt technical, managerial and financial reforms to make it economically viable, accountable and transparent. Moreover, a progressive subsidy reform is also imperative.

Electricity is currently heavily subsidized in Algeria. Households pay the equivalent of $0.038 per kWh for electricity, that is, one seventh of the price paid in the United Kingdom, and businesses pay $0.033.[39] These low prices, amounting to around one third of the production cost, are made possible by subsidies.

There are also additional indirect subsidies, through subsidies to fossil fuels, which are relevant here given that the country's electricity comes primarily from fossil fuels. Unfortunately, these subsidies, alongside Algeria's regressive tax system, do not benefit the needy classes as much as they enrich business and capital owners in the country. A just subsidy reform is thus a political and economic imperative. This reform needs to be progressive in terms of which sections of society are subsidized: this should not include the richest classes and capitalist groups (who thereby increase their profits); rather, it should seek to lessen the plight of the most vulnerable in society.[40]

Algeria is currently facing a double socioeconomic and political crisis. The mass protest movement *Hirak* that started in February 2019 and lasted for more than a year was a serious political challenge to the Algerian regime. In addition to the very negative economic repercussions of the COVID-19 pandemic, this makes any blanket removal of energy subsidies in the near future politically untenable, as well as unjust for the millions of Algerians that have been pauperized over the last few years.

In response to the fall in global oil prices in 2020, which the IMF estimates will shrink the economy by six percent and drop the country's foreign reserves from $62 billion to $47 billion by the end of 2020,[41] the government limited its commitment on social spending with a 30 percent cut. Nevertheless, the country saw a significant budget deficit, reaching 18.4 percent of GDP in 2021.[42] On top of increased poverty levels, hundreds of thousands of jobs have been lost, including in the precarious informal sector. According to government data, 500,000 jobs were lost in 2020 alone.[43,44] At the time of writing, the Algerian economy is still under stress, though the war in Ukraine might prove to be a blessing to Algeria's rulers, as oil and gas prices are once again soaring.

Integrating renewable electricity in the governmental subsidy schemes may be an option for promoting the deployment of renewable energy.

However, the existing global neoliberal climate policy paradigm has proven inefficient in disincentivizing fossil fuels through a carbon pricing model and by encouraging low carbon investment through subsidies and favoured contractual structures. This policy paradigm sees governments as protectors of private players' capital, which prohibits them from addressing social and environmental challenges. With its massive investment deficit and techno-logical inefficiency, privatization market approaches have so far failed to bring about the energy transition that is so essential in Algeria.[45] What is needed is a strong political commitment towards the energy transition within the Algerian public energy sector, with the controlled inclusion of private sector actors, alongside more participative, transparent and demo-cratic governance of state-owned enterprises (SOEs).

ALGERIA'S URGENT NEED FOR A JUST ENERGY TRANSITION

The oil boom of the 2000s enhanced the country's fiscal balance and allowed for massive investments. However, the sharp drop in oil prices since June 2014, including during the COVID pandemic – although now reversed due to Russia's invasion of Ukraine – has cast doubt on the country's economic and energy strategy. Export income covered only 67 percent of total imports in 2015, while the revenue regulation fund and foreign exchange reserves have been steadily declining, from $121.9 billion in October 2016 to $42 billion by March 2021.[46]

The Algerian economy is vulnerable to fluctuations in international markets, due to its excessive reliance on hydrocarbon revenues, which cannot be a solid foundation on which to build long term economic plans. This instability is underlined by recent economic performance, which has been marked by downturns and upswings. The COVID-19 pandemic exac-erbated the previously existing economic crisis in the country and has been an additional threat to Algerian policy-makers in terms of building a diversified economy and implementing an energy transition, leading to a growing budgetary deficit, economic contraction, and a reduction in public spending. All of these have resulted in difficulties in finding money to finance the energy transition. However, after these travails, oil prices soared in the fourth quarter of 2021, following the global post COVID economic upswing: Brent crude prices averaged $71 per barrel in 2021 (peaking at $86), having started the year at $50 per barrel. Hydrocarbon export revenues increased from $20 billion to $34.5 billion during this period. Thereafter, the invasion of Ukraine drove up oil prices even further, briefly surpassing $123 per barrel for Brent crude at the start of the war.[47]

While this may appear to be good news – at least for Algeria's rulers – such high prices may be a hindrance to a green energy transition as they may entrench the extractivist and rentier mentality, and may provoke a push for more drilling for fossil fuels, especially in the current context in which the EU is seeking to end its reliance on Russian gas by diversifying its sources.[48] There are some worrying signs that this is the direction that is indeed being taken by the Algerian ruling classes: they have agreed to pump more gas to Italy, and are now considering exploring and exploiting new fields, with the help of European countries and companies.[49]

The country's long term needs necessitate a quick transformation, and a double transition – both an economic and an energy transition. The economic transition involves moving away from a mainly fossil fuel-based economy and towards a more diversified economy based on sustainable industrial and agricultural activities, as well as transitioning from being a net importer of products to being an industrialized producer. The energy transition, on the other hand, involves a shift away from the burning of fossil fuels to the use of more sustainable energy carriers via the deployment of green technologies, and an examination of how and by whom energy should be used to maximize social justice and wellbeing.

According to the International Energy Agency's net zero 2050 roadmap, decarbonization in accordance with the Paris Agreement does not allow for investment in new oil and gas fields. As things currently stand, by 2030 it is predicted that countries will have surpassed by more than 20 percent the maximum emissions possible if the Paris Agreement's 1.5°C goal is to be sustained. Furthermore, oil firms control trillions of dollars in assets that, according to the Intergovernmental Panel on Climate Change (IPCC), they will never be able to burn if humans are to continue to survive on this planet. Numerous hidden risks in the oil business are now increasingly becoming apparent to private investors and public financial institutions. Stabilizing prices, phasing out production and seeking more alternative commodities, such as blue and green hydrogen, are thus top imperatives for hydrocarbon producers.

The urgency of the climate crisis necessitates the participation of the oil and gas industry, governments, financial institutions, as well as the active involvement of workers and trade unions in taking firm decisions and making a strong commitment to turn them into action. The development of, and support to, the hydrocarbon sector must come to an end, with a fast but equitable strategy to phase out existing and planned oil production while assisting oil dependent countries, communities and workers around the world in a smooth transition.[50]

Unfortunately, it is not clear that Algeria is planning to phase out oil production in the near future. As indicated above, due to the rising demand from Europe, Algeria is currently seeking to develop further its oil and gas industry by attracting investors to explore new oil and gas fields. This strategy enhances the risk of lock-in and stranded assets in the country, and entails the risk that a chaotic fall in oil demand will lead to economic collapse for Algeria, due to its complete reliance on fossil fuel exports. Such a collapse would involve a loss of government support for public services, a fall in subsidized essential products, a weakening (if not outright collapse) of national firms, and hence privatization, which would be a conduit for neoliberal capitalism and neocolonialism. Few Algerians could afford to live decently in the face of such extreme effects.

Due to Russia's war in Ukraine, the current opportunity presented by the oil price increase offers a good moment to commit significant funding to Algeria's renewable energy deployment and to the energy transition. The government should immediately prioritize energy transition and economic diversification, in order to protect national sovereignty and to provide a secure future for Algerians in the long run. This needs to be accompanied by climate grants from the global North, as part of a climate repair effort, to help Algeria to adapt to climate change impacts and to transition to renewable energy. Looking further ahead, exporting green hydrogen and other renewable energies could allow the country to maintain its position as a reliable energy exporter in the post hydrocarbon age. This scenario will be viable only if renewables become a much bigger portion of the country's energy mix. Otherwise, the growth of infrastructure exclusively dedicated to hydrogen production for export, for example, will impede the country's transformation, exacerbate its energy problems and perpetuate its subordinate position within the global economy.

Several policy instruments can be adopted to manage a fair phase-out of oil and gas in Algeria.[51] The Algerian government should develop bottom-up policy tools to protect workers, their families, and the communities that will be impacted by the decommissioning of oil and gas industries. A public entity should be formed to identify oil and gas workers and communities who will be directly harmed by the transition. Financial assistance, requalification and re-entry into the green energy market are all required.

Green skills programmes are needed to enable affected workers to access new opportunities in the job market.

Protecting communities from any negative effects of the energy transition requires democratic governance, and more bottom-up and participatory mechanisms. Educating the public about the dangers of climate change and the need for an energy transition, through schools, mosques and various educative and religious channels, would be useful in instilling a more responsible attitude towards the world. Youth should be mobilized and solidarity should be built with Algerian communities resisting and adapting to climate change through the international global climate justice movement. Setting up grassroots commissions and a dedicated ministry (or at least a division within it) to manage the just transition could be an effective strategy. Learning from other experiences across the world will be vital: the ongoing coal phase-out in numerous nations provides valuable lessons on managing just oil and gas phase-outs.[52] These experiences show the importance of the following:

- Setting ambitious targets that are consistent with the Paris Agreement through a clear process, with different funding, implementation and energy cooperation mechanisms, and making clear the needed timeframe for enacting the recommendations in legislation.
- Adopting an inclusive approach, with the representation of women, youth and underrepresented minorities.
- Maintaining a balance of power among various actors, and recognizing the centrality of human rights claims, including people's right to energy, decent livelihoods and a liveable environment, but also the rights to land or other resources of communities that may be impacted by renewable projects.
- Guaranteeing transparent decision-making structures and institutional processes while providing opportunities for confidential deliberation.
- Considering structural changes, rather than having only a narrow focus on economic implications. It is important to identify and engage with the affected communities, and to consider the so far neglected gender implications of a phase-out of oil and gas.
- Stressing the significance of the government's ambitious climate policies and commitment to achieving a just transition by assisting affected regions and communities.

- Mobilizing international financial, technical and managerial assistance to support a just transition that leave no one behind.

Finally, to ensure the long-term viability of its business, the state-owned oil corporation Sonatrach should conduct a thorough risk analysis of any future oil investment, and should seriously consider renewable energy business opportunities.

CONCLUSION

Algeria's situation and concerns reflect those of several other oil and gas rich countries when faced with the necessary global energy transition. Algeria has wasted numerous opportunities to embark on an energy transition, due to the ruling elite's unwillingness to prioritize this until recently. In the current global economic crisis, the country will struggle to finance an energy transition while simultaneously avoiding a degradation of socioeconomic conditions, including through maintaining the subsidized electricity supply to the population. Indeed, social peace depends on low cost energy.

In addition to these domestic issues, the global energy transition may put extra strain on the country. Apart from the direct impact on the national oil and gas company, its employees and their families, a global energy transformation may lead to a fall in oil export returns, economic collapse and destitution for many Algerians.

Global neocolonial energy transition trends, particularly EU-centric green export policies, are focused primarily on supporting a green transition at the European level, at the expense of cheap resources and undervalued labour in the global South. Prioritizing the export of renewable energies and green hydrogen, for example, would impede Algeria's transition since a strategy directed towards export would then take precedence over addressing local energy and economic challenges. If the limited funds allocated to Algeria's energy transition are used to build and/or renovate infrastructure for export, this will hamper the country's ability to meet its local energy transformation demands. Rather than focusing on exports, it would be more logical and just to prioritize the local energy transition needs, as opposed, for example, to exporting green hydrogen and consuming fossil fuels at home.

A just transition requires going beyond a switch to green energy technologies, to also include ensuring the protection of oil and gas employees and their families, as well as the provision of necessary skills for their integration into the green labour market. It also requires ensuring affordable and reliable energy access for everyone, and a recovery from all of extractivism's negative consequences on the economy, society, politics and the environment. Most importantly, to avoid the transition being merely a transition from fossil "brown" extractivism to renewable "green" extractivism, there is a need to protect the land and resource rights of communities living near renewable energy sites.

While the public sector's involvement in the energy transition needs to be consolidated, public sector institutions must also be managed better, and must be more transparent and accountable. Ultimately, there is a need for better policies and mechanisms as regards evenly distributing the national wealth and closing the gap between social classes, as well as a more democratized energy sector that allows Algerians to actively participate in determining their own future and finding genuine solutions to their energy problems. First and foremost, Algeria should commit to climate initiatives and should set high emissions reduction goals, and should reject any future investments that are not in accordance with the Paris Agreement. Thereafter, the next step must be the thorough preparation of the fossil fuel phase-out and a progressive decarbonization of the energy sector. To avert a dramatic collapse of Algeria's economy, a planned and fair phase-out of the fossil fuel industry will require international cooperation. Last, but not least, effective public communication regarding current and future energy issues, as well as the economic risks facing the oil and gas sector, is needed, to promote Algerian working people's readiness to accept and cooperate in the necessary measures to alleviate the economic hazards involved in the energy transition.

NOTES

1. Hannah Ritchie, Max Roser and Pablo Rosado, "CO$_2$ and Greenhouse Gas Emissions," *Our World in Data*, 2020, accessed May 30, 2022, https://ourworldindata.org/co2-and-other-greenhouse-gas-emissions.
2. Lars Kamer, "Africa's Share in Global Carbon Dioxide (CO$_2$) Emissions from 2000 to 2020," *Statista*, August 2021, accessed May 30,2022, https://www.statista.com/statistics/1287508/africa-share-in-global-co2-emissions.
3. IPCC, *Global Warming of 1.5°C. An IPCC Special Report on the Impacts of Global Warming of 1.5°C above Pre-Industrial Levels and Related Global Greenhouse Gas Emission Pathways, in the Context of Strengthening the Global Response to the Threat of Climate Change, Sustainable Development, and Efforts to Eradicate Poverty* (Cambridge University Press, Cambridge, 2018), https://doi.org/10.1017/9781009157940.

4. The Information Architects of *Encyclopaedia Britannica*, "Algeria Facts," *Britannica*, accessed April 25, 2022, https://www.britannica.com/facts/Algeria.

5. John Entelis, "Sonatrach: The Political Economy of an Algerian State Institution," *Oil and Governance: State-Owned Enterprises and the World Energy Supply* (Cambridge University Press, Cambridge, 2011), pp. 557-98, doi:10.1017/CBO9780511784057.016.

6. Roger Albinyana and Aurèlia Mañé-Estrada, "Energy Interdependence. The Linkage of the Political Economy of Algeria's Natural Gas with that of the Western Mediterranean Region. A Methodological Approach," *Revista UNISCI, UNISCI journal*, No. 47 (May 2018), pp. 29-64, https://doi.org/10.31439/UNISCI-2.

7. Ahmed Bouraiou et al., "Status of Renewable Energy Potential and Utilization In Algeria", *Journal of Cleaner Production*, No. 246, 2020, p. 119011, https://doi.org/10.1016/j.jclepro.2019.119011.

8. *Encyclopaedia Britannica*, "Algeria Facts."

9. Entelis, "Sonatrach".

10. "Algeria GDP Per Capita," Trading Economics, accessed April 25, 2022, https://tradingeconomics.com/algeria/gdp-per-capita.

11. Younes Zahraoui, M. Reyasudin Basir Khan, et al., "Current Status, Scenario, and Prospective of Renewable Energy in Algeria: A Review," *Energies* 14, No. 9, p. 2354, 2021, https://doi.org/10.3390/en14092354.

12. Michael Hochberg, *Algeria Charts a Path for Renewable Energy Sector Development*, Middle East Institute, 2020, accessed April 25, 2021, https://tinyurl.com/539mzevv.

13. Mostefa Ouki, *Algerian Gas in Transition: Domestic Transformation and Changing Gas Export Potential*, The Oxford Institute for Energy Studies, October 2019, https://doi.org/10.26889/9781784671457.

14. Hamza Hamouchene, *Extractivism and Resistance in North Africa*, Transnational Institute, 2019, www.tni.org/en/ExtractivismNorthAfrica.

15. "Eni: Agreement with Sonatrach up to 9 Billion Cubic Meters of Gas", *The Limited Times*, April 11, 2022, accessed June 30, 2022, https://tinyurl.com/48hutpt4.

16. Antonio Hill, Hervé Lado, Thomas Scurfield and Amir Shafaie, *Europe's Demand for Africa's Gas: Toward More Responsible Engagement in a Just Energy Transition*, Natural Resource Governance Institute, 2020, https://tinyurl.com/yruymafz.

17. Ali Aissaoui, *Algerian Gas: Troubling Trends, Troubled Policies*, Oxford Institute for Energy Studies, May 2016, https://tinyurl.com/7y2byypf.

18. Sahnounea, F., Belhamela, M., Zelmatb and M., Kerbachic, R., "Climate Change In Algeria: Vulnerability and Strategy of Mitigation and Adaptation," *Energy Procedia*, 36, 2013, https://doi.org/10.1016/j.egypro.2013.07.145.

19. Soumission Portal, UNFCCC, "Contribution prévue déterminée au niveau national CPDN – Algérie," September 3, 2015, https://tinyurl.com/2p8rxhfp.

20. Commissariat aux Energies Renouvelables et à l'Efficacité Energétique, *Transition Energétique en Algérie: Leçons, état des lieux et perspectives pour un développement accéléré des energies renouvelables*, 2020.

21. Dennis Kumetat, "Managing the Transition: An Analysis of Renewable Energy Policies in Resource-Rich Arab States with a Comparative Focus on the United Arab Emirates and Algeria," (PhD dissertation, London School of Economics and Political Science, 2012), pp. 1-380, http://etheses.lse.ac.uk/id/eprint/623.

22. "Algeria," U.S. Energy Information Administration, accessed May 2, 2022, www.eia.gov.

23. "Algeria Looks to Solar Energy to Meet Growing Electricity Demand," *Fanack*, March 8, 2019, accessed April 25, 2021, https://tinyurl.com/5ecrs6dm.

24. Dalila Hamiti and Sultana Bouzadi-daoud, "La stratégie algérienne de transition énergétique conformément au programme de développement des énergies renouvelables et d ' efficacité énergétique : état des lieux et perspectives de développement," *Journal of Contemporary Business and Economic Studies*, pp. 4, 2, 2021.

25. Ali Smai and Mohamed Lamine Zahi, "Les potentialites de l'algerie en energies renouvelables," *Recherches économiques et managériale*, June 2016, p. 27, https://www.asjp.cerist.dz/en/downArticle/101/10/1/47030

26. Zahraoui, Basir Khan, and al., "Current status."

27. Commissariat aux Energies Renouvelables et à l'Efficacité Energétique (CEREFE). *Transition Energétique en Algérie*, 2020. https://www.cerefe.gov.dz/wp-content/uploads/2022/02/Rapport_CEREFE_TE-2020-4.pdf, p. 36

28. Zahraoui, Basir Khan, et al. "Current Status."

29. CEREFE (2020) *Transition énergétique en Algérie*.

30. Thomas Herman, "Algeria – Launch of the Solar 1,000 MW Call for Tender," *Africa Notes*, Herbert Smith Freehills, January 17, 2022, accessed April 28, 2022, https://tinyurl.com/yc7wknpt.

31. CEREFE, *Transition énergétique en Algérie*.

32. "En Algérie, les incendies qui ont ravagé la Kabylie sont éteints," *Le Monde*, August 18, 2021, accessed May 30, 2022, https://tinyurl.com/3w37mvra.

33. Andreas Franke, "COP26: Western Nations Pledge USD8.5 Billion for South African Coal Phase-Out," *S&P Global Commodity Insights*, November 3, 2021, accessed July 1, 2022, https://tinyurl.com/2xhv4n3a.

34. "Nos Membres," Cluster Solar Energy, http://www.clustersolaire-algeria.com/liste-des-membres-du-ces/.

35. Hamza Hamouchene and Brahim Rouabah, "The Political Economy of Regime Survival: Algeria in the Context of the African and Arab Uprisings," *Review of African Political Economy*, 43, pp. 150, 668-80, August 8, 2016, doi: 10.1080/03056244.2016.1213714.

36. Hamouchene, "Algeria, an Immense Bazaar: The Politics and Economic Consequences of Infitah," *Jadaliyya*, January 30, 2013, https://tinyurl.com/msfprduu.

37. CEREFE, *Transition énergétique en Algérie*.

38. Philippe de Richoufftz, "Algérie: Contenu local et énergies renouvelables," *Alexander & Partner*, Novembre 2021, accessed April 29, 2022, https://tinyurl.com/3kmdcpcx.

39. Laura El-Katiri and Bassam Fattouh, "A Brief Political Economy of Energy Subsidies in the Middle East and North Africa," *International Development Policy*, 7, 2017.

40. Tim Gould and Zakia Adam, *Low Fuel Prices Provide a Historic Opportunity to Phase Out Fossil Fuel Consumption Subsidies*, IEA, June 2, 2020.

41. "Algeria on the Brink as Pandemic and Low Oil Price Take Their Toll," *Financial Times*, accessed July 1, 2022, https://tinyurl.com/2p92w3x8.

42. Peterson Ozil, "COVID-19 in Africa: Socio-Economic Impact, Policy Response and Opportunities," *International Journal of Sociology and Social Policy*, May 2020, doi: 10.1108/IJSSP-05-2020-0171.

43. "Impact de la pandémie en Algérie: Plus de 500.000 emplois perdus," *Algerie Eco*, June 22, 2022, accessed July 1, 2022, https://tinyurl.com/nmxdhena.

44. "Where We Work," Algeria Overview, World Bank, accessed July 1, 2022, https://www.worldbank.org/en/country/algeria/overview.

45. Sean Sweeney, John Treat and Daniel Chavez, *Energy Transition or Energy Expansion?* Transnational Institute, October 22, 2021, accessed May 30, 2022, https://www.tni.org/en/publication/energy-transition-or-energy-expansion.

46. Francisco Serrano, *Higher Oil Prices are Giving Algeria's Regime Breathing Room*, Middle East Institute, May 25, 2022, accessed May 30, 2022, https://tinyurl.com/3xc9s57v.

47. Mark McCarthy Akrofia and Sarpong Hammond Antwib, "COVID-19 Energy Sector Responses in Africa: A Review of Preliminary Government Interventions," *Energy Research & Social Science*, 68, p. 101681, 2020, doi: 10.1016/j.erss.2020.101681.

48. "REPowerEU: Affordable, Secure and Sustainable Energy for Europe," European Commission, accessed July 1, 2022, https://tinyurl.com/mpj5nxjc.

49. Monika Bolliger, "Könnten Sie morgen Gas nach Deutschland liefern, Herr Arkab?" *Der Spiegel*, June 19, 2022, accessed June 1, 2022, https://tinyurl.com/ycyrpche.

50. Sivan Kartha and Greg Muttitt, "Equity, Climate Justice and Fossil Fuel Extraction: Principles for a Managed Phase Out," *Climate Policy*, 20:8, September 6, 2022, https://doi.org/10.1080/14693062.2020.1763900.

51. Hanna Brauers and Danae Fenner, "Comparing Coal Commissions: What to Learn for Future Fossil Phase-Outs?" *Coal Transitions*, accessed May 2, 2022, https://tinyurl.com/4jemwamu.

52. *Ibid.*

13

Unjust Transitions: The Gulf States' Role in the "Sustainability Shift" in The Middle East and North Africa

Christian Henderson

In recent years the Gulf states' ability to deal with climate change has been subject to speculation in the Western media. In some cases, the very survival of these countries has been called into question. According to one article in the UK's *Guardian*, the region is likely to face an "apocalypse" in the near future as a result of temperature increases and rising sea levels.[1] The piece paints a picture of states that are shaped by hostile environments, with fragile societies that will be pushed over the edge by the climate crisis. In addition to the challenges of climate change, the article also intimates that a decline in demand for oil and gas will be an additional cause of the demise of the Gulf states, due to their dependence on the export of hydrocarbons.

Aside from their dramatic tone, reports of this kind have serious analytical shortcomings. They tend to assume that the Gulf states are passive actors in the politics of climate change. Rather than a source of power, their control of 30-40 percent of proven oil reserves is framed as a vulnerability, and it is implied that the increased use of renewables will mean these countries become superfluous as the global economy makes a transition to green forms of energy. Based on a perception of shared environmental and climatic conditions, this framing also surmises that the Gulf region is in the same boat as other parts of the Middle East and North Africa (MENA) when it comes to the threat of the climate crisis and the challenges of the energy transition.

This chapter contradicts these assumptions. It shows how, rather than being powerless producers, the Gulf countries are working to ensure they remain at the centre of the global energy regime. This entails the formulation of a dualistic policy: one that allows them to benefit from both fossil

fuels and renewable energies. The Gulf Cooperation Council (GCC) countries are intent on extracting, producing and selling oil and gas, and their
downstream by-products, for as long as there is demand.[2] At the same time,
they are also gaining a foothold in renewable energy markets and in the
development of other fuels, such as hydrogen, and they are using their
capital to invest in wind and solar farms across the MENA region. In contrast
to the assumption that the Gulf countries are subject to the same socioeco-
logical dangers as other MENA countries, this chapter also illustrates that
some of the GCC states are investing in infrastructure that will offer some
protection from crises. This will give them a capacity to manage food, water
and energy that far surpasses that of other countries in the region, and in
doing so may offer some defence from environmental disruption.

Comprehending these dynamics is essential to understanding the
contours of a just transition in the MENA region. Energy flows, extraction
and development in this area have been characterised by historical patterns
of domination of the global South by the North. The colonial period led to
the subordinate integration of many regional societies within the global
economy. The economies of North Africa, for example, were defined by the
extraction of agrarian commodities and natural resources, a legacy that continues to this day.[3] What should be understood, however, is that this
hierarchy also has a regional manifestation. The emerging political and
economic power of the Gulf states is creating a highly polarized regional
dynamic. The capital of the GCC is invested in the formal economies of
some of the most populous Arab countries: the Gulf is one of the largest
sources of foreign capital in countries such as Jordan, Egypt and Sudan.[4] At
the same time, the Gulf countries are also playing a role in superintending
the internal politics of these states: their aid and investment is propping up
their leaderships, allowing them to weather economic storms and repress
internal political dissent. As a result, the power of the Gulf states stands in
the way of the social and democratic progress on which a just energy transition depends. Equal access to clean energy and other resources, such as food
and water, and forms of restitution, such as climate reparations, requires a
political transformation as much as environmental-technical innovation.

This highly polarized regional dynamic also has global implications. One
political goal of the GCC countries is to ensure that the rising social concerns
about the grim realities of the climate crisis do not result in government regulation that interferes with the demand for fossil fuels and leads to a
devaluation of fossil fuel endowments. This is an objective that is shared
with other companies, markets and ruling classes in the global economy. In

this sense, a successful strategy for a just transition should take into account the Gulf's role in such alliances, and the outcome of their influence in the global economy. The Gulf states' power is manifested in their investment in global markets, advertising, sports and various institutions, such as the forthcoming UN Climate Conference in the United Arab Emirates.

A GREEN ENERGY SHIFT IN THE GULF?

In recent years the buzzwords of "sustainability" and the "green economy" have been used in the Gulf states as much as they have been anywhere else. The GCC countries are keen to portray themselves as eager participants in environmental change.[5] This is most apparent in Saudi Arabia, the UAE and Qatar, the three countries that are the focus of this chapter. These countries have promoted their investment in renewable energy and have publicized a programme of environmental modernization, including plans for "decarbonized oil and gas", a circular economy, vertical farming and an array of technology-based solutions.[6,7] However, these conceptions obfuscate an actuality that is very far from the principle and practice of environmental sustainability. In reality, these countries have no intention of curbing their oil production and have articulated their commitment to expanding production for as long as there is demand. In this sense, the Gulf's position is completely aligned with that of most other hydrocarbon exporters and oil companies.

This position has been explicitly stated by Gulf officials. In the summer of 2021, the Saudi energy minister, Prince Abdulaziz bin Salman Al Saud, communicated it with crystal clarity. According to a *Bloomberg* report, at a private meeting the prince commented on his country's intention to continue producing and selling oil no matter what. "We are still going to be the last man standing," he said, "and every molecule of hydrocarbon will come out."[8] This sentiment has also been echoed by other officials in the region. In 2022, the UAE minister of state for climate and food security, Mariam al-Mheiri, stated that "for as long as the world needs oil and gas, we're going to give it to them".[9] This intention to protect the value of hydrocarbon assets and meet demand is reflected in the plans of every single Gulf state to ramp up its production of oil and gas.[10]

In light of this unwavering commitment to oil and gas, how do renewables fit into the energy policies of the Gulf states? First, it should be underscored that present progress in the transition to renewable energy within the Gulf states remains very slow. In 2019, the UAE had the largest production of renewable energy within its energy mix compared to all other GCC states, with a figure of 0.67 percent of the country's total national energy consumption.[11] This is far lower than many other non-GCC countries.[12] However, some Gulf countries have said that they intend to change this. The UAE has announced a commitment to meet 50 percent of its electricity demand with "clean energy" by 2050, using a combination of renewables, nuclear and "clean coal".[13] Saudi Arabia intends to achieve the same target by 2030.[14]

These are highly ambitious policies, and they should be treated with some scepticism. Making such declarations allows these countries to present the appearance of pursuing environmental sustainability. The commitment to the transition to renewable energy is thus part of a broader apparent commitment to environmental sustainability, which is also manifested in public exhibitions, such as Dubai's Expo 2020, which was pervaded with narratives about sustainability.[15] Narratives about ecological consciousness also underpin major developments, such as Neom, the new futuristic city that is planned for Saudi Arabia's Red Sea coast. According to the promotional material, Neom will be a "blueprint for tomorrow in which humanity progresses without compromise to the health of the planet".[16] In some cases these public relations campaigns result in pronouncements that are patently false. The organizers of the 2022 World Cup in Qatar claimed that it was the first carbon-neutral tournament in history, an assertion that was quickly debunked by journalists and activists.[17]

Irrespective of the questionable and superficial nature of these claims, this hyperbolic greenwashing serves an important purpose. It helps to obfuscate the reality of the Gulf states' function as major producers of oil and gas in the global economy. It allows these countries to maintain their legitimacy on the international stage and ensure that they are central actors in debates over energy politics. On the one hand, the commitment to oil and gas will ensure that the GCC states will retain their control over energy markets, manifested in the leading role of Saudi Arabia, the UAE, Kuwait and Qatar in the Organization of the Petroleum Exporting Countries (OPEC). On the other hand, the image of sustainability and environmental consciousness portrays the Gulf states as important stakeholders in renewable energy markets and a lower-carbon future. One example is the next United Nations Climate Change Conference of Parties, COP28, in 2023, which will be held in Dubai. These global climate summits, which have been taking place for three decades, are intended to lead to an international agreement that will result in the reduction of greenhouse gas emissions and thereby curb climate change. However, in the UAE the negotiations at the COP 28 will be presided over by the head of the Abu Dhabi National Oil Company (ADNOC), a move that one activist described as "putting the fox in charge of the henhouse".[18] This obviously presents a contradiction, but it is one that characterizes sustainability politics everywhere.

Aside from politics, however, it is likely that the Gulf states will eventually take steps to increase the level of renewables in their domestic energy mix. They may not attain the rapid transition that they have pledged to achieve, but renewable energy is likely to take hold in the heart of world oil extraction. In order to understand this, we need to look deeper into the configuration of the region's energy economy and the requirements of a social metabolism in a hot and arid ecology.[19] These countries have very high levels of domestic power consumption. Saudi Arabia, the UAE and Qatar have some of the highest levels of electricity consumption per capita in the world,[20] and all of the GCC states have consumption per capita that is higher than the average for high-income countries. One cause of this high usage is the domestic consumption of energy for air conditioning, a demand that has been exacerbated by subsidized energy, although this support is now being rolled back by many GCC governments. Another cause of this demand arises from the production of desalinated water, which accounts for the majority of domestic water consumption in most Gulf states. The desalination of water is a highly energy-intensive process. In Saudi Arabia, for example, it accounts for around 20 percent of energy consumption.[21] One estimate suggests that desalination plants in the Gulf states account for 0.2 percent of the world's electricity consumption.[22] As a result of economic and demographic growth, this energy demand has expanded in recent years. In Saudi Arabia, for example, consumption of energy has more than doubled from 1,335 terawatt hours (TWh) in 2000 to 3,007 TWh in 2021.[23] Similar increases can be observed elsewhere in the region.

This huge level of energy consumption is becoming a costly hindrance for the Gulf economies. Electricity in the Gulf countries is provided mostly by oil- and gas-fired power stations. As a result of the increasing domestic demand, increasing amounts of oil are being diverted away from export to global consumers, who pay market rates. The internal demand for oil shows no sign of abating, and some estimates suggest that the domestic consumption of oil could continue to increase by as much as five percent a year.[24] One study suggests that by 2030 the internal consumption of oil in Saudi Arabia could match the amount that is exported.[25] These trends are spurring the expansion of renewable energy production in the Gulf states. In these countries the green energy shift is actually impelled by the need to retain oil for export: it is motivated by a commitment to fiscal sustainability, rather than environmental concerns.

A NEW MARKET

Aside from the need to reconfigure domestic energy production, the Gulf states also view renewables and fuels such as hydrogen as a new market opportunity. Green energy is an investment asset for the GCC's surplus capital. The sector is relatively low-risk: it receives support from development finance institutions and guarantees from host governments. As a result, Gulf conglomerates are active in the sector. New energy companies have emerged that often receive a degree of state backing and funding. One example is Masdar in the UAE. Owned by the state of Abu Dhabi, Masdar initially became known for its plan to build a city in Abu Dhabi that would be based on the principle of sustainability and that would utilize renewable energy.[26] The company also has a large investment arm that owns around $20 billion in renewable energy assets in a number of markets across the world.[27] Another case is ACWA, which is partly owned by the Saudi state. This company, which has a global presence, owns $75 billion in assets, but of these assets only a minority are in the renewable energy category.[28]

These firms are very active in the MENA region. Economies such as those of Morocco, Jordan and Egypt are accessible to Gulf companies as a result of close bilateral relations. Renewable acquisitions are often included in packages of state-led Gulf aid and investments, thereby ensuring that ventures receive backing at the highest level. This is part of the trend of the Gulf states' expanding influence over the politics and economics of the region. It is a pattern that is concomitant with investment in other sectors, such as food production and infrastructure, and also direct government aid to regional allies. The clearest example of this is Egypt: it is estimated that between 2014 and 2016 Saudi Arabia, the UAE and Kuwait gave President Abdelfattah al-Sisi's government around $30 billion in aid. This played a central role in enabling his rule and stabilizing the country in the counter-revolutionary phase, following the 2011 revolution. This flow of money was central to the restoration of authoritarian rule in what is the most populous Arab country.

One example of state-to-state support for the renewables sector was evident at COP27, held in Sharm al-Sheikh in November 2022. Sheikh Mohammed bin Zayed, the president of the UAE, and Abdelfattah al-Sisi both personally attended the signing of an agreement between Masdar and

Infinity, the largest renewables company in Egypt, for a wind farm that would be the largest of its type in the country.[29] Another example is a deal that was signed between the governments of the UAE, Egypt and Jordan in 2022, named the Industrial Partnership for Sustainable Economic Growth. This covers food, fertilizers, textiles, pharmaceuticals, minerals and petrochemicals.[30] The agreement also includes plans to enhance renewable energy production.

One dimension of these deals is the role of financing from development banks. Institutions such as the World Bank, the European Bank for Reconstruction and Development and the African Development Bank have financed projects in which Gulf states have invested. The involvement of both governments and these international institutions creates powerful stakeholders in these projects and de-risks them. This type of support has allowed Gulf investors to become major stakeholders in the renewable energy policies of some governments in the MENA region. Gulf capital has entrenched itself in the future of energy in the region, ensuring that it can syphon off profits from the transition to renewables.

One clear example of the powerful combination of government and institutional stakeholders is the Ouarzazate Solar Complex in Morocco, which is one of the largest concentrated solar power plants in the world. This project is funded by a consortium that includes Saudi Arabia's ACWA, the Moroccan Agency for Solar Energy and TSK, a Spanish company. Other backers include the World Bank and several other development banks. Another example is the investment by AMEA Power, a UAE company, in a wind farm and solar energy plant in Egypt. These projects are being implemented in partnership with the Sumitomo Corporation, and are financed by the International Finance Corporation, the Dutch Entrepreneurial Development Bank and the Japan International Cooperation Agency.[31]

An additional example of the extent to which money from the GCC is embedding itself in the future of renewables and resource governance in the region is a proposal that has been signed by the UAE, Israel and Jordan. These three states have agreed on a plan for UAE's Masdar to invest in a solar energy facility in Jordan that will sell all of its electricity to Israel. In return, Israel will sell desalinated water to Jordan.[32] If it proceeds, the deal will illustrate how UAE capital and Israeli technology can gain a greater foothold within the region. This deal will also normalize and deepen Israel's occupation of Palestinian territories and the system of apartheid that it imposes on the Palestinian population. It is an illustration of how these types of projects can have highly uneven results. Power from a solar farm constructed on Jordanian territory will be diverted to the Israeli market. Networks of water and electricity production will be delivered to wealthier consumers, while excluding deprived populations who are subjugated by military occupation.

In conjunction with solar and wind, hydrogen may play a role in the energy transition, as an alternative fuel/energy carrier.[33] Several Gulf countries, such as Saudi Arabia, Qatar, Oman and the UAE, are unveiling projects that will cater to a growing global demand for hydrogen. Whether these projects will produce "green" hydrogen (from renewables), "blue" hydrogen (from gas with carbon capture) or "grey" hydrogen (from fossil fuels without carbon capture) remains to be seen. It is difficult to ascertain to what extent the final product will be a zero-carbon or low-carbon fuel. The competitive advantage of these countries is natural gas: by using this fuel they will be able to produce hydrogen at a far cheaper cost than using renewable power and huge amounts of desalinated water (which would require more energy consumption). Green hydrogen will cost as much as 11 times more than natural gas, five times more than grey hydrogen and two times more than blue hydrogen.[34] With this considered, the details of these plans are vague and it is possible that the classification of the categories of hydrogen could be blurred, making it difficult to determine whether these fuels genuinely have low or zero carbon emissions.

Gulf investors are also acquiring foreign assets in the hydrogen sector. Egypt is seeking to become a hub for the production of green (and blue) hydrogen, and GCC companies are seeking to capitalize on these plans. For example, Masdar has signed a proposal to invest in two green hydrogen sites in Egypt, one on the Mediterranean coast and another in the Suez Canal Economic Zone at Ain Sukhna on the Red Sea coast.[35] The agreement also includes a plan to produce green ammonia, which can be used to make "carbon-neutral" fertilisers. Other Gulf companies are also investing in Egypt's strategy to become a hub for the production of green hydrogen. Financing for an Egyptian company involved in this plan has also been provided by the European Bank for Reconstruction and Development. Thus, these projects have been de-risked by finance from the Gulf and Europe.[36]

Whether these plans are feasible and realistic remains to be seen, but the focus on hydrogen has strong political overtones. Hydrogen is touted as a panacea within energy markets. It is seen as a means to reduce consumption of fossil fuels, and this has taken on an added urgency since Russia's invasion of Ukraine in 2022, which has forced many European governments to seek alternatives to their dependency on Russian gas exports. Should these plans be realized, they will result in the expansion of renewable energy projects (solar and wind), with Gulf and Western capital investing in state-led projects that will be integrated into European energy networks. From the perspective of Gulf producers, one possible motive for this policy is the role of gas in the production of hydrogen. Growing hydrogen markets provide a hedge that allows the Gulf economies to partake in the energy transition but also maintain the value of their gas reserves.

A REGION OF INEQUALITY

How are the Gulf states using their hydrocarbon revenues to safeguard their future in light of the risks of climate change? The resources and capital of the Gulf states place them at the top of the regional political and economic hierarchy, which is characterized by increasing polarization. There is chasmic inequality between the poor countries of the region and the wealthy ones. For example, GDP per capita in Yemen is $701, while in the UAE it is $44,315.[37] Other examples of this differential can be found elsewhere in the region: GDP per capita in Syria is $533, while in Qatar it is $66,000.[38] As a result of this imbalance the MENA countries do not share the same prospects in terms of the effects of climate change. The political and economic power of the Gulf states means they have more capacity to manage the problems of a warming climate. This ability contrasts with that of other countries in the region, such as Yemen, Lebanon and Syria, which are suffering from a combination of economic collapse, suffocating public debt, conflict and internal instability.

The status of the Gulf states' food security is one example of this regional inequality. The GCC states are highly reliant on food imports, and between 80 and 90 percent of commodities are imported. This creates a vulnerability to geopolitical disruption that could affect logistics and supply chains. The Gulf countries have used their capital to mitigate this risk. They have invested heavily in transport and storage infrastructure. This means that they can import food from several different global locales, thus ensuring that they have a diversified source of commodities. Gulf countries import food from all global regions, and they have also acquired land in North Africa, the Black Sea region, the US and Latin America.[39] They have also established large food processing, poultry and dairy operations. These facilities serve Gulf markets and provide some self-sufficiency, but they still require the import of raw commodities, such as cattle feed. More recently, the Gulf states have begun investing in agri-tech capabilities that allow them to grow food in fully controlled indoor environments.[40] These projects are energy intensive and they benefit from a subsidized supply of electricity and other inputs from Gulf governments.[41]

They are also a form of ecological modernization and constitute an attempt to gain greater control over the social and environmental relations of food production, which provides benefits in a warming climate. The absence of domestic agriculture creates a reliance on imports, but it also lessens the direct exposure to climate change. Societies that depend heavily on smallholder agriculture as a source of income and employment are more vulnerable to the effects of climate breakdown. Agriculture in Yemen, Egypt and Morocco accounts for between 20 and 35 percent of employment, while in the Gulf states it accounts for less than five percent.[42] The Gulf states are not immune from this danger altogether, as climate breakdown may compromise production in regions from which they source commodities, but their buying power and diversified network of supply chains reduces their exposure – at least for the time being. The use of oil revenues to finance food imports is another example of the way in which these states remain highly dependent on the export of oil and gas. It is a matter of existential importance for them.

The regional inequality is also manifested in Gulf investment in agribusiness abroad. There is sometimes a tendency in development literature to assume that cooperation and regional investment flows are a means to address the food security needs of MENA states. These flows are presented as a means to resolve the food insecurity of the Gulf states and at the same time invest in the agriculture sectors of poorer Arab economies.[43] However, the reality of Gulf investment in agriculture contradicts this interpretation. The acquisition of large areas of land in Egypt, Sudan and Ethiopia constitute plantations that consume water and other resources for food that is exported directly back to the Gulf states. One of the most common crops on these enclosures is alfalfa, a cattle feed that is used in the large dairy operations that have been established in the Gulf states.[44] These calories are removed from economies that have high levels of food insecurity and a history of famine. In Sudan, for example, Gulf investors have acquired more than 500,000 hectares of land, often in prime agricultural areas that are adjacent to the Nile – land that is often claimed by small farmers.[45] These farms produce grain and cattle feed that is exported back to the GCC economies, even as the Sudanese population continues to experience food insecurity, with 12 million people, out of a total population of 44 million, currently estimated to face acute food insecurity.[46] Half a million children are estimated to suffer from severe acute malnutrition in the country. These kinds of large-scale land acquisitions are often described as "land-grab-bing" and pose well-documented threats to the rights, livelihoods and health of local people.[47] Such land grabs and export-oriented large-scale agricultural production weaken the food sovereignty of countries like Sudan.

Another dimension of the regional inequality is storage capacity for food grains, which creates a buffer against price rises and supply shocks. This is particularly important for Arab countries, given their reliance on imported food and the potential for climate and market shocks. The Gulf states have invested heavily in grain silos and food stores, and this infrastructure has been included in port and airport projects in these countries. As a result, their storage capacity far outstrips that of other countries in the region. For example, Saudi Arabia has a grain storage capacity of around 3.5 million tonnes, for a population of 35 million,[48] whereas Egypt's grain storage capac-ity is around 3.4 million tonnes, for a population three times larger than Saudi Arabia's, at around 105 million people.[49] Qatar's storage capacity is around 250,000 tonnes, for a population of 2.6 million,[50] while Yemen has a similar capacity, for a population of 30 million. This contrast also exists in other comparisons with countries around the region, particularly with those societies that have been hit by war and disasters. For example, the devastating Beirut port explosion in August 2020 destroyed the harbour's 100,000-tonne grain silos.

In addition to food silos, the Gulf states are also investing in other forms of infrastructure that will provide a means to manage their essential resources in the face of the effects of climate change. Saudi Arabia, Qatar and the UAE have recently completed water storage facilities that will guarantee supply. In some cases, these facilities are some of the biggest in the world: Qatar's 6.5 million cubic metre water tank is sufficient for seven days of national consumption.[51] The construction of this infrastructure illustrates how the Gulf states are securitizing their social metabolism: water and food storage capacity can offer resilience in the face of conflict, climate crisis and logistical disruption. This provides an insight into the divergence of the region's developmental trajectories: the capacity to deal with climate change, environmental stress and their potential shocks is highly uneven.

THE GULF AND THE JUST TRANSITION

The social and economic principles that are inherent to a just transition are at odds with the strategies that have been described in the preceding pages.

By investing in renewable energy, agribusiness and infrastructure upgrades, the Gulf states are pursuing a capital-intensive and technological programme of environmental modernization. This involves techno-fixes and accumulation by dispossession in the name of "sustainability". These methods are primarily motivated by profit and security considerations, with a commitment to environmental sustainability being a secondary concern. This approach places little or no emphasis on equality, justice and universal basic needs. It is premised on the idea that environmental sustainability is a technocratic issue, one that can be dis-embedded from the highly political questions of the distribution of wealth and resources, consumption, and the extraction of profit.

This is a propensity that has regional ramifications. As described above, the influence of the Gulf states is patent in renewable energy investments in economies such as those of Egypt, Tunisia, Morocco and Jordan. Through their investments, state-led GCC conglomerates are entrenching themselves within the regional renewable energy transition. However, this influence is also present on a broader level across the region. Aid and investment from the Gulf states helps underpin a number of Arab governments, such as those of Egypt, Jordan and Tunisia: the Gulf states provide loans that finance these governments and prop them up. In addition to financing, the Gulf states also steer regional politics in other ways. Saudi Arabia and the UAE launched a military intervention in Yemen, and Qatar and Saudi Arabia have supported reactionary proxies in Syria. These interventions foreclose the democratic space that is required for a truly just transition; they hinder the emergence of social movements that can demand a more equitable and sustainable use of national resources. Furthermore, as discussed above, the use of large areas of land for renewable energy production and agribusiness enclosures often relies on the dispossession of other land users, appropriation that is achieved through authoritarian and repressive forms of governance. For a just transition to be achieved within many countries in the Arab area, the question of social and environmental justice must consider this regional dimension. The path of revolutionary and social change cannot be understood as involving only struggles that are determined by class conflict at the national level: the weight of Gulf influence within the regional political economy must also be included in the equation.

Such obstacles to a just transition can also be observed at a global level. The Gulf states have a presence within the politics of climate change and they use their resources to launder the image of the oil-based economy. This is manifested in the greenwashing and sustainability branding that is at work within these countries and is apparent in the appointment of an oil executive as president of COP28. This marketing is also evident in investments by the Gulf states in high-profile assets in the West. The clearest example is football teams, and some of the largest football clubs in Europe are owned by Gulf countries or have signed advertising deals with Gulf airlines and entities. Gulf states' ownership of clubs such as Paris Saint Germain, Barcelona, Newcastle and Manchester City launders the reputation of these states and internalizes their oil and gas revenues within these emblems of working-class pride and identity; it is an effort to sustain the familiarity of fossil fuels through culture, and ensure their ongoing demand in the global market.

The Gulf states are not alone in their intention to safeguard a political climate that continues to accept the carbon emissions from oil and gas. Their commitment to fossil fuels aligns with global capital; they share this objective with multinationals, financial markets and states. The Gulf states are indispensable to the hegemony of these structures, because of their export of oil and gas and through their capital, which is invested across the global economy. This will ensure that the Gulf states remain a locus of imperial power for some time. Furthermore, the growing energy demands of emerging economies in Asia will ensure that the Gulf states retain this relevance. With this considered, attempts to bring about a just transition within the societies of the Middle East will have to confront this alliance of national ruling classes, the Gulf states and global capital.

However, despite their power, a number of uncertainties are on the horizon for the Gulf states. As with all societies, those of the Gulf states are not immune from the realities of climate change. Their economic reliance on oil and gas means that they must diversify their economies in order to pay for the rising cost of their food imports, energy production and water consumption. Rising temperatures may affect food yields globally and cause shocks to international commodity chains, and this disruption could hit these economies. On a regional level, their ability to shore up an authoritarian alliance, on which their accumulation and extraction of food partly rests, could also be tested. The pressures that led to the Arab revolutions of 2010 and 2011 have not been resolved; deep structural reconfiguration is still needed. It is too early to foresee how these quandaries will develop but the Gulf states are not immune from popular calls for democracy, equity and redistribution that define the just transition.

NOTES

1. Patrick Wintour, "'Apocalypse Soon': Reluctant Middle East Forced to Open Eyes to Climate Crisis", *The Guardian*, October 29, 2021, https://tinyurl.com/yckkm49d.

2. The GCC is a political and economic union of six countries on the Arabian Peninsula: Saudi Arabia, the United Arab Emirates, Qatar, Kuwait, Bahrain and Oman. The union was founded in 1981. With the possible exception of Kuwait, democratic representation in these countries is minimal.

3. Two examples of this are cotton in Egypt and wine in Algeria, two crops that dominated these economies during the colonial period. See: Giulia Meloni and Johan Swinnen, "The Rise and Fall of the World's Largest Wine Exporter – And its Institutional Legacy", *Journal of Wine Economics* 9, Vol. 1(2014), pp. 3-33, https://doi.org/10.1017/jwe.2014.3; Aaron G. Jakes, *Egypt's Occupation: Colonial Economism and the Crises of Capitalism,* (Stanford University Press, Stanford, 2020).

4. Adam Hanieh, *The Lineages of Revolt*, (Haymarket Books, Chicago, 2013); Armelle Choplin and Leïla Vignal, "Gulf Investments in the Middle East: Linking Places, Shaping a Region", in Leïla (ed.), *The Transnational Middle East,* (Routledge, London, 2016), https://doi.org/10.4324/9781315535654; Karen E. Young, *The Economic Statecraft of the Gulf Arab States: Deploying Aid, Investment and Development Across the MENAP,* (I.B. Tauris, London, New York, Dublin, 2022), https://doi.org/10.5040/9780755646692; Christian Henderson, "The Rise of Arab Gulf Agro-Capital: Continuity and Change in the Corporate Food Regime", *The Journal of Peasant Studies* 49, No. 5 (2021), pp. 1079-100, https://doi.org/10.1080/03066150.2021.1888723.

5. Tobias Zumbraegel, *Political Power and Environmental Sustainability in Gulf Monarchies,* (Palgrave Macmillan, 1st ed, London, 2022).

6. "As long as the world needs oil and gas, we're going to give it to them, says UAE minister", *CNBC*, Video, 03:41, September 27, 2022, https://tinyurl.com/28e53fyn.

7. The meaning of decarbonized oil and gas is vague but it refers to technological innovations such as carbon capture and the electrification of production.

8. Javier Blas, "The Saudi Prince of Oil Vows to Drill 'Every Last Molecule'," *Bloomberg*, July 22, 2021, https://tinyurl.com/jsdapt42.

9. "As long as the world needs oil and gas", CNBC.

10. It should be made clear that this is a decision that has been made by the ruling classes of the Gulf states. One question that requires further research is the extent to which this is supported by ordinary people in the region. As a result of the restrictions on public freedom of speech in many countries in the region, identifying and understanding the degree of environmentalism within the GCC states is difficult, but that is not to say that it does not exist.

11. "Renewable Energy Consumption (% of Total Final Energy Consumption)", the World Bank database, https://data.worldbank.org/indicator/EG.FEC.RNEW.ZS

12. For example, renewable energy accounts for 12.2 percent of consumption in Tunisia (the highest in North Africa). In Jordan this figure is 8.17 percent. See: World Bank, "Renewable Energy Consumption".

13. "*UAE Energy Strategy 2050*," About the UAE, The UAE Government portal, accessed February 16, 2023, https://tinyurl.com/342hybwp.

14. "Energy & Sustainability," About the Kingdom, Vision 2030 – Kingdom of Saudi Arabia, www.vision2030.gov.sa/thekingdom/explore/energy.

15. Natalie Koch, "Sustainability Spectacle and 'Post-Oil' Greening Initiatives", *Environmental Politics*, 2022, https://doi.org/10.1080/09644016.2022.2127481.

16. Merlyn Thomas and Vibeke Venema, "Neom: What's the Green Truth Behind a Planned Eco-City in the Saudi Desert?," *BBC News*, February 22, 2022, https://www.bbc.com/news/blogs-trending-59601335.

17. Gilles Dufrasne et al., *Poor Tackling: Yellow Card for 2022 FIFA World Cup's Carbon Neutrality Claim*, Carbon Market Watch, May 2022, https://tinyurl.com/mr2jshsf.

18. Sam Meredith, "UAE Sparks Furious Backlash by Appointing Abu Dhabi Oil Chief as President of COP28 Climate Summit," *CNBC*, January 12, 2023, https://tinyurl.com/zmsv7fpv.

19. The term social metabolism refers to a society's need for a constant flow of material and energy for its economy and continuance.

20. "Per Capita Electricity Generation, 2022", Charts and Explorers, *Our World in Data*, https://tinyurl.com/ydtzrw3j.

21. Ayhan Demirbas, Ayman A. Hashem and Ahmed A. Bakhsh, "The Cost Analysis of Electric Power Generation in Saudi Arabia", *Energy Sources, Part B: Economics, Planning, and Policy* 12, No. 6 (2017), pp. 591-96, https://doi.org/10.1080/15567249.2016.1248874.

22. Naser Al Wasmi, "Demand for Desalinated Water Puts Pressure on Gulf Ecosystems", *The National News*, February 2, 2017, https://tinyurl.com/3nyhytju.

23. Hannah Ritchie, Max Roser and Pablo Rosado, "Saudi Arabia: Energy Country Profile", Energy, 2022, *Our World in Data*, https://ourworldindata.org/energy/country/saudi-arabia.

24. Dermot Gately, Nourah Al-Yousef and Hamad M.H. Al-Sheikh, "The Rapid Growth of Domestic Oil Consumption in Saudi Arabia and the Opportunity Cost of Oil Exports Foregone," *Energy Policy* 47, (2012), pp. 57-68, https://doi.org/10.1016/j.enpol.2012.04.011.

25. *Ibid.*

26. Masdar city was started in 2006 and it remains under construction. Some reports suggest it will be completed in 2030.

27. "Masdar-led Consortium Signs Deal to Develop Suez Canal Green Hydrogen Project," *Reuters*, November 16, 2022. https://tinyurl.com/yd4w9yv2.

28. "Saudi ACWA Power's Assets Expected to Reach $230bn by 2030: CEO," *Arab News*, November 3, 2022, https://www.arabnews.com/node/2193036/business-economy.

29. "UAE President, Egyptian Counterpart Witness Signing of Agreement to Develop One Of World's Largest Onshore Wind Projects in Egypt", *Masdar*, February 16, 2023, https://tinyurl.com/3sas3s4b.

30. Mohammed Abu Zeid, "UAE, Egypt and Jordan Draft Agreements for Renewable Energy Projects", *Arab News*, May 29, 2022, https://arab.news/v65bs.
31. Andy Sambidge, "Dubai Firm Plans $1.1bn Clean Energy Investment in Egypt", *Arabian Gulf Business Insight*, December 1, 2022, https://tinyurl.com/5n8h645p.
32. Riedel and Sachs, "Israel, Jordan, and the UAE".
33. At the point of use, green hydrogen has no greenhouse emissions and its consumption emits only water. However, its status as a zero-carbon fuel depends on how the production process is powered; for hydrogen to be classed as a green fuel it has to be produced with renewable energy. The production of hydrogen with natural gas is classed as blue or grey, depending on the level of CO_2 emissions.
34. Michael Barnard, *Assessing EU plans to Import Hydrogen From North Africa: The Cases of Morocco, Algeria and Egypt*, Transnational Institute and Corporate Europe Observatory, May 17, 2022, https://tinyurl.com/swyzvsc7.
35. "Masdar-led Consortium Signs Deal to Develop Suez Canal Green Hydrogen Project", *Reuters*, November 16, 2022. https://tinyurl.com/yd4w9yv2.
36. Nibal Zgheib, *EBRD Strengthens Egypt's Construction and Utilities Sectors*, European Bank for Reconstruction and Development, November 14, 2019, https://tinyurl.com/yvvpx5ef.
37. World Bank, GDP per capita-2021, https://data.worldbank.org/indicator/NY.GDP.PCAP.CD.
38. *Ibid.*
39. Eckart Woertz, "Wither the Self-Sufficiency Illusion? Food Security in Arab Gulf States And The Impact of COVID-19", *Food Security 12*, (2020), pp. 757-60, https://doi.org/10.1007/s12571-020-01081-4.
40. Henderson, "The Power of Food Security," *Globalizations*, https://doi.org/10.1080/14747731.2022.2075616.
41. Oxford Business Group, "Agri-Tech & Food Security in the GCC : COVID-19 Response Report", 2022, https://aoad.org/GCC_Agritech_CRR.pdf.
42. World Bank, "Employment in Agriculture (% Of Total Employment) (Modeled ILO Estimate)-2019", https://data.worldbank.org/indicator/SL.AGR.EMPL.ZS.
43. El Zein et al., "Health and Ecological Sustainability in the Arab World: A Matter of Survival", *The Lancet 383*, No. 9915 (104):458-76, https://doi.org/10.1016/S0140-6736(13)62338-7.
44. Henderson, "Land Grabs Reexamined".
45. "Map", Land Matrix, 2023, https://landmatrix.org/map.
46. Save the Children, "One in Four People Face Severe Hunger in Sudan as Food Crisis Deepens – Sudan", June 22, https://tinyurl.com/49a79rt6.
47. Transnational Institute, *The global land grab: A Primer*, 2013, www.tni.org/en/publication/the-global-land-grab.
48. "Saudis Grains Storage Capacity Rises by 37% with 2 New Silos," *Zawya*, September 19, 2021, https://tinyurl.com/2p9dnpny.
49. "Inside Egypt's Plan to Tackle the Wheat Crisis", *Middle East Eye*, March 16, 2022, https://tinyurl.com/mt8pf8es.
50. "Arrival of First Vessel at Hamad Port's Strategic Food Security Facilities Terminal," *Qatar News Agency*, August 13, 2022, https://tinyurl.com/ycypebyu.
51. "Water Security Mega Reservoirs", Projects, Arcadis, www.arcadis.com/en/projects/middle-east/qatar/water-security-mega-reservoirs.

About the Contributors

Mohamed Salah Abdelrahman is an environmental researcher, political ecologist and community mobilizer with more than 10 years of experience in conducting studies on environmental and resource extraction issues, such as hydro dam projects and gold mining. He is interested in the correlation between environmental policies, questions of justice and market drivers of conflict in different communities around Sudan. Among many other publications, his first book *The Social and Environmental Price of Gold Mining* (in Arabic) was published in 2018. Mohamed holds a master's degree in environmental science from the University of Khartoum.

Joanna Allan is an academic in the department of Geography and Environmental Sciences at Northumbria University. She is also an activist with Western Sahara Campaign UK and Western Sahara Resource Watch.

Asmaa Amin is a Jordanian writer and researcher interested in energy and environmental questions. She has been working in the sector of renewable energy since 2016. She holds a Master's degree in business administration and a Bachelor's degree in electrical engineering. She has published articles and studies on various websites and journals.

Razaz H. Basheir is a researcher focusing on questions of urban infrastructure in general, and energy infrastructure in particular. She has previously worked in power generation projects, from design to implementation. Razaz has an educational background in mechanical engineering and recently finished her master's studies in Southern Urbanism at the African Center for Cities – University of Cape Town. Razaz is currently a researcher at Innovation, Science and Technology Think Tank for People-Centered Development (ISTinaD) in Khartoum, and an editor at Amar, an online socio-technical publication.

Chafik Ben Rouine is the Co-Founder and President of the Tunisian Observatory of Economy.

Imane Boukhatem is an energy policy researcher and a sustainability advocate. She has represented the Middle East and North Africa (MENA) region as a climate ambassador at the Global Youth Climate Network and Student Energy and in global youth-led organizations that are empowering young people to mitigate climate change and to accelerate the transition to a sustainable and equitable energy future. Imane's PhD research focuses on the potential role of green hydrogen in the energy transition in Algeria.

Mohamed Gad is an economic journalist who, since 2004, has specialized in the Egyptian economy. His work is published in local and international newspapers, as well as on independent news websites. He has published academic work with independent research centres and human rights organizations, and has edited four books on the Egyptian economy published by Dar al-Maraya.

Hamza Hamouchene is a London-based Algerian researcher-activist, commentator and a founding member of Algeria Solidarity Campaign (ASC), Environmental Justice North Africa (EJNA) and the North African Food Sovereignty Network (Siyada). He is currently the Arab region Programme Coordinator at the Transnational Institute (TNI). His work is focused on issues of extractivism, resources, land and food sovereignty as well as climate, environmental, and energy justice in the Arab region. He is the author/editor of three other books: "*The Arab Uprisings: A decade of struggles*" (2022), "*The Struggle for Energy Democracy in the Maghreb*" (2017) and "*The Coming Revolution to North Africa: The Struggle for Climate Justice*" (2015). He also contributed chapters to various books including *The Oxford Handbook of Economic Imperialism* (2022), *The Routledge Essential Guide to Critical Development Studies* (2021), *Fanon Today: Reason and Revolt of the Wretched of the Earth* (2021), "A Region in Revolt: Mapping the Recent Uprisings in North Africa and West Asia" (2020), *The Palgrave Encyclopaedia of Imperialism and Anti-Imperialism* (2016) and "Voices of Liberation: Frantz Fanon" (2014).

Adam Hanieh is a Professor of Political Economy and Global Development at IAIS, University of Exeter, and Joint Chair at the Institute of International and Area Studies (IIAS) at Tsinghua University, Beijing, China. He published four books that explore different aspects of the Middle East region. His most recent book, *Money, Markets, and Monarchies: The Gulf Cooperation Council and the Political Economy of the Contemporary Middle East,* was published by Cambridge University Press in 2018, and was awarded the 2019 British International Studies Association International Political Economy Group Book Prize and the 2019 Political Economy Book Prize of the Arab Studies Institute.

Christian Henderson is assistant professor of International Relations and Modern Middle East Studies at the Leiden Institute for Area Studies in Leiden University. His research focuses on food systems in the Gulf states and the political ecology of the Middle East and North Africa. He holds a PhD in Development Studies from the School of Oriental and African Studies. His work has been published in the *Journal of Peasant Studies*, *Environment and Planning A*, *Globalizations*, and *New Political Economy*.

Hamza Lakhal is a PHD student in Anthropology at Durham University. He is a Sahrawi poet and pro-Western Sahara independence activist.

Mahmoud Lemaadel is an independent researcher and media activist. He is also the co-founder of the local media and human rights platform; Nushatta Foundation for Media and Human Rights that operates in the Moroccan-occupied Western Sahara and the Sahrawi refugee camps in the south-west of Algeria.

Jawad Moustakbal is the National Coordinator in Morocco for the International Honors Programme "Climate Change: The Politics of Food, Water, and Energy" at the School of International Training (SIT) in Vermont, USA. He has worked as a project manager for several companies, including OCP, the Moroccan state phosphate company. Jawad is also an activist who works for social and climate justice. He is a member of the National Secretariat of ATTAC/CADTM Morocco, and a member of the shared Secretariat of the International Committee for the Abolition of Illegitimate Debts.

Saker El Nour is a visiting Postdoctoral Fellow at the International Research Group on Authoritarianism and Counter-Strategies of the Rosa Luxembourg Foundation and the Center for Middle Eastern and North African Politics, Freie Universität Berlin. His scholarly interests include political ecology, rural sociology, rural social movements, and agri-environmental politics, with a focus on Arab countries. He is co-founder of the North Africa and Middle East Network for a Just Transition (RÉSEAU TANMO).

Karen Rignall is a cultural anthropologist and associate professor at the University of Kentucky (US). Her research examines the politics of land access, rurality, and natural resource governance in Morocco's pre-Saharan oases and the Appalachian US. She has conducted ethnographic fieldwork and multi-disciplinary collaborations, with a current focus on supporting grassroots networks rooted in rural communities and working towards energy and economic transition.

Flavie Roche is an intern at the Tunisian Observatory of Economy.

Katie Sandwell is a Programme Coordinator at the Transnational Institute (TNI). She holds a BA in Philosophy and a Master of Environmental Studies degree from York University, Canada focusing on food sovereignty and local food movements. She has lived in Canada, Germany, the United Kingdom, and the Netherlands and has been involved with a range of food, community, environmental, and social justice organizations. Her research focuses on Just Transition, human rights, corporate concentration, food sovereignty, and oceans.

Manal Shqair is a Palestinian climate activist, researcher, and the international advocacy officer of Stop the Wall Campaign, a grassroots organization based in Palestine. Currently, Manal is doing her PhD in sociology at Queen Margaret University, Scotland. In her PhD thesis she examines the role of Palestinian semi-nomadic women's everyday practices of sumud (steadfastness) in maintaining and reinforcing group solidarity and in enabling popular resistance to disrupt Israeli settler colonial dispossession intertwined with patriarchy and capitalism.

Index

Abu Dhabi
  ADNOC (Abu Dhabi National Oil
    Company), 3, 236, 245, 279, 280
  Masdar, 280
ADNOC (Abu Dhabi National Oil
  Company), 245, 279
AFD (Agence Française de
  Développement), 32, 168
AfDB (African Development Bank), 32,
  168, 227
Algeria, 15, 252–74
  agriculture, 111, 112, 114, 116, 123
  climate breakdown, 1
  emissions, 109, 121
  fossil fuels, 4, 20, 35–7, 45, 201,
    252–74
  IMF, 114
  independence, 112
  Morocco, 42, 54, 61
  pipelines, 39, 42
  refugees, 2, 50, 51, 61
  Tunisia, 38, 42, 201
Asia, 19, 52, 235–51, 287

Bahrain, 67, 69, 74, 236, 247, 248
Bolivia, 31, 119
BOT (build-operate-transfer), 157, 160
BP, 235, 237, 239, 242
BRI (Belt and Road Initiative), 244, 245

Canada, 11, 236, 238
CEGCO (Central Electricity Generation
  Company), 177, 178
Chevron, 75, 235, 239
Chile, 31, 40
China, 6, 19, 40, 209, 235, 236, 243, 264
  loans, 139
  oil, 239–41, 242, 244, 245
  Sinopec, 242, 245
CO_2 (carbon dioxide), 42, 252

COP (United Nation Climate
  Conference of the Parties), 3, 4
  COP7, 3
  COP15, 3
  COP18, 3
  COP21, 146
  COP22, 3, 35, 54
  COP26, 89, 262
  COP27, 3, 4, 37, 68, 69, 236, 247, 248,
    280
  COP28, 3, 69, 236, 247, 248, 277, 278,
    287
Copenhagen, 3, 120
COVID-19 pandemic, 6, 29, 38, 43, 99,
  117, 185, 194, 202, 208, 265, 266
CSP (concentrated solar power), 32, 91,
  93, 94, 222, 227, 259
CSPRON (Committee for the
  Protection of Natural Resources in
  Western Sahara), 58
CSR (corporate social responsibility),
  89, 99, 100
Cyprus, 76, 77

DCFTA (Deep and Comprehensive Free
  Trade Agreement), 206, 207
Desertec, 40–43, 205, 214
Dii (Desertec Industrial Initiative), 41,
  52, 93, 206
DRC (Democratic Republic of Congo),
  31, 40, 136

EBRD (European Bank for
  Reconstruction and Development),
  3, 281, 282
EDCO (Electricity Distribution
  Company), 177, 178
EEHC (Egyptian Electricity Holding
  Company), 163–7
EEM (Énergie Électrique du Maroc),
  221

EEM (Énergie Éolienne du Maroc),
   225–6
Egypt, 1, 15, 37, 39, 276, 280
   agriculture, 112, 113, 114, 116, 117,
      123, 124, 126, 136, 137, 284, 285
   climate change, 121
   COP27, 3, 37, 68, 69, 247
   emissions, 109, 252
   energy, 74, 77, 157–74, 281, 286
   energy transition, 45, 111, 172
   hydrogen, 282
   Israel, 67, 68, 75, 76
   Jordan, 175, 179, 180, 181, 183, 184,
      187, 196–7
   liberalization, 19, 158, 171
   privatization, 19
   refugees, 2
   Sudan, 135, 145
EIB (European Investment Bank), 31,
   32
EMRC (Energy and Minerals
   Regulatory Commission), 177,
   178
Enefit (Estonian National Energy
   Company), 176, 185
ENI, 37, 76, 208, 239, 254
ENLT-NewMed, 69, 72, 74–7, 78
   ENLT (Enlight Green Energy), 69, 75
   NewMed, 69, 74, 76
EPC-F (Engineering, Procurement,
   Construction and Finance ESC:
   Economic and Social Council),
   157, 165
Estonia, 176, 185
EU (European Union)
   agriculture, 18, 53, 117, 118
   climate transition, 5, 43, 254, 270
   energy, 2, 4, 36, 37, 43, 75, 76, 206,
      248, 267
   fish, 53
   hydrogen, 39, 40
   Israel, 75, 76, 77, 79
   subsidies, 207
Europe, 6, 7, 10, 30, 37–42, 45, 52, 54,
      70, 76, 92, 110, 120, 121, 205, 207,
      236, 237, 240–44, 246, 254–5, 267,
      268, 282
ExxonMobil, 235, 237, 238, 239, 242

France, 9, 36, 209, 225, 228, 239, 252
   Total, 36, 41, 76, 235, 239

Germany, 32, 40, 53, 95, 206, 209, 210,
      227, 228
   Lahmeyer International, 138
   Siemens, 35, 53, 157, 225, 226, 228
GHG (greenhouse gas), 29, 53, 109, 110,
      119, 121, 141, 201, 252, 254, 278
GIZ (German Agency for International
      Cooperation), 5, 206, 210, 256
global North, 18, 20, 30, 52, 109, 116,
      117, 118, 119, 120, 204, 262, 268
global South, 3, 13, 15, 21, 31, 37, 38,
      43, 44, 72, 77, 90, 109, 116, 119–20,
      147, 149, 151, 204, 235, 254, 262,
      264, 270, 276

hydroelectric power. See hydropower
hydrogen, 4, 15, 17, 38–43, 247, 248,
      276, 280, 282, 283
   green hydrogen, 93, 267, 268, 270,
      282
hydropower, 134, 140, 146, 150, 226

IFIs (international financial
      institutions), 31, 163, 170, 221
ILO (International Labour
      Organization), 111, 117
IMF (International Monetary Fund)
   colonialism, 31
   elites, 2–3
   energy, 18, 19, 142, 159, 169, 175,
      177–8, 185, 227
   inequality, 5
   liberalization, 19, 114, 158, 159
   loans, 158, 159, 169, 175, 176, 227
   oil prices, 265
   privatization, 19, 114, 177–8, 179
   SAP, 31, 221
   subsidies, 179
IPCC (Intergovernmental Panel on
      Climate Change), 109, 252, 267,
      271
Iraq, 1, 21, 197, 240, 245
   OPEC, 238
   refugees, 2, 69
   United States, 179

Israel, 4, 17, 67–87
  Jordan, 183–4, 196, 281–2
  LNG, 37, 39, 176, 181, 189
  Palestine, 67–87, 281
Italy, 37, 165, 209, 267
  ENI, 37, 76, 208, 239, 254

Japan, 209, 227, 228, 246, 281
JNF (Jewish National Fund) (Israel),
    70, 79
Jordan, 1, 15, 67, 74, 276, 280, 281
  COP27, 68, 69
  energy, 18, 19, 73, 74 ,75, 76,
    175–200, 286
  refugees, 2
  water, 70–72, 77

Khaya, Sultana, 57–8
Kuwait, 1, 178, 236, 238, 244, 245, 248,
    278, 180
Kyoto Protocol (aka Kyoto Accord), 3,
    256

Latin America, 98, 238, 283
Lebanon, 2, 70, 75–6, 197, 283
Leviathan field, 73, 75, 76, 78, 176, 183,
    184
Libya, 21, 42
LNG (liquefied natural gas), 37, 180,
    245

Maghreb, 9, 17, 123, 124, 125
Malta, 38, 205
Masdar, 32, 68, 72, 225, 248, 280, 281,
    282
Masen (Moroccan Agency for
    Sustainable Energy), 32, 35, 99,
    224, 227–8
Mauritania, 50, 51
MENA (Middle East and North Africa)
  climate change, 2, 29
  climate transition, 13, 89, 275–90
  energy, 6, 40, 52, 93, 206
  greenwashing, 72
Morocco, 15
  Abraham Accords, 67
  agriculture, 111, 112, 115, 116, 123,
    284

climate breakdown, 1, 2
  COP22, 3
  emissions, 109, 121
  energy dependency, 220–32
  energy transition, 17, 19, 32, 45,
    88–108, 111, 229–30
  environmental destruction, 31
  extractivism, 17, 49–66
  greenwashing, 61
  hydrogen, 40
  independence, 50, 220, 221
  inequality, 220–32
  Israel, 67, 69, 74, 76
  mining, 88–108
  pastoralism, 33
  pipelines, 42
  privatization, 5, 38
  refugees, 2, 17, 34, 49–66
  renewable energy, 17, 35, 49, 88–108,
    203, 221–3, 280, 281, 286
  Saharawi peoples, 2, 17, 34, 49–66
  Soualiyate women, 34
  Western Sahara, 49–66
  workers' exploitation, 31

NEPCO (National Electric Power
    Company) (Jordan), 177, 183, 185,
    195
NOCs (national oil companies), 235–51
Noor Midelt project, 32, 225, 227, 228
Norway, 210, 254

OCP (Office Chérifien des Phosphates),
    35, 53, 293
Oman, 1, 67, 69, 74, 236, 244, 248, 282
ONE (Office National de l'Électricité)
    (Morocco), 221, 223, 224, 226, 227
OPEC (Organization of Petroleum
    Exporting Countries), 238, 242,
    278

Palestine, 2, 15, 17, 67–87, 281
Paris Climate Agreement (2015), 3, 12,
    60, 111, 120–21, 146, 201, 247, 267,
    269, 271
PNEREE (National Programme for the
    Development of Renewable Energy

and Energy Efficiency) (Algeria),
258–9, 263
POLISARIO (Popular Front for the
Liberation of Saguia El Hamra and
Río de Oro), 50–52, 54, 59
PPPs (public–private partnerships), 5,
201, 203, 205, 207, 208, 214–15,
220, 223, 229
PST (Plan Solaire Tunisien), 203–4
PV (photovoltaic) energy, 32, 35, 207,
208, 209, 210, 212–13, 215, 259,
260, 263

Qatar, 4, 15, 283, 286
COP18, 3
energy, 4, 76, 236, 244, 245, 247, 248,
279
hydrogen, 282
LNG, 37, 39, 180
migrant workers, 2
OPEC, 278
renewable energy, 277
water, 285

Russia, 53, 76, 235, 238, 240, 244
gas, 4, 37, 39, 254, 267
Ukraine, 36, 77, 266, 268, 282

SABIC (Saudi Basic Industries
Corporation), 243, 244
SADR (Sahrawi Arab Democratic
Republic), 51, 53, 60–61, 62
Sahara, 10, 15, 17, 29, 34–5, 36, 38,
49–66, 252, 260
Said, Edward, 9–10
SAMIR (Société Anonyme Marocaine
de l'Industrie du Raffinage), 221,
222, 225
SAP (Structural Adjustment
Programme), 31, 44, 115, 221
Saudi Arabia, 1, 15
ACWA Power, 35, 38, 93, 99, 225,
280, 281
agriculture, 140, 285
aid, 280
Aramco, 235, 241, 242, 245, 246, 247,
248
China, 242

COP22, 35
COP27, 69
energy transition, 278, 281
ENLT, 74
fossil fuels, 4, 93, 221, 235–51, 277,
279
hydrogen, 40, 282
Israel, 67, 69
land acquisition, 139–40
pipelines, 245
SABIC, 243, 244
Syria, 286
Yemen, 286
Shell, 41, 235, 237, 239, 242
Sinopec, 242, 245
solar energy, 34, 69, 146, 147, 168, 187,
189, 226, 257, 258, 261, 263, 264,
282, 283
solar farms, 40, 61, 276
solar power plants, 30, 146, 164, 167,
209, 258
South Africa, 11, 40, 45, 65, 252, 262
South America, 11, 240
South Korea, 146, 246, 248
Spain, 50, 51, 120, 209, 226
STEG (Tunisian Company of Electricity
and Gas), 201, 204, 207–9, 214
Sudan, 15, 16, 67, 134–54, 276, 284, 285
refugees, 2
Syria, 1, 2, 21, 68, 69, 70, 71, 72, 77, 283,
286

Tamar field, 75, 78, 183
Tanzania, 14, 142
Total, 36, 41, 76, 235, 239
Tunisia, 15
agriculture, 2, 45, 110, 111, 112, 113,
115, 123
Algeria, 38, 42
climate breakdown, 1, 2, 121
emissions, 109, 121, 201
energy transition, 5, 19, 38, 201–19
ETAP, 208
fish, 2
independence, 112, 201
inequality, 110, 113
pipelines, 42
privatization, 5, 38, 201–19

refugees, 2
renewable energy, 5, 19, 201–19, 286
    STEG, 201, 204, 207–9, 214

UAE (United Arab Emirates), 4, 15, 236,
        281, 283, 285
    Abraham Accords (2020), 67
    COP27, 68, 69, 247
    COP28 3, 69, 247, 277, 278–9
    Dubai, 3, 143, 278
    electricity consumption, 279
    ENLT-NewMed, 69, 74, 75
    ethylene, 243
    hydrocarbon, 277
    hydrogen, 248, 282
    Israel, 68, 69, 75, 76, 196
    Jordan, 68, 69, 196
    Masdar, 32, 68–9, 72, 225, 248, 280,
        281, 282
    migrant workers in, 2
    petrochemicals, 243, 244
    Project Prosperity, 68–9
    renewable energy, 69, 277–9
    Taqa, 225
    Yemen, 286
UGTT (General Union of Tunisian
        Workers), 208, 214
UK (United Kingdom), 9, 38, 205, 235,
        237, 265, 275
    BP, 235, 237, 239, 242
    Nur Energy, 38, 205
    Shell, 41, 235, 237, 239, 242
Ukraine, 4, 6, 36, 38, 39, 75, 76, 77, 254,
        265, 266, 267, 268, 282
UN (United Nations)
    COP, 3, 4
    COP26, 89, 262
    COP27, 3, 4, 37, 68, 69, 236, 247, 248,
        280
    COP28, 3, 69, 236, 247, 248, 277, 278,
        287
    decolonization, 50
    IPCC report, 2
    Israel, 74

Resolution 194, 68
Resolution 1514 (XV), 51
Sudan, 142
UNFCCC (United Nations Framework
    Convention on Climate Change),
    3, 120, 256
USA (United States of America), 2, 10,
    11, 75, 97, 149, 170, 171, 236, 243,
    283
    Appalachian coalfields, 90, 100
    Chevron, 75, 235, 239
    ExxonMobil, 235, 237, 238, 239, 242
    Iraq, 179
    LNG, 37
    oil, 237, 238, 240
    USAID (United States Agency for
        International Development), 5

wind energy, 34, 55, 69, 146, 147, 168,
    187, 189, 213, 223, 226, 257, 258,
    261, 282, 283
    wind energy companies, 34, 35, 53,
        57, 225
    wind farms, 30, 34, 40, 52, 53, 61, 74,
        75, 209, 276, 281
    wind power plants, 146, 164, 167,
        207, 209, 258
    wind turbines, 60, 74, 203, 209, 210,
        213–14
World Bank, 2, 144–9
    colonialism, 31
    Egypt, 161, 166, 168
    energy transition, 5, 32
    Jordan, 177–9, 185
    liberalization, 18, 19, 114, 160, 168,
        169
    loans, 158, 176, 281
    Morocco, 33, 221, 227
    privatization, 18, 19, 114, 177–9
    subsidies, 168–70, 171, 177–9
    Sudan, 138, 142, 143
    Tunisia, 115

Yemen, 1, 2, 21, 283, 285, 286

The Pluto Press Newsletter

Hello friend of Pluto!

Want to stay on top of the best radical books
we publish?

Then sign up to be the first to hear about our
new books, as well as special events,
podcasts and videos.

You'll also get 50% off your first order with us
when you sign up.

Come and join us!

Go to bit.ly/PlutoNewsletter

Thanks to our Patreon subscriber:

Ciaran Kane

Who has shown generosity and
comradeship in support of our publishing.

Check out the other perks you get by subscribing
to our Patreon – visit patreon.com/plutopress.

Subscriptions start from £3 a month.